Chalcogenide Glasses for Infrared Optics

Chalcogenide Glasses for Infrared Optics

Dr. A. Ray Hilton, Sr.
Chairman of the Board and Technical Director
Amorphous Materials, Inc.
Garland, Texas

New York Chicago San Francisco
Lisbon London Madrid Mexico City
Milan New Delhi San Juan
Seoul Singapore Sydney Toronto

The McGraw·Hill Companies

Cataloging-in-Publication Data is on file with the Library of Congress

Copyright © 2010 by The McGraw-Hill Companies, Inc. All rights reserved. Printed in the United States of America. Except as permitted under the United States Copyright Act of 1976, no part of this publication may be reproduced or distributed in any form or by any means, or stored in a data base or retrieval system, without the prior written permission of the publisher.

1 2 3 4 5 6 7 8 9 0   DOC/DOC   0 1 5 4 3 2 1 0 9

ISBN 978-0-07-159697-8
MHID 0-07-159697-6

Sponsoring Editor Taisuke Soda	**Project Manager** Gita Raman, Glyph International	**Indexer** Edwin Durbin
Editing Supervisor Stephen M. Smith	**Copy Editor** Patti Scott	**Art Director, Cover** Jeff Weeks
Production Supervisor Richard C. Ruzycka	**Proofreader** Shivani Arora, Glyph International	**Composition** Glyph International
Acquisitions Coordinator Michael Mulcahy		

Printed and bound by RR Donnelley.

McGraw-Hill books are available at special quantity discounts to use as premiums and sales promotions, or for use in corporate training programs. To contact a representative, please e-mail us at bulksales@mcgraw-hill.com.

This book is printed on acid-free paper.

Information contained in this work has been obtained by The McGraw-Hill Companies, Inc. ("McGraw-Hill") from sources believed to be reliable. However, neither McGraw-Hill nor its authors guarantee the accuracy or completeness of any information published herein, and neither McGraw-Hill nor its authors shall be responsible for any errors, omissions, or damages arising out of use of this information. This work is published with the understanding that McGraw-Hill and its authors are supplying information but are not attempting to render engineering or other professional services. If such services are required, the assistance of an appropriate professional should be sought.

About the Author

Dr. A. Ray Hilton, Sr., is known worldwide through publications describing his efforts in chalcogenide glasses. He serves as Chairman of the Board and Technical Director of Amorphous Materials, Inc. (AMI). Under Dr. Hilton's direction and guidance, a process was developed to compound and cast high-purity homogeneous plates of chalcogenide glasses carried out under high vacuum in high-purity quartz containers. Thousands of glass blanks required for government FLIR systems produced in the 1980s for the Army were supplied by AMI. AMI is currently developing new glass compositions for lenses for use in inexpensive infrared cameras.

Contents

Acknowledgments xi
Introduction xv

1 **Transmission of Light by Solids** 1
   1.1 Solids 1
   1.2 Beginning of Transmission of Light—
       An Electronic Transition 2
   1.3 Long-Wavelength Cutoff 5
   1.4 Extrinsic Loss within the Band, Impurities,
       Scatter, and Quality 10
   1.5 Optical Constants and Dispersion
       due to Strong Absorption 12
       References 14

2 **Chalcogenide Glasses** 17
   2.1 Historical Development 17
   2.2 The Periodic Table and Glass Formation . . . 21
   2.3 Evaluating Possible Glass Forming Systems . . . 29
   2.4 Qualitative Evaluation of Compositions
       for Development 35
   2.5 General Physical Properties of
       Chalcogenide Glasses 40
       2.5.1 Softening Points and Hardness 40
       2.5.2 Thermal Coefficients of Expansion . . . 41
       2.5.3 Density 41
       2.5.4 Molar Refraction 42
       2.5.5 Electrical Properties 47
       2.5.6 Physical Strength 47
       2.5.7 Softening Points 48
   2.6 Chemical Bonding in Chalcogenide Glasses . . . 48
       2.6.1 Composition Location in the
            Glass Forming Diagram 48
       2.6.2 Molecular Vibrations of Constituent
            Atoms 50
       2.6.3 Mass Spectrometric Investigation
            of Bonding in the Glasses 55
       2.6.4 X-ray Radial Distribution Analysis of
            Chalcogenide Glasses 57
       2.6.5 Conclusions from the TI Exploratory
            Programs of 1962 to 1965 59

viii   Contents

    2.7 Chalcogenide Glasses Containing Transition Elements 60
    2.8 Discussion of Results 66
    References 69

3 Glass Production 71
    3.1 Reactants 71
    3.2 Compounding Methods 73
    3.3 Compounding with Reactant Purification ... 74
    3.4 Open Casting Methods 77
    3.5 Purification, Compounding, Casting—One Closed Operation 84
    3.6 Summary 86
    References 87

4 Characterization of Glass Properties 89
    4.1 Thermal Expansion, Glass Transition Temperature, and Softening Point 89
    4.2 Transmission, Precise Refractive Index, and Thermal Change in Refractive Index 94
    4.3 Physical Properties Important for Optical Use 108
        4.3.1 Hardness 108
        4.3.2 Young's Modulus, Shear Modulus, and Poisson's Ratio 109
        4.3.3 Rupture Modulus 110
        4.3.4 Thermal Conductivity 112
        4.3.5 Electrical Resistance 113
    4.4 Resistance to Chemical Attack 114
    4.5 Final Production Procedure 114
    References 118

5 Conventional Lens Fabrication and Spherical Surfaces 119
    5.1 Lens Blank Preparation 119
    5.2 Generation of Spherical Surfaces 121
    5.3 Polishing 122
    5.4 Testing 123
    5.5 Antireflection Coatings 126
    Reference 129

6 Unconventional Lens Fabrication, Aspheric Surfaces, and Kinos 131
    6.1 Optical Designs 131
    6.2 Diamond Turning 132
    6.3 Slump Molding 133
    6.4 Precision Molding 133

	6.5 Volume Production	146
	6.6 Problem of Refractive Index Change When Pressure Molding	148
	References	151
7	**Glass Processes for Other Applications**	**153**
	7.1 AMI as Supplier of Chalcogenide Glasses for IR Fibers	153
	7.2 AMI Fiber Drawing Process	158
	7.3 Chemical Applications of AMI IR Fiber	168
	7.3.1 Fiber Summary	173
	7.4 Extrusion of Chalcogenide Glasses	174
	7.4.1 Glass Extrusion Summary	178
	References	178
8	**IR Imaging Bundles Made from Chalcogenide Glass Fibers**	**181**
	8.1 The Stacked Ribbon Method	181
	8.2 IR Imaging Bundles of 1-m Length	184
	8.3 Goals of the Navy SBIR 10-m IR Imaging Bundle Program	191
	8.4 The Navy Phase II 27-Month Program	192
	8.4.1 The 1-m C2 Imaging Bundles	192
	8.4.2 AMI Glass Clad Fibers	194
	8.4.3 AMI Production of High-Purity Arsenic Trisulfide Glass	194
	8.4.4 The 50 Percent Transmission Goal	196
	8.4.5 Formation of Bundles on the 10-m Drum	199
	8.4.6 Optical Evaluation of 10-m Imaging Bundles	204
	8.5 Summary	209
	References	210
9	**AMI Infrared Crystalline Materials**	**211**
	9.1 Cadmium Telluride	211
	9.2 Previous Work at TI	212
	9.2.1 Conclusions Concerning This Effort	214
	9.3 AMI DARPA-Funded Large Plate Process	215
	9.3.1 Conclusions	221
	9.4 Vacuum Float Zoned Silicon Detector Material	222
	9.5 Silicon as an Infrared Optical Material	225
	9.6 Single-Crystal Silicon Fibers	228
	9.7 Gallium Arsenide as an Infrared Optical Material	230

	9.8	Production of GaAs at AMI	231
	9.9	Horizontal Bridgman Production of GaAs Plates at AMI	233
		References	243
10	**Early Work at Texas Instruments**		245
	10.1	First Job	245
	10.2	Infrared Applications to Materials	245
	10.3	Optical Interference and Film Thickness	247
	10.4	The Infrared Scan Technique for Epitaxial Film Thickness	248
	10.5	Elliptical Polarization of Light on Reflection	252
	10.6	Measuring the Elliptical Polarization Angles ψ and Δ	255
	10.7	Ellipsometers Used at TI	259
	10.8	Infrared Ellipsometry	259
	10.9	The TI Automatic Ellipsometer System	264
	10.10	Summary	269
		References	270
	Index		271

Acknowledgments

When you read this book describing my 50 years' experience, you will find the applied research process is not flawless. Bad choices and false starts I made are identified. But when it came to choosing with whom I was going to share my life, I was dead on. My wife, Madora Pauline Bull Hilton, of 58 years has been in my corner, on my side every step of the way, through the years in college, followed by service in the Air Force, and graduate school while raising three children on the G.I. Bill. A few people influenced me as I pursued this path in life. We had help from John Beckham, the business manager of the chemistry department, and I taught freshman labs under Dr. Tom Burkhalter and finally received a research fellowship via Dr. Albert Jache, my senior adviser, who also taught me the love of research. We finally finished in 1959, there was one more oil company job in Houston, and then it was off to a good-paying job at Texas Instruments in Dallas in 1960.

Texas Instruments was a great place to work. The colleagues who helped me most were Charlie Jones, Harold Hafner, and Dr. George Cronin. Technicians were Jimmie Parker and Joyce Jones. In 1974, after 14 years in the TI CRL, 5 years as a senior scientist, I left to manage the infrared glass production in the EO Division. The production of the glasses had become very important. I soon realized there was a need for a second source of the glass, a unique opportunity for me. Like many, I had always wanted to run my own company. In 1977 when I told Madora what I wanted to do, she said go ahead, since it has always been your dream. I will join the company, she said, but I will still need time to be with my children and grandchildren.

So I left TI in March 1977, borrowed some money from a bank, using land we owned as collateral, and accepted stockholders. We started in a small rented building. Our first employee was Glen Whaley, a master glass blower from TI. His son, Greg Whaley, and our oldest son, Ray Hilton, Jr., worked part-time while still going to school. Glen's friend Mitchell Jones was our first technician. Our oldest daughter, Gail Hanna, soon joined us as a technician followed by James McCord from TI. AMI has been in operation 32 years, and our son is now president. Greg Whaley is vice president and director

Acknowledgments

for sales and contracts. Our daughter, Gail Hanna, is our antireflection coating specialist. Madora is retired and spends time with her 12 grandchildren, all of whom live here.

I would also like to acknowledge the following individuals who contributed to the work reported in the designated chapter(s):

	Chapters
Cam Allen, System Design, TI	10
Max Amon, Optical Design, LMCO	6
Dr. R. J. Archer, Ellipsometry, Hewlett-Packard	10
Dr. Werner Beyen, laboratory director, TI	10
Dr. Steve Boldish, AMI consultant	9
Maurice Brau, SC Materials, TI	2, 6
Max Bryant, Mass Spectrometry, TI	2
Dr. Tom Burkhalter, laboratory director, TI	10
Ed Carr, coatings consultant, AMI	5, 6, 8
Ron Child, glass blower, TI	10
Igeno Lombardo Codman, Johnson & Johnson	7
Dr. George Cronin, SC Materials, TI/AMI	7, 9
Gene Daniels, computer programmer, TI	10
Jim Davidson, Thermalscan	8
Dr. Robert Dobrot, X-ray Structure, TI	2
Gerald Ferguson, AMI Navy Representative	7
Dr. Jacob Fraden, Thermoscan	7
Carl Fritz, grandson, IR Refractometer, AMI	7
Dr. Tommy George, Mass Spectrometry, TI	2
Amy Graham, IR&D, LMCO	6
Harold Hafner, Glass Science, TI	3, 9
Dr. John Hall, Optical Design, NVL	9
Gail Hanna, daughter, coating technician, AMI	5, 6, 8
Bob Harp, senior technician, TI	9
Dr. Jim Harrington, professor, Rutgers University	7
Dr. Don Hayes, president, Micro Fab	3
Ray Hilton, Jr., president, AMI	3, 5–9
Bob Icovazzi, Optics Production, LMCO	9
Dr. Pete Johnson, Ceramics, TI	10

Acknowledgments

Charlie Jones, Materials, TI	2, 3, 9, 10
Dr. Eric Jones, Computer Science, TI	10
Mitchell Jones, senior technician, AMI	3, 9
Edward Knerr, IRAD program, LMCO	6
Dr. Heinz Krebs, professor, Stuttgart University	2
John Lawson, Optics Design, CBC America Optics	6
Rich LeBlanc, IR&D, LMCO, AMI director	6
Tom Loretz, Glass Science, AMI consultant	7, 8
Al Lyon, Optical Design, LMCO	9
James McCord, senior technician, AMI	3, 6–9
Dr. Peter Melling, president, Remspec	8
Paul Modlin, Advantek Engineering	8
Robert Patterson, TI 1173 Glass Composition	4
Charles Ratliff, Computer Science, TI	10
Dr. Mike Rechtin, Materials Science, TI	3
Dr. Allen Rineberg, Physics, TI	10
Dr. Grady Roberts, Materials Science, TI	2
Nate Rosebrooks, Sancliff Equipment	7
Dr. Laird Shearer, Physics, Lasers, TI	10
Doug Sinclair, senior technician, TI	10
Horace Spain, production engineer, Exxon	10
Dr. Larry Swink, X-ray Structure, TI/AMI	3
Jan Terlouw, vice president, CBC America Optics	6
William Thompson, consultant for optics, AMI	4, 6, 8
Ronald Timm, senior technician, AMI	6
Dale Welt, senior technician, AMI	9
Dub Westfall, senior technician, TI	10
Glen Whaley, vice president, AMI	3, 7–10
Dr. Donald Wierauch, TI 20 Glass Composition	2
Gary Wiese, Optical Design, LMCO	8
Dan Woody, IR&D, LMCO	6

Introduction

The purpose of this book is to describe the technology developed over 50 years to utilize chalcogenide glasses as infrared optical materials. Chalcogenide glasses are based on the chalcogen elements sulfur, selenium, and tellurium excluding oxygen, the first member of the family. The name is a misnomer since chalcogen is from the Greek meaning chalk former and oxygen is the only member of the family that forms chalk. All its compounds are called oxides. Methods used to identify qualitatively chalcogenide glass compositions with promise to become useful infrared optical materials are discussed. Once identified, the optical and related physical properties must be measured quantitatively. The method best suited for the production of homogeneous glass in high purity and quantity must then be developed. Thus, a great deal of effort is required before a glass composition is considered by optical designers ready for use in an infrared system. For this reason, only a few glass compositions have been fully developed and used in quantity over the years.

Infrared light by definition is light with a wavelength greater than the sensitivity region of the human eye, 4000 to 8000 Å. For infrared discussions, the more commonly used terms are 0.4 to 0.8 μm with μm being the abbreviation for micrometers. Of special importance are materials useful for infrared imaging systems designed to respond to infrared energy transmitted through the atmosphere. Figure I.1 illustrates infrared light absorption in the air at sea level due to water vapor, carbon dioxide gas, and ozone. The bottom illustration is the resultant total for the three gas molecules. Notice there are two windows indicated where energy is transmitted well, from 3 to 5 μm (hot window) and about 7 to 14 μm (thermal window). The window is called thermal since the peak of emitted radiation from a body at room temperature, about 300 K, occurs in this window. Thermal imaging of a living subject is based on emitted radiation, which is transmitted in this atmospheric window. The hot window refers to the fact that heated objects emit at the shorter wavelengths in this range. Examples might be the tailpipe of a jetplane or a missile exhaust.

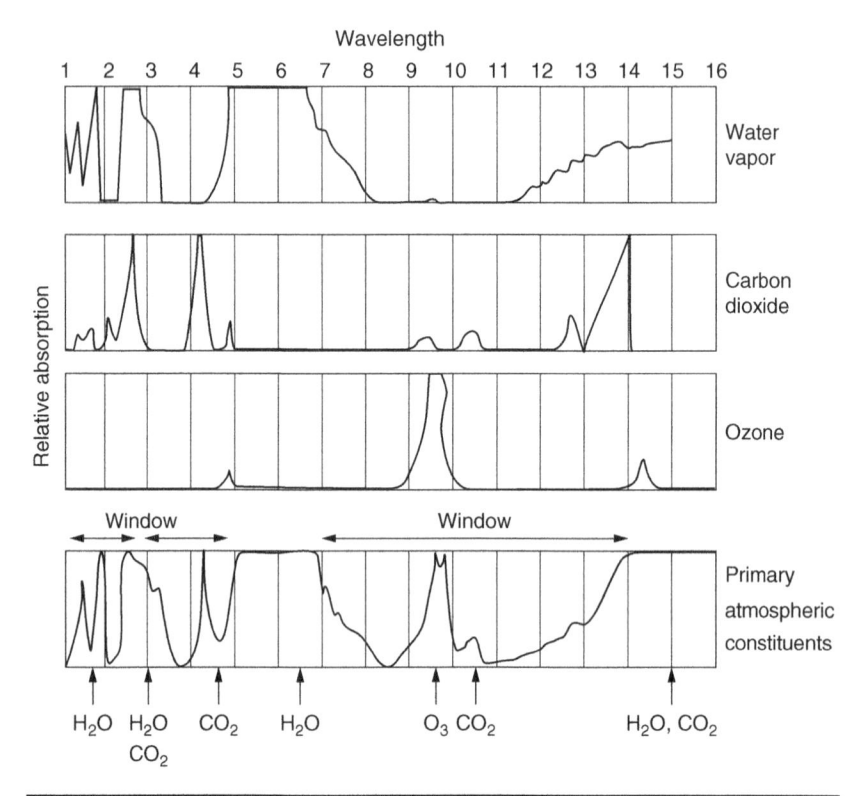

FIGURE I.1 Infrared absorption bands of primary atmospheric constituents for average conditions at sea level.

Over the years, materials used in infrared systems have included alkali halides, alkaline earth halides, melt formed semiconductors, vapor grown fine-grain polycrystalline semiconductors, and chalcogenide glasses. Each of the crystal materials has some advantages and some disadvantages that will be discussed toward the end of this book. However, this book will concentrate chiefly on chalcogenide glasses. After 17 years at Texas Instruments (TI), the author left in 1977 to found Amorphous Materials (AMI), a small company dedicated to producing infrared transmitting glasses for use in infrared optical systems. The company is still active in developing new glass compositions for new applications.

Some crystalline materials were produced at AMI. The production of vacuum float zoned silicon, gallium arsenide, and cadmium telluride, all useful in infrared technology, will be described. Most of the early glass work reported here was carried out at TI in government-sponsored programs as indicated in the references. Discussions of glasses developed at AMI and their applications will be given. Some results of infrared techniques applied to semiconductors at TI will be described. Glasses have a major advantage over crystalline materials in that they can be easily cast, molded, extruded, and drawn into fiber. Such processes generally cannot be applied to crystalline materials but were applied to chalcogenide glasses. Also, the composition of a glass can be changed within limits to enhance properties important to an application. Very little such latitude exists with crystals. The ratios among constituent atoms of most crystalline materials are fixed. Examples of how the composition of some glasses was changed at AMI to enhance a property will be discussed.

Comments to the readers who are students: The author considers himself a physical chemist. Chemistry is an applied science and mostly empirical. Tools used while conducting a research project have changed immensely since 1948 when this author started out, from slide rules and burets to computers, infrared FTIR spectrophotometers, Raman instruments, electron microscopes, and differential thermal analysis (DTA) for glasses. Lasers were not even invented until 1959. There was no material sciences school, only chemistry and physics. Chemical structural theories have changed greatly based on results from the new instruments and techniques. The language of science is constantly changing, reflecting people's increased understanding, which improves their descriptions. Not nature! Nature never changes. Avoid having a preconceived solution to a problem before you start. Let nature guide you through the results of your experiments. Always remember, it is the investigation that is important, not the investigator. It is not important to be right at the start—only at the finish.

Chalcogenide Glasses for Infrared Optics

CHAPTER 1
Transmission of Light by Solids

1.1 Solids

In nature, material exists as gas, liquid, or solid. Gas atoms or molecules are free to move within the confines of their container. Liquids move to fill the shape of their container while solids are rigid in shape. There is about a 1000-fold increase in density in going from gas to the condensed state of liquid or solid. The atoms or molecules come much closer together as a liquid and closer still as a solid. The dense solid may have a precise three-dimensional spatial arrangement for the atoms making up the solid that covers thousands of neighbors in all three directions. When the long-range order is perfect, the solid may be referred to as single-crystal. Or the order may be maintained over limited atomic distances and be referred to as polycrystalline. The atoms or molecules of liquids are free to move within their arrangements continuously in any direction and are said to have no long-range three-dimensional order. Order found is that of nearest neighbors or second nearest neighbors or even more, but not long range in the structural sense. Depending on elemental composition, when the atoms or molecules of a solid come close together, they begin to share their electronic bonding states, which results in formation of an energy structure for the solid. When excited, bonding valence electrons are elevated into a higher conduction band state and are free to travel through the solid as charge if an electric field is applied. The energy difference between the valence state and the free conduction state is called the bandgap of the solid. The vacancies left in the valence band are called holes and can constitute charge flow moving in the opposite direction to the field. The band structure is well developed and precise in crystalline solids with good crystalline perfection. Liquids and amorphous solid glasses are condensed states but without long-range three-dimensional orders. A glass is referred to as a disordered solid. The energy band structure may exist, but the energy level states in the band structure are not nearly as precise as in a crystalline solid.

1.2 Beginning of Transmission of Light—An Electronic Transition

Generally speaking, infrared optical materials are insulators or semiconductors as judged by their bandgaps and resistivity. Photons of light corresponding to energy greater than the bandgap of the solid are strongly absorbed at the surface. As the wavelength is increased and the photon energy decreased below the bandgap, light is transmitted through the solid. The beginning of light transmission of a solid occurs at the wavelength that corresponds to the bandgap energy. The absorption of the photon is a very strong, quantized electronic transition. One may think of this energy as representing the average ionization energy for the primary chemical bonds formed between the atoms that make up the solid. If the required ionization energy is large enough, transmission begins in the ultraviolet region of the spectrum, as in the case for alkali halides or alkaline earth halides. Then the solid appears water-clear or colorless. If it occurs in the visible region, the solid appears colored. If the absorption edge occurs in the infrared region, the solid appears metallic because all visible light is strongly absorbed and reflected.

The use of infrared spectroscopy as an analytical tool to identify and measure concentrations of organic compounds began in the late 1940s. Instruments, crude by today's standards, used salt prisms to disperse the light, salt windows for the instrument, and cells to contain the samples being analyzed. Petroleum refineries used the infrared-based technology for quality control in their laboratories. The bandgaps for both the alkali halides and the alkaline earth halides occur in the ultraviolet region and were not a factor in their infrared use. Most of these ionic solids are soft, weak, and hygroscopic, making them unsuitable for use outside of the laboratory.

The semiconductor revolution began in the early 1950s at Bell Telephone Laboratories when Gordon Teal et al.[1] developed the ability to grow high-purity germanium in single-crystal form. The result was the germanium transistor. Later in the 1950s, Gordon Teal joined Texas Instruments and under his direction accomplished the same feat for silicon, resulting in the silicon transistor. Both germanium and silicon found use as infrared optical materials and as infrared light detectors. Germanium windows and lenses became the optical material standard for the industry due to their wide transmission, 2 to 20 µm, with very little change in refractive index (low dispersion) and good physical properties. Silicon found use as a missile dome material due to its superior physical properties such as strength and hardness. The transmission range was 2 to 14 µm again with little refractive index change and physical properties superior to those of germanium. However, silicon has a lattice absorption at 9 µm in the middle of the most desirable 8- to 14-µm atmospheric window. Generally, the bandgaps of semiconductors decrease with increasing atomic mass as illustrated in Fig. 1.1. The plots show the bandgaps for the group IV elements, the II–VI and III–V compound semiconductors as a function of their molecular weights. A similar relationship may be established for chalcogenide glasses. An average molecular weight for glasses may be calculated by multiplying the percentage of each constituent atom by its atomic mass and adding to get the total.

The application of chalcogenide glasses as infrared materials began in 1950 when R. Frerichs[2,3] rediscovered arsenic trisulfide glass. Good infrared transmission had been reported previously[4] in 1870. W. A. Fraser and J. Jerger[5] continued the development of the glass into a product at Servo Corporation in 1953. Devices were developed

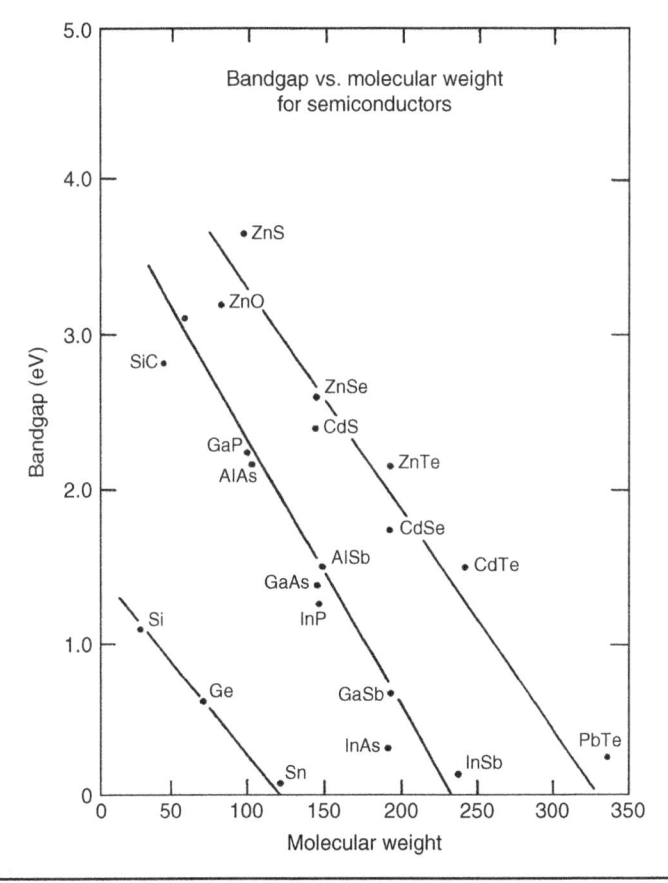

FIGURE 1.1 Bandgaps of IV, II–VI, and III–V crystalline compound semiconductors as a function of their molecular weights.

using arsenic trisulfide glass lenses to sense overheated bearings on railroad cars. In further efforts, many chalcogenide glass compositions were discovered and evaluated at Servo[6] Corporation. Arsenic trisulfide glass is a red glass. Other sulfur glass compositions may show visible light transmission. Most of the glasses based on selenium or containing tellurium are opaque to the visible region and show metallic luster.

As stated before, glasses have energy band structure similar to that of crystalline materials employing the same elements. The chalcogenide glasses are electronic conductors with free electrons and holes as are their crystalline counterparts. However, their disordered nature leads to very low carrier mobility. Glasses based on sulfur and/or selenium are very high in resistivity, almost semi-insulators. Glasses containing tellurium have a more metallic nature and may have fairly low resistivity. For selenium glasses, except those containing sulfur, transmission begins at their absorption edge, which for the most part is 1 µm or more in wavelength. An example of the beginning of transmission for a chalcogenide glass is shown in Fig. 1.2. The glass is an arsenic selenide glass designated Amtir 4 by AMI, and the

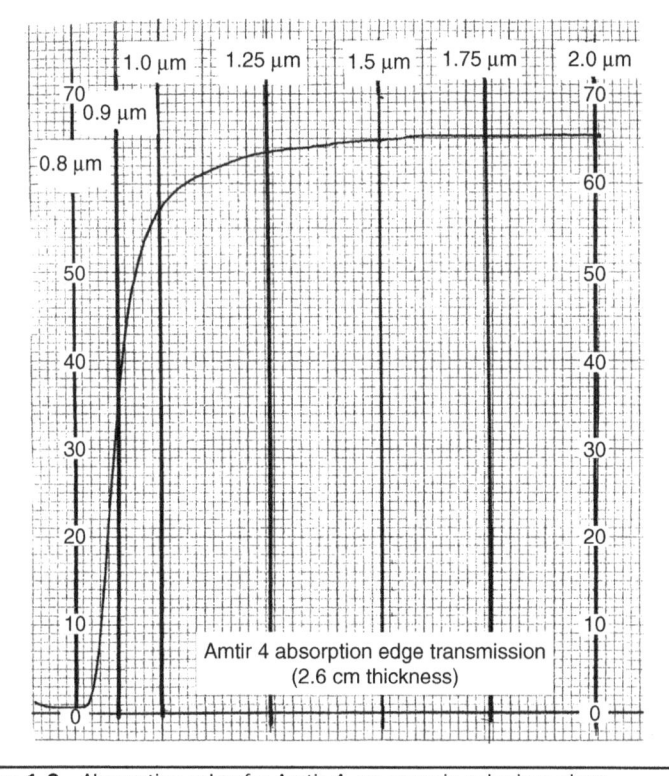

FIGURE 1.2 Absorption edge for Amtir 4, an arsenic-selenium glass.

true transmission is about 65 percent at 1.5 µm, decreasing rapidly to 0 percent at about 0.84 µm. The shape of the curve changes with thickness which in this case is 2.6 cm. A thicker sample would hit zero sooner at a longer wavelength due to internal absorption. The 65 percent transmission would be essentially the same for a thinner piece since the absorption in that wavelength is very slight. The limit of 65 percent is due to Fresnel reflection loss, which is determined by the refractive index of the glass. The optical constants for the glass or any solid are discussed later.

1.3 Long-Wavelength Cutoff

The long-wavelength limit for an infrared optical material is usually determined by a multiphonon lattice band, a combination band, or some vibrational absorption involving constituent atoms of the solid. A qualitative understanding of the factors involved in determining the long-wavelength cutoff for materials may be obtained by considering the expression for the frequency of a simple free diatomic vibration

$$f_o = \frac{1}{2\pi}\left(\frac{k}{\mu}\right)^{1/2}$$

where f_o = fundamental frequency for vibration between atoms of elements A and B
k = force constant for chemical bond between atoms A and B
μ = reduced mass for elements A and B, from

$$\frac{1}{\mu} = \frac{1}{ma} + \frac{1}{mb}$$

where ma and mb are the atomic masses of elements A and B

When infrared energy with a wavelength corresponding to the frequency of vibration is absorbed by the molecular pair, the pair is raised to a higher vibrational energy level. Energy is increased in the solid. The absorption strength depends upon the ionic character between atoms A and B. If the atoms are the same, purely covalent, as in silicon-silicon or germanium-germanium, the absorption is weak or nonexistent. A purely covalent bond means the negative and positive centers of charge between the atoms coincide—there is no separation. Linus Pauling[7] developed the concept of electronegativity values for each atom. If the atoms are different, in a Pauling electronegativity sense, there is some separation of charge between the atoms, some ionic character. Separation of charge constitutes an electric moment or dipole in the chemical bond. The dipole couples with the electric field of the infrared light, allowing energy to be transferred from the light to the molecule. For crystalline solids, the absorption may be very intense which leads to the presence of a strong infrared reflection peaks often referred to as a *Reststrahlen* band. Examples will be shown in later discussions for ionic solids, crystalline semiconductors.

In general, one may say that good physical properties are the result of strong chemical bonds formed between atoms of low atomic mass. The combination of strong bonds (large k) and small atomic mass (small μ) leads to high vibrational frequencies that fall in the wavelength region of interest. Oxide materials are not useful in the longer infrared wavelengths for this reason. Generally speaking, it will become apparent that infrared optical materials transparent out to 14 µm do not have good physical properties.

Fair predictions may be made concerning the vibrational frequencies for crystalline compounds. Figure 1.3 shows a plot of the transverse optical mode wavelengths for III–V compound semiconductors, group IV elemental solids, II–VI crystalline semiconductors, and alkali halides[8] as a function of molecular weight. Figure 1.4 shows the fact that a multiphonon process, a multiple of the fundamental, transmission to 14 µm requires a solid made from heavy atoms. A good

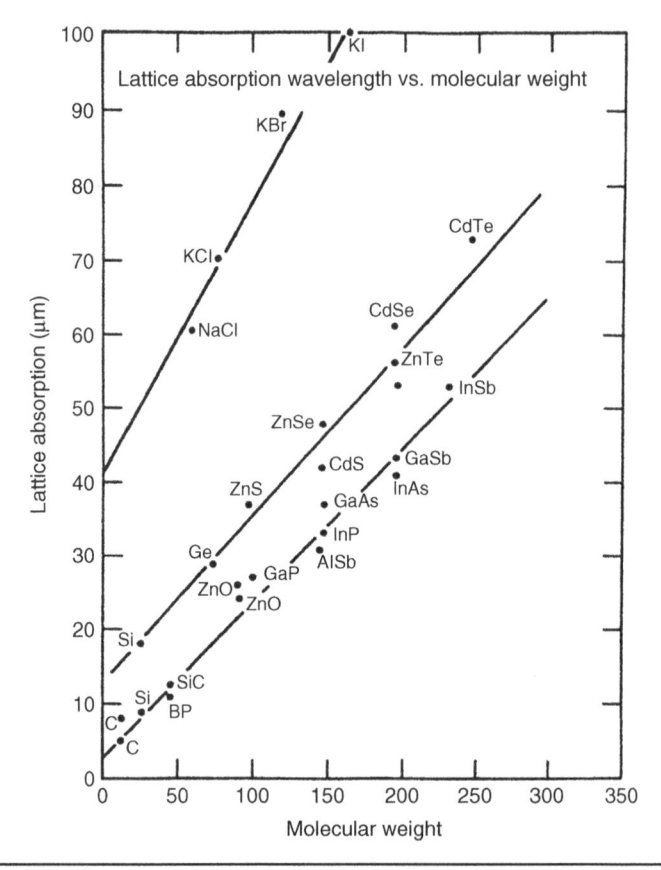

FIGURE 1.3 The wavelength corresponding to the transverse optical mode frequency for group IV semiconductors, II–VI and III–V compound semiconductors as a function of their molecular weights.

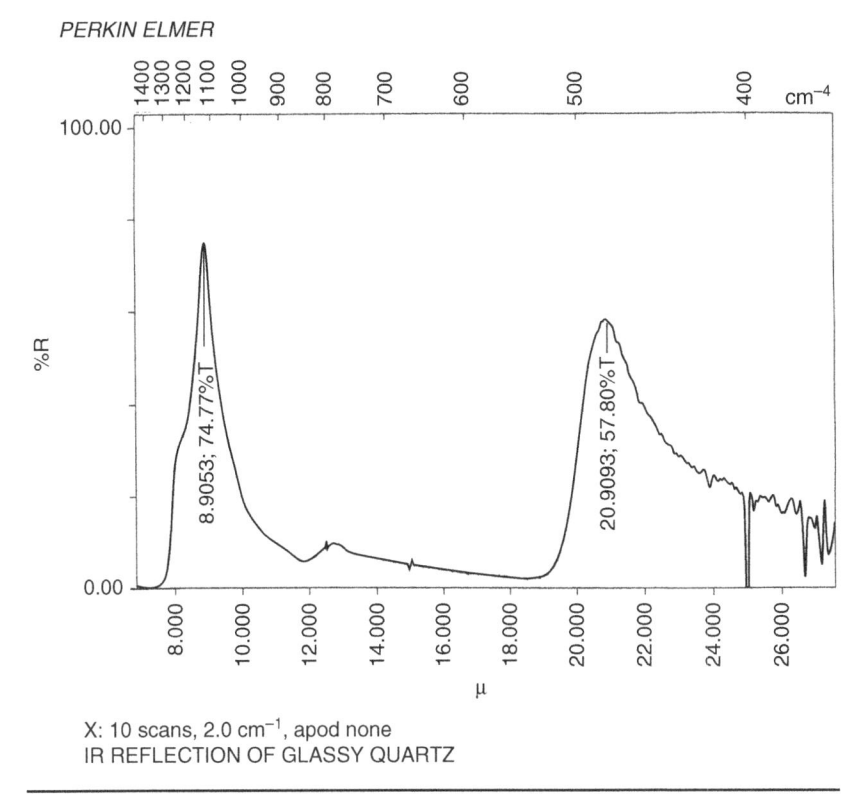

FIGURE 1.4 Far infrared reflection spectra of glassy quartz.

rule of thumb to follow[9] is that the cutoff occurs at twice the frequency of the highest longitudinal optical mode.

Predictions of lattice mode frequencies can be made to a fair degree of accuracy using a number of empirical methods. For example, the lattice structure for a crystal compound AB can be predicted from Pauling electronegativity differences for the binary AB and the principal quantum numbers of their bonding electrons. These concepts were first proposed by Mooser and Pearson.[10] The concepts were enlarged by Parthe.[11] A lattice mode treatment using empirical force constant tables,[12] of which there are many, should lead to a fairly accurate prediction of the long-wavelength cutoff for a hypothetical solid.

The spectroscopic selection rules for active vibrations in crystalline solids rely on symmetry of the crystal cell. A crystal has long-range order in the spatial arrangement of the atoms relative to one another. As a consequence, not all vibrational modes are active due to symmetry considerations. For glasses, the structure is molecular with no long-range order. There is no symmetry. All modes are active. It should be mentioned that the elemental semiconductors germanium and silicon have zero Pauling electronegativity differences, so their fundamental lattice modes are not spectroscopically active. In both materials, weak absorption by higher-order lattice vibration modes is observed.

An example of far infrared Reststrahlen-type reflection bands in glasses is shown in Fig. 1.4. The infrared reflection for glassy quartz is measured using the AMI Perkin Elmer FTIR spectrophotometer. Note the very strong band at about 20 µm followed by another strong band at 9 µm, about one-half the wavelength of the other. Note the reflection peaks are 75 and 58 percent, really quite strong. Keep in mind that the degree of ionic character in the silicon-oxygen bond is considerable in comparison to those of the selenium-based covalent glasses. The second band stops the infrared transmission of glassy quartz, although it had already been stopped by the inpurity of water, which absorbs strongly at 2.9 µm.

As mentioned earlier, infrared materials transmitting to 14 µm are formed from heavier elements and do not generally have good physical properties. One physical property important for producing lenses from optical materials is surface hardness. Figure 1.5 shows a

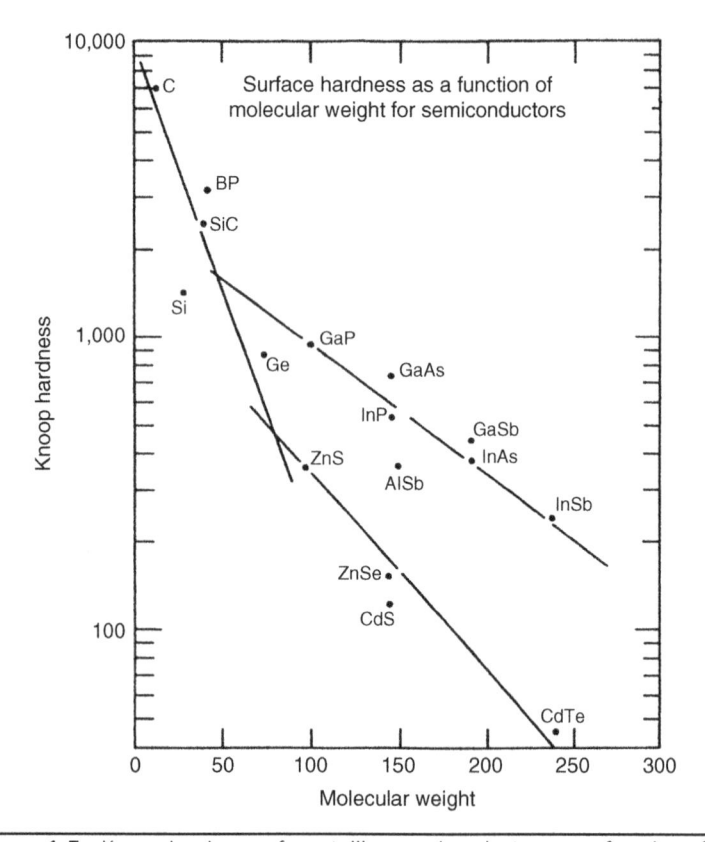

Figure 1.5 Knoop hardness of crystalline semiconductors as a function of their molecular weights.

plot of Knoop surface hardness for crystalline semiconductors as a function of molecular weight. Some of these materials are useful as infrared (IR) optical materials. Note the low Knoop hardness values with increasing mass.

To summarize, the transmission range of a solid is determined by the bandgap of the material on the short-wavelength side and a lattice-type absorption band involving constituent atoms on the long-wavelength side. Both quantities are qualitatively predictable from the location of the elements in the periodic table, as will be discussed later. An example of Reststrahlen-like reflection bands for some chalcogenide glasses is shown in Fig. 1.6.

The percent reflections for As_2S_3 glass, Ge_2S_3 glass, a Si-As-Te glass, and a Ge-P-Te glass, all measured in the far infrared using the Perkin Elmer 301 infrared spectrophotometer at TI, are shown. Note the percent reflection for these glasses is much smaller than that shown by glassy quartz. The fundamental frequencies for the As—S bond, the Ge—S bond, the Si—Te bond and the Ge—Te bond in the glasses are all near the peak of their respective reflections. Their long wavelength cutoff then is about one-half the wavelength of their peaks.

Figure 1.7 depicts the transmission range of glasses based on sulfur, selenium, or tellurium. We see the sulfur-based glasses showing some visible light transmission but cutting off after 10 μm. Notice transmission rates for sulfur-based glasses are the highest because the Fresnel loss is less due to their lower refractive index.

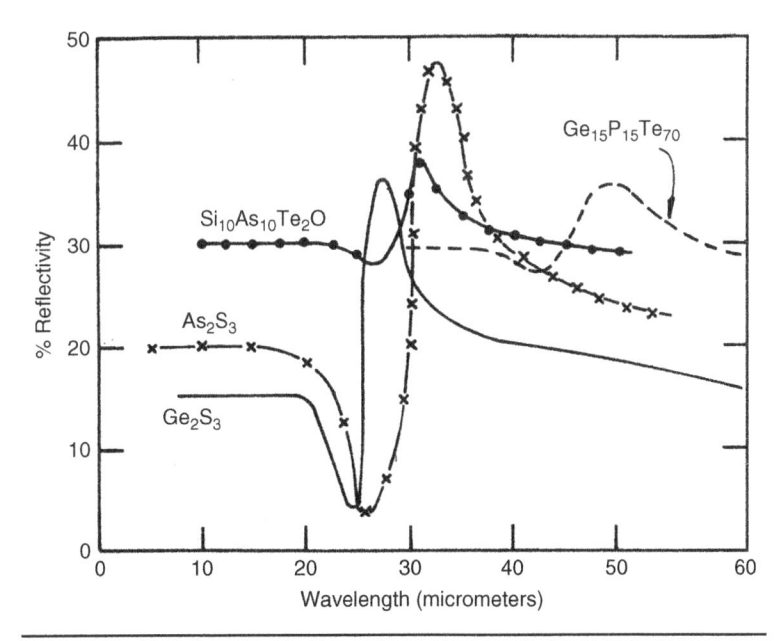

FIGURE 1.6 Far infrared reflection spectra of some chalcogenide glasses.

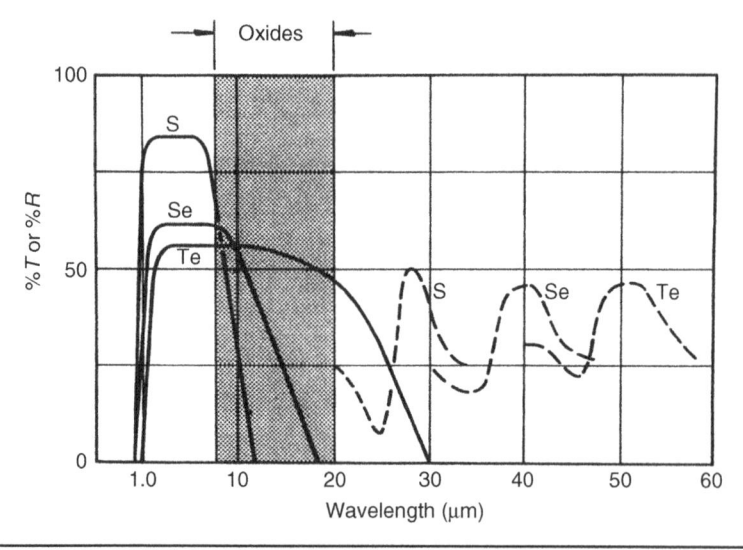

FIGURE 1.7 Pictorial representation of the transmission range for glasses based on sulfur, selenium, or tellurium.

The selenium-based glasses start transmitting at about 1 μm and start cutting off after about 12 μm. The tellurium glasses start transmitting at about 2 μm and cutting off about 20 to 30 μm. Tellurium glasses have the highest index and are the hardest to make without crystallizing. This depiction is for only one chalcogen in the composition. Mixed chalcogen glasses such as sulfur-selenium or selenium-tellurium will be somewhat different with regard to transmission, index, and tendency to crystallize.

1.4 Extrinsic Loss within the Band, Impurities, Scatter, and Quality

Electronic, vibrational, or physical defects related to purity or method of preparation may affect the performance of a material within the transmission range. Thus, in a general sense, the effects are considered extrinsic, not intrinsic, to the solid. Impurity atoms in crystalline semiconductors may be electrically active in the host material, leading to charge carriers in excess of intrinsic levels. The free carriers with high mobility classically in semiconductors may absorb infrared radiation in proportion to the infrared wavelength squared.[13] Inclusion of scattering terms may lead to a dependence greater than wavelength squared. Intervalence band transitions may produce absorption bands in P-type materials such as germanium and gallium arsenide. Examples of such effects are described in standard texts such as the one by Moss.[14] The solution for low carrier mobility materials such as found in chalcogenide glasses leads to an expression independent of wavelength.[12] Free carrier absorption for melt-formed chalcogenide glasses used for optical applications has not been observed but has been reported[15] for highly conducting chalcogenide glasses containing tellurium used in electronic devices.

The presence of impurity atoms such as oxygen, water, and carbon bonded to constituent atoms leads to localized molecular absorption bands. Classic examples are the 9-μm absorption in silicon due to the Si—O bond,[16,17] the 11.6-μm absorption in germanium due to the Ge—O[17] bond, and a Si—C absorption for C_{12} in silicon at 16.5 μm.[18] It is interesting to note the fundamental for pure Si—C occurs at 12.6 μm.[19] Molecules such as water (H_2O), hydrogen sulfide (H_2S), or hydrogen selenide (H_2Se) occur often in sulfur or selenium glasses. The impurity molecules couple (in a weak bond) to the positive element in the glass composition. An arsenic-selenide glass has only one absorption band due to H_2Se at about 4.6 μm while an arsenic-germanium-selenium glass has two absorption bands, one for arsenic at 4.5 μm and one for germanium at 4.9 μm. A sulfur glass will have an absorption for water at 2.9 μm and one for hydrogen sulfide at 4 μm. Silicon-oxygen (Si—O) absorption will occur at about 9.5 μm in glasses if present as an impurity. Hydrocarbons present during the compounding process for selenium glasses lead to the formation of H_2Se due to the reaction of liquid selenium with a hydrocarbon.

A laboratory method used to generate small amounts of H_2Se is to melt paraffin and selenium together. Localized absorption due to low-level impurities in crystals or glasses produces narrow sharp bands useful for determining the impurity concentration. Crystal materials are often grown from a melt in near-perfect lattice form. There is little chance of bubbles or particulate inclusions. Vapor-grown polycrystalline materials may contain particulate matter from the gases or vapors used in the growth. Glasses are formed from melts as well. The melts may be in containers opened to the atmosphere or sealed in evacuated chambers. Glasses are usually mixed by some form of agitation of the melt to ensure complete reaction of the components and to form a homogeneous melt from which the solid forms. Bubbles may form during the process. Often, at the end of the compounding mixing process, agitation will be stopped and a period of time is allowed for the bubbles to rise to the surface and be eliminated. Bubbles act as hollow spheres in the glass, scattering the light. Most optical glass specifications allow bubbles with no larger than 0.002-in diameter and only a few. Inert particulate matter may also be present in the glass due to contaminate in the reactants. Such particles are small and absorb light. Some glass compositions produce crystallites during processing. Crystallites may be distinguished from particles on examination because they generally transmit light and crystalline facets. Crystallites, as bubbles, cause transmission loss.

There is usually a slight difference in the refractive index of the crystallite from the glass from which it formed. Because of this difference, light is reflected (scattered) at each surface interface and is a transmission loss. The scattering particles may range from the microscopic to the macroscopic. The intensity of light scattered at right angles to the incident light depends upon the particle size relative to the wavelength of light and upon the difference between the particle and the surrounding medium.[15] Such would be the case for large-grain polycrystalline materials. As particles become very small relative to the wavelength, the intensity drops off dramatically, as the inverse of the wavelength to the fourth power.

Another source of loss of transmitted light and optical distortion is striae in glass. During the glass process, local variations in composition or density produce regions where the refractive index is different from that of the whole. The beam can be diverted in direction in these local regions, harming the integrity of the transmitted image or reducing the intensity of light reaching the detector plane. The supplier must develop a reliable process to supply high-purity, homogeneous glass, free of bubbles, particles, striae, and crystallites.

1.5 Optical Constants and Dispersion due to Strong Absorption

If a beam of light in air incident on a flat surface at an angle of O_1 from the normal and the refracted beam is O_2 from the normal, the refractive index may be calculated[20] from $N_1 \sin O_1 = N_2 \sin O_2$. Since N_1 for air is ~1.000,

$$N_2 = \frac{\sin O_1}{\sin O_2}$$

The refractive index of an optical material is not a constant, and how it varies with wavelength is perhaps the most important parameter for its use by an optical designer. Already we have seen that Fresnel reflection losses decrease transmission for a material. The Fresnel reflection coefficient R can be calculated[20] from the expression

$$R = \frac{(N-1)^2}{(N+1)^2}$$

Again, the value of R is not a constant since N changes with wavelength. The greater the value of N, the higher the value of R. Thus far, we have accounted for the Fresnel reflection losses only from transmission. Losses due to absorption must be measured and calculated from the expression

$$T = \frac{(1-R)^2 e^{-\alpha x}}{1 - R^2 e^{-2\alpha x}}$$

where T = transmission I/I_0
 R = reflectivity
 α = bulk absorption coefficient, cm^{-1}
 X = sample thickness, cm

Optical designers desire index values to five numbers for imaging systems. How these values are obtained will be discussed in a later section. Absorption values may be calculated from measured transmission by using the sample thickness and precise index values. The full transmission expression may be programmed at wavelength points with thickness and transmission as variables. Another method often used incorporates two samples of different thickness with the change in transmission used in the solution for the absorption value.

All discussions thus far pertain to the transparent region where absorption is low. In this region the index is taken as a simple number. However, the index and the absorption coefficient are interdependent. That is, the index is really a complex number and should be written[20] as

$$N = n - ik$$

where n = real part of index
 k = imaginary part, the absorption coefficient $\alpha = 4\pi k/\lambda$
 α = bulk absorption coefficient, cm^{-1}
 λ = wavelength, cm

In the transparent region, the value of k is so small it can be ignored. However, in strong absorption wavelength (big k) regions such as at the electronic absorption edge or where lattice-type absorption between constituent atoms occurs, the real part of the index changes dramatically as k increases. The optical constants are interdependent. The effect of these two regions, one on each end of the transmission region, carries over into the transparent region as a major factor in the dispersion or change in index with wavelength. The other major factor is change in temperature. As a solid expands or contracts, the number of atoms per cubic centimeters changes. The density changes. The index reflects the mass of the atoms in the solid and the number of atoms per cubic centimeter.

The variation of the refractive index with wavelength in the transparent range may be depicted with the use of two dispersion curves, one at each end, shown in Fig. 1.8. The first depicts the change in reflectivity that occurs at the electronic absorption edge for a solid. At the short-wavelength side the reflectivity first falls as the absorption k increases. Then reflectivity increases as k rises to a maximum. On the long-wavelength side of the peak, k falls to a negligible value, and the reflectivity falls and levels out at a value slightly greater than before. Thus, the refractive index has increased. The real part of the index n first falls as k increases and then as k declines returns to a value greater

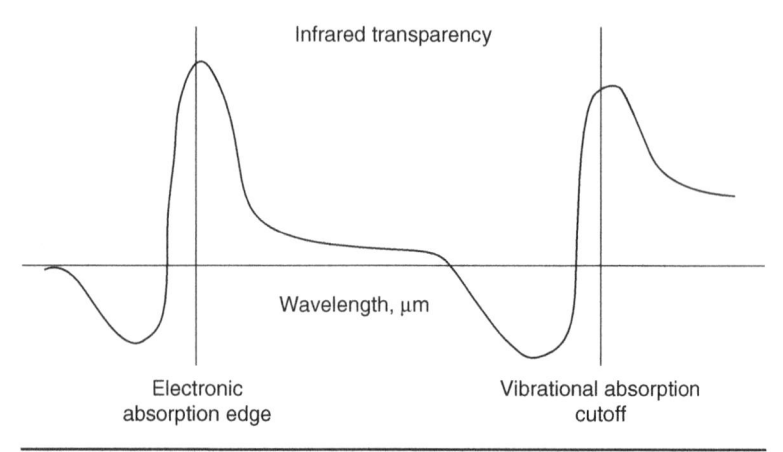

FIGURE 1.8 Two dispersion curves separated in wavelength depicting the change in refractive index due to the strong absorption at the electronic edge and near the lattice-type absorption. Their interactions affect the refractive index in the transparent region.

than before the influence of strong k. On the long-wavelength low-energy end, we have a dispersion curve for the lattice-type absorption between constituent atoms. As before, in the short-wavelength side of the strong absorption, the reflectivity and index fall below level, rebound to a peak close to the maximum absorption (k), and finally declines to a level above the previous level as k becomes negligible.

The index increases on the long-wavelength side of the electronic edge absorption and decreases in the short-wavelength side of the cutoff absorption due to constituent atoms. The overall effect in the transparent region depends upon how close in wavelength the two effects are to each other. When the solid is heated, both absorptions broaden and shift in wavelength, contributing to a thermal change in index. The other factor is volume expansion. For materials like the alkali halides, the absorption edge and lattice cutoff are widely separated in wavelength. Hence, the volume expansion prevails, and there are negative changes in index with temperature as the number of atoms per cubic centimeter declines.

References

1. G. K. Teal, M. Sparks, and E. Buehler, *Phys. Rev.* 81, 637 (1951).
2. R. Frerichs, *Phys. Rev.* 78, 643 (1950).
3. Rudolph Frerichs, *J. Opt. Soc. Am.* 43, 1153 (1953).
4. C. Schultz-Sellack, *Ann. Physik* 139, 162 (1870).
5. W. A. Fraser and J. Jerger, *J. Opt. Soc. Am.* 43, 332 (1953).
6. J. Jerger, Jr., and R. Sherwood, "Investigate the Properties of Glasses Transmitting in the 3 to 5 and 8 to 14 Micron Windows." Servo Corp., Hicksville. L.I.N.Y. Final Tech. Report, Contract No. Nonr 4212(00), August 1964.
7. Linus Pauling, *The Nature of the Chemical Bond*, Cornell University Press, Ithica, N.Y., 1948.

8. E. Burstein, "The Intrinsic Infrared and Lattice Vibrational Spectra of Cubic Diatomic Crystal," *Lattice Dynamics, Proceeding of the International Conference,* Copenhagen, 1963.
9. William B. White, "Infrared Transmitting Materials," Report No.1, Contract No. 14-67-0385-005, Office of Naval Research, December 1968.
10. E. Mooser and W. B. Pearson, *Acta Crystallogr.* 12, 1015 (1959).
11. E. Parthe, *Chemical Structure of Tetrahedral Structures,* Gordon and Breach Science Publishers, New York, 1964.
12. G. R. Somajula, *J. Chem. Phys.* 28, 814 (1958).
13. See standard SC texts such as *Semiconductors* by R. A. Smith, Cambridge, MA, 1959.
14. T. S. Moss, *Optical Properties of Semiconductors,* Academic Press, New York, 1959.
15. D. L. Mitchell, S. G. Bishop, and P. C. Taylor, *J. Non. Cryst. Sol.* 8–10, 231 (1972).
16. W. Kaiser and P. H. Keck, *J. Appl. Phys.* 28, 882 (1957).
17. W. Kaiser, P. H. Keck, and C. F. Lang, *Phys. Rev.* 101, 1264 (1956).
18. R. C. Newman and J. B. Willis, *J. Phys. Chem. Solids* 26, 373 (1965).
19. W. G. Spitzer, D. Kleinman, and D. Walch, *Phys. Rev.* 113, 127 (1959).
20. R. W. Ditchburn, *Light,* Dover Publications, Mineola, N.Y., 1991.

CHAPTER 2
Chalcogenide Glasses

2.1 Historical Development

The investigation of chalcogenide glasses as optical materials in infrared systems began with the rediscovery of arsenic trisulfide glass[1,2] by R. Frerichs in 1950. Good transmission for arsenic trisulfide had been reported in 1870.[3] Development of the glass as a practical optical material was continued by W. A. Fraser and J. Jerger[4] in 1953 at Servo Corporation. During the 1950–1970 period, the glass was made in ton quantities by American Optical and Servo in the United States and by Barr and Stroud in the United Kingdom along with several others in Europe. The glass was used for commercial devices. As an example, devices that detected overheated bearings on railroad cars were made and marketed by Servo Corporation. Hot objects could be detected at this time by radiation transmitted through the 3- to 5-μm atmospheric window where arsenic trisulfide glass was transparent. However, the need for other chalcogenide glass compositions capable of transmitting longer wavelengths arose with the concept of passive thermal optical systems.

Jerger, Billian, and Sherwood[5-7] extended their investigation of arsenic glasses containing selenium and tellurium, and later adding germanium as a third constituent. The goal was to use chalcogen elements heavier than sulfur to extend long-wavelength transmission to cover the 8- to 12-μm window and at the same time improve physical properties. In parallel, Russian work at the Ioffee Institute in Leningrad under the direction of Boris Kolomiets was reported in 1959.[8] Work along the same line was begun in the United Kingdom by Nielsen and Savage[9-11] as well. The Royal Radar work led to limited production of chalcogenide glass at the British Drug House Laboratories. The U.K. results are summarized in a recent book by Savage.[12]

Work at Texas Instruments (TI) began as an outgrowth of the thermoelectric materials program. The glass forming region for the silicon-arsenic-tellurium system was mapped out by Hilton and Brau.[13] This development led to an exploratory DARPA-ONR program at TI, funded from 1962 to 1965.[14] The ultimate goal of the program was to find infrared transmitting chalcogenide glasses with physical properties comparable to those of oxide optical glasses and a softening point of 500°C. An attempt was made to achieve this goal by incorporating transition element titanium in the composition.[15] Further work[16] was funded on sulfur-based glasses in 1973 to 1974. The exploratory programs resulted in eight chalcogenide glass U.S. patents. Three papers were published in an international journal detailing the results.[17] As a result of publishing the three papers, the author was invited to present a paper at the Fourth All-Union Symposium on the Vitreous Chalcogenide Semiconductors held in May 1967 in Leningrad (now called St. Petersburg). While at the meeting, the author met Valentina Kokorina, head of the Russian chalcogenide glass production facility. Valentina Kokorina, now retired, recently published a book detailing the Russian effort.[18] Representing the West also was Prof. Douglas McKenzie of the University of California, Los Angeles, a well-known glass scientist. From the University of Edinburgh in Scotland was Prof. Alan Owen, head of the Electrical Engineering Department. Another from the West at the meeting was Stanley Ovshinsky, founder of Electron Energy Conversion Devices in Troy, Michigan, who presented his first paper advocating switching devices based on the electronic properties of chalcogenide glasses. Similar work in the United States was reported earlier by A. David Pearson[19] of Bell Telephone Laboratories. Thus started a great worldwide effort to investigate chalcogenide glasses and their electronic properties. The purpose was to pursue a new family of inexpensive electronic devices based on amorphous semiconductors. The effort in this field far exceeded the effort directed toward optical applications. Some of the results of the efforts in the United States were reported in a symposium in 1969[20] and in another symposium in 1971.[21] Generally, the glasses used could be made conducting with applied voltage resulting in the formation of a crystalline filament path. Such glasses were not suitable for optical use and are not discussed in this book. Our discussion concentrates on efforts in the United States to develop infrared optical materials, specifically, chalcogenide glasses from groups IVA, VA, and VIA from the periodic table.

In 1966, the Air Force decided to fund a Research Program on Infrared Optical Materials at Central Research Laboratories at TI. There was a six-month delay in funding. During this time, Robert Patterson mapped out the glass forming region for the germanium-antimony-selenium system. A U.S. patent[22] was granted covering the best composition selected, TI 1173.

When the author returned to the United States in 1967 and reported the existence of a chalcogenide glass production facility in Russia the Air Force decided to fund a Development Program at TI with a goal of establishing a similar glass production facility. Funding for this purpose by the Air Force covered the years 1967 through 1974.

Also in 1967, a Materials Advisory Panel[23] was formed headed by Norbert Kreidl, a noted glass scientist and former director of research at Bausch and Lomb. The author was a member of the panel. One of the recommendations of the panel was that a program be funded to develop zinc selenide as an infrared optical material. The Air Force decided to fund a Raytheon program headed by Jim Pappis.[24] The development at Raytheon of the chemical vapor deposition method to produce plates of small-grain zinc selenide and zinc sulfide has been a very important advancement in the production of infrared optical elements.

In 1967, the Air Force Production Development Program at TI was transferred to the Semiconductor Production Division under the direction of Charlie Jones, a colleague of the author from the Central Research Laboratory. Joining the effort was Harold Hafner who had just joined TI. His background was in glass science as he had served as deputy research director at Bausch and Lomb under Norbert Kreidl. Hafner made many important contributions including a glass casting process and a glass tempering process. The program concentrated the efforts on a glass from the germanium-arsenic-selenium system. TI results[25] agreed with the conclusions of the Russian, U.K., and Servo efforts that the germanium-arsenic-selenium system produced the best glasses for infrared system applications. Don Weirauch at TI conducted a crystallization study on the germanium-arsenic-selenium family of glasses and identified a composition in which crystallites would not form. The composition was fully characterized and produced at Texas Instruments as TI 20 glass. No patent could be granted for the glass since the glass system was reported in several places in the literature. One of the accomplishments of the program was to increase the rupture modulus of TI 20 from 3000 to 6000 psi by tempering. In 1972, a window of TI 20 glass was cast 12 in × 24 in × 0.5 in, polished flat and parallel and antireflection-coated.[26] The window was installed in a U.S. Navy F4 in front of its FLIR system. The plane on returning to Florida flew into a hailstorm, which shattered the window and damaged the FLIR. Thus started the rain erosion testing of infrared optical materials at the facility at Wright-Patterson Air Force Base.

Attempts were made at TI to develop a business selling infrared glass outside the company. Brochures and advertising were employed. Little in sales resulted. The program had reached a production stage and was moved to the Electro-Optics Division, the user division. Production of TI 20 was stopped as well as sales of chalcogenide glasses outside the company.

In 1972–1973, TI had to produce the glass in quantity and quality for the Navy P3 system. Their yields were so low that glass blanks began to cost more than the system. In 1974, the author left TI Central

Research, moved to the TI EO Division, and took over the glass production program. His efforts succeeded and yields improved, leading to decreased cost. Standard production became 7-kg plates 12 in × 12 in of TI 1173.

In the late 1960s and early 1970s, passive 8- to 12-μm systems began to be produced in small numbers, mostly for the Air Force or Navy. The numbers were in the hundreds, corresponding to the number of airplanes. Notably was the system for the B52 made by Hughes Aircraft. The potential numbers reached thousands when the customer became the Army. The Army embraced a *common module* approach whereby different companies could all build from the same design. There would be three programs: the ANTAS for the Infantry, the Tank Thermal Sight, and TADS PNVS for the helicopter. All companies could compete, and the winner's design would be adopted as the common module. Since the wavelength range was wide, color correction would be necessary which meant that a second optical material would be needed to go along with germanium, by then the industry standard. The TI design used TI 1173 with germanium. TI won the first two programs. The Army became aware of the fact that TI was the only source of the glass used in their common modules. The third program, TADS PNVS, had not been decided. It became obvious to the author that a second source of glass would be needed. Early in 1977, the author gave notice to TI that he planned to leave the company. He was called in and told by TI that he would be allowed to produce the glass. The Army was told that there would soon be a second source. However, the Army awarded the TADS PNVS contract to Martin Marietta in 1977. Thus, for two of the programs the Army was committed to a design that used a patented glass as the second material, not available to competitors. The alternate glass, TI 20, was no longer in production at TI. Clearly a second source of glass was needed to ensure the success of the Army common module approach.

The author left TI in 1977 and in May founded Amorphous Materials (AMI). Very soon he found TI had changed its mind and would not license Amorphous Materials to produce TI 1173 glass. The decision was made to produce the TI 20 germanium-arsenic-selenium glass at AMI because its physical and optical properties had been fully studied and published in a government report.[25] Besides, TI 20 was a better glass than TI 1173. The glass was not and could not be patented. Next, AMI had to convince the Army that producing this glass would save the concept of the common module approach. It took over a year to persuade the Army to help AMI. The effort was supported through a three-year Manufacturing Methods and Technology Program funded by the Army through the Night Vision Laboratory at Ft. Belvoir. Amorphous Materials renamed the glass composition Amtir 1 and went on to supply Magnavox, Kollsman Instruments,

Hughes, Martin Marietta, and Westinghouse in several programs. Amtir 1 is still the major FLIR glass produced by AMI. Standard production is 9-kg plates 8-in diameter. AMI has produced 35 tons of Amtir 1 glass from 1978 to 2007.

Later, in 1991 after the TI 1173 patent expired, the Night Vision Laboratory provided a letter contract to qualify AMI as a second source of the glass which AMI renamed Amtir 3. It is interesting to note that in the United States during the period from 1950 to the present, only three glass compositions have been produced in ton quantities: arsenic trisulfide, TI 1173 (Amtir 3), and TI 20 (Amtir 1). No other widely used new compositions have emerged. The reason in part is due to the considerable effort required to identify, produce, and characterize a new glass composition to the state that optical designers, system designers, and corporate management are willing to use it in a new system. Even if a new, better glass emerged, there would be great reluctance to redesign a system once it is in production.

2.2 The Periodic Table and Glass Formation

Previously, it was pointed out that efforts to find and develop chalcogenide glasses for infrared systems were most successful using elements from the IVA, VA, and VIA groups of the periodic table. This statement is to point out that the periodic table is not an inexhaustible supply of elemental combinations that should be investigated. The three groups named have fueled investigations of many systems: binary, ternary, or those containing even more elements. A review of other materials transparent in the infrared but from different elemental families may help to explain why some elements are favored more than others. Figure 2.1 presents a shortened version of the periodic table of the elements. Outlined are the families of the elements from which infrared optical materials are formed.

As indicated in the chart, the alkali halides form from the IA alkali metal elements Li, Na, K, Rb, and Cs in combination with the group VIIA halogens F, Cl, Br, and I. The alkaline earth halides form from the IIA alkaline earth metal elements Be, Mg, Ca, Sr, and Ba in combination with the VIIA halogen elements F, Cl, Br, and I. Notice also in Fig. 2.1 that a change in Pauling electronegativity[27] is indicated as one moves up or down in the chart or from left to right. On the left, the value decreases going from lighter alkali and alkaline earth elements to heavier. Thus, Cs and Ba have the lowest values for the two families. Conversely, Pauling electronegativity is indicated to be increasing with atomic number going across the chart from IA elements to VIIA elements. At the same time on the right of the chart, Pauling electronegativity increases going up the chart from the heavier halogens to the lighter elements. Fluorine in the top right corner has the highest value, 4.0, one unit higher than its next row element Cl.

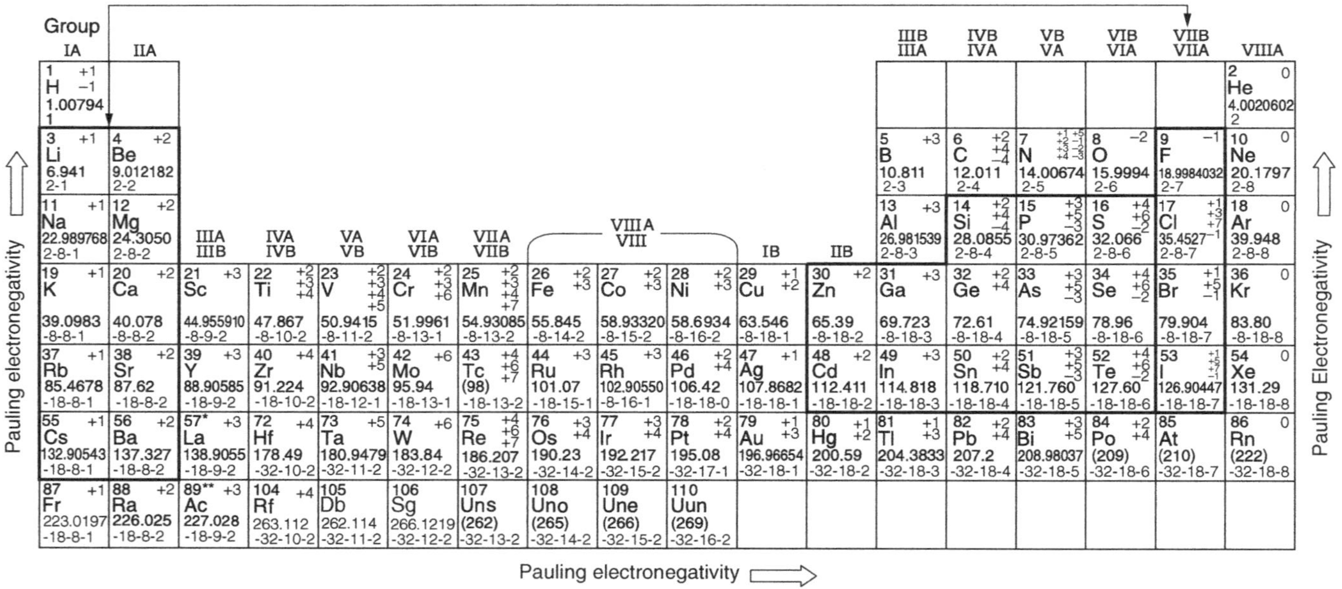

Alkali halides, IA elements Li, Na, K, Rb, Cs with VIIA elements F, Cl, Br, I

Alkaline earth halides, IIA elements Be, Mg, Ca, Sr, Ba with VIIA elements F, Cl, Br, I

II–VI crystalline compounds, IIB elements Zn and Cd with VIA elements S, Se, Te

Elemental crystalline semiconductors, IVA Si, Ge, Sn

Compound crystalline semiconductors, IIIA elements Ga, and In, with VA elements P, As, Sb

Chalcogenide glasses, IVA elements Si, Ge, Sn along with VA elements P, As, Sb combined with VIA elements S, Se, Te

FIGURE 2.1 Designation of the elements from which infrared optical materials are formed.

Chalcogenide Glasses

A purely covalent bond, such as exists in amorphous selenium or crystalline silicon, has an even distribution of the bonding electrons between the two atoms, zero percent ionic character. The positive and negative centers for the atom pair coincide midway between. A large value of electronegativity indicates a negative element which tends to attract and hold the bonding electrons closer and away from the positive element. A low electronegativity indicates a positive element which tends to furnish the bonding electrons to the other atom. The positive and negative centers for the atom pair do not coincide. The bond has ionic character. Pauling electronegativity[27] values for the important elemental families already mentioned are listed below:

Pauling Electronegativity Elemental Values					
IA	IIA	IVA	VA	VIA	VIIA
Li 1.0	Be 1.5	C 2.5	N 3.0	O 3.5	F 4.0
Na 0.9	Mg 1.2	Si 1.8	P 2.1	S 2.5	Cl 3.0
K 0.8	Ca 1.0	Ge 1.8	As 2.0	Se 2.4	Br 2.8
Rb 0.8	Sr 1.0	Sn 1.8	Sb 1.9	Te 2.1	I 2.5
Cs 0.7	Ba 0.9	Pb 1.8	Bi 1.9	Po 2.0	At 2.2

One should notice the values for oxygen and sulfur are a full unit different which is significant in the fact that oxides and other chalcogenides are very different from one another. Oxygen is a gas at room temperature while sulfur, selenium, and tellurium are solids and in amorphous forms made up of chains and rings in a polymerlike structure. Nitrogen and fluorine are also gases. First-row elements are not important to our chalcogenide glass formation discussion. First, we will look at the percent ionic character in the chemical bonds formed between the elements of the alkali halides and the alkaline earth halides. Using X_A and X_B as the electronegativity values for elements A and B, in the following table percent ionic character for the A-B bond is found from $X_A - X_B = \Delta$.

Δ	0.1	0.2	0.3	0.4	0.5	0.6	0.7	0.8	0.9	1.0	1.1	1.2
Percent Ionic	0.5	1	2	4	6	9	12	15	19	22	26	30

Δ	.13	1.4	1.5	1.6	1.7	1.8	1.9	2.0	2.1	2.2	2.3	2.4
Percent Ionic	34	39	43	47	51	55	59	63	67	70	74	76

Δ	2.5	2.6	2.7	2.8	2.9	3.0	3.1	3.2
Percent Ionic	79	82	84	86	88	89	91	92

Applying these numbers to the alkali halides, we find their bonds average more than 50 percent ionic character. For the often used alkali halide NaCl, Δ is 2.1 which corresponds to 67 percent ionic character. The same procedure applied to alkaline earth halides averages to about 50 percent. For the often used alkaline earth halide CaF_2, Δ is 3.0 which corresponds to 89 percent ionic character. Both families are broadband in transmission from the ultraviolet and some into the far infrared. Their refractive indexes are relatively low while their expansion coefficients are large, leading to negative changes in index with temperature. Some can be grown in large single-crystal form but are soft and weak and may cleave. Their melts are not viscous. Most are readily attacked by water. Alkali halides are used mostly in the laboratory. A major exception among the alkaline earth halides for system use is fine-grain polycrystalline magnesium fluoride. The point is that these materials are ionic crystalline solids and as such are not useful in night vision systems used in the field. The same treatment applied to the combinations in the IVA, VA, and VIA elements illustrates the covalent nature of the chalcogenide infrared glasses.

IVA Atom Pairs	Δ	Percent Ionic	VA Atom Pairs	Δ	Percent Ionic
Si 1.8–O 3.5	1.7	51	P 2.1–O 3.5	1.4	39
Si 1.8–S 2.5	0.7	12	P 2.1–S 2.5	0.4	4
Si 1.8–Se 2.4	0.6	9	P 2.1–Se 2.4	0.3	2
Si 1.8–Te 2.1	0.3	2	P 2.1–Te 2.1	0	0
Ge 1.7–O 3.5	18	55	As 2.0–O 3.5	1.5	43
Ge 1.7–S 2.5	0.8	15	As 2.0–S 2.5	0.5	6
Ge 1.7–Se 2.4	0.7	12	As 2.0–Se 2.4	0.4	4
Ge 1.7–Te 2.1	0.4	4	As 2.0–Te 2.1	0.1	0.5
Sn 1.7–O 3.5	1.8	55	Sb 1.9–O 3.5	1.6	47
Sn 1.7–S 2.5	0.8	15	Sb 1.9–S 2.5	0.6	9
Sn 1.7–Se 2.4	0.7	12	Sb 1.9–Se 2.4	0.5	6
Sn 1.7–Te 2.1	0.4	4	Sb 1.9–Te 2.1	0.2	1
Pb 1.9–O 3.5	1.6	47	Bi 1.9–O 3.5	1.6	47
Pb 1.9–S 2.5	0.6	9	Bi 1.9–S 2.5	0.6	9
Pb 1.9–Se 2.4	0.5	6	Bi 1.9–Se 2.4	0.5	6
Pb 1.9–Te 2.1	0.2	1	Bi 1.9–Te 2.1	0.2	1

Excluding oxygen bonds, the IVA elements bonded with VIA elements average 8 percent ionic character, and the VA bonds with VIA elements average only 4 percent. One may ask why ionic character is important in infrared chalcogenide glasses. Covalent bonding has very specific bonding angle requirements to nearest neighbors which resist being changed by force.

Myuller[28] points out that the deformation of covalently bonded substances in the liquid state during viscous flow requires much more energy than the deformation of materials bonded ionically. Thus, covalent melts are viscous. The bonding requirements for ionic and metallic substances are not rigid with respect to bonding angles. Thus, metallic and ionic melts are not viscous and freeze into a solid when cooled to their melting points.

Glass melts are different. Figure 2.2 shows a thermal expansion analyzer[29] (TEA) curve for AMI C1 glass measured using a dilatometer. A sample about 2 in long is heated at a controlled rate, and the change in length is measured and plotted as a function of temperature. Note the slope is typical of the expansion of a solid. As the temperature increases, the slope at some point begins to gradually change, taking on a steeper slope more typical of a liquid. The intersection of the two slope lines is called the glass transition temperature T_g for the glass. For this instrument, T_g is also defined in terms of viscosity, $T_g \sim 10^{13}$ poise. Had this been a crystalline material, the expansion curve would have ended abruptly at the melting point of the solid as it turned into a nonviscous liquid. As the measurement of the glass continues, a point is reached where, under the conditions of the instrument, expansion stops and the sample begins to contract. This point is called the dilatometric softening point T_d of the glass. For the conditions of

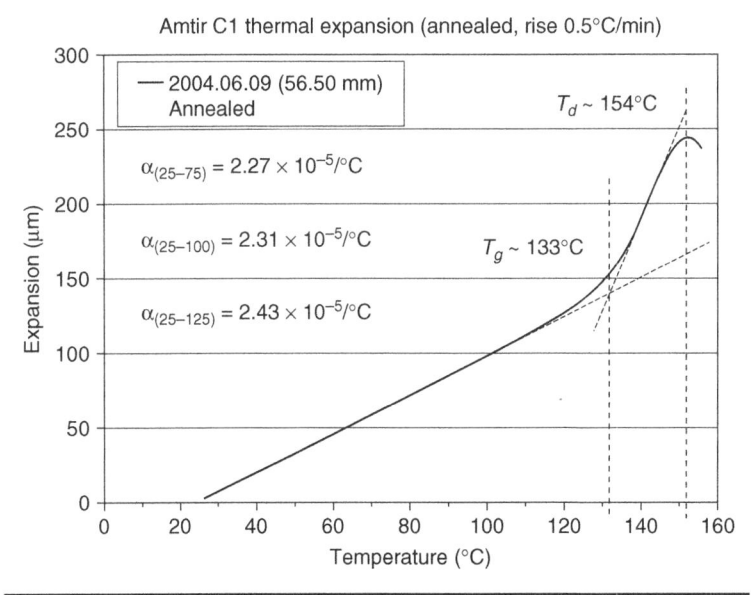

Amtir C1 thermal expansion (annealed, rise 0.5°C/min)

— 2004.06.09 (56.50 mm) Annealed

$\alpha_{(25-75)} = 2.27 \times 10^{-5}/°C$

$\alpha_{(25-100)} = 2.31 \times 10^{-5}/°C$

$\alpha_{(25-125)} = 2.43 \times 10^{-5}/°C$

$T_d \sim 154°C$

$T_g \sim 133°C$

FIGURE 2.2 Measurement of thermal expansion and glass transition temperature for AMI C1 glass.

this instrument, T_d is also defined in viscosity terms as $10^{9.5}$ poise. If the glass temperature curve is approached from the other direction, i.e., cooling from above the softening point toward room temperature, the shape will be slightly different depending upon the rate of cooling (quench). As the glass is cooled into the T_g range, the slope then decreases and begins to level out to a value typical of a solid. Different rates of cooling produce a family of curves slightly separated from one another. The slight differences are reflected in the density and thus the resultant refractive index. For this reason, the thermal history of a glass is important. To stabilize the refractive index from batch to batch, the quenching process and the annealing procedure must be established and consistently followed when the glass is produced. Nucleation and growth of small crystallites in a glass melt do occur depending upon composition and temperature. The process requires movement of atoms to come together to form molecules of exact composition ratios. Such processes are much more difficult in viscous melts and occur slowly if at all. As the temperature increases, the viscosity decreases, making the crystallization process possible. Thus, the crystallization process is time- and temperature-dependent. Figure 2.3 is a differential thermal analysis (DTA) plot[29] for AMI C1 glass that shows difference in temperature for a fine powder glass sample referenced against a fine powder of standard aluminum oxide powder when the samples are heated at a controlled rate. Two regions are designated. The first is at a temperature right above the T_g that dips below the reference line, an endothermic process. This change is

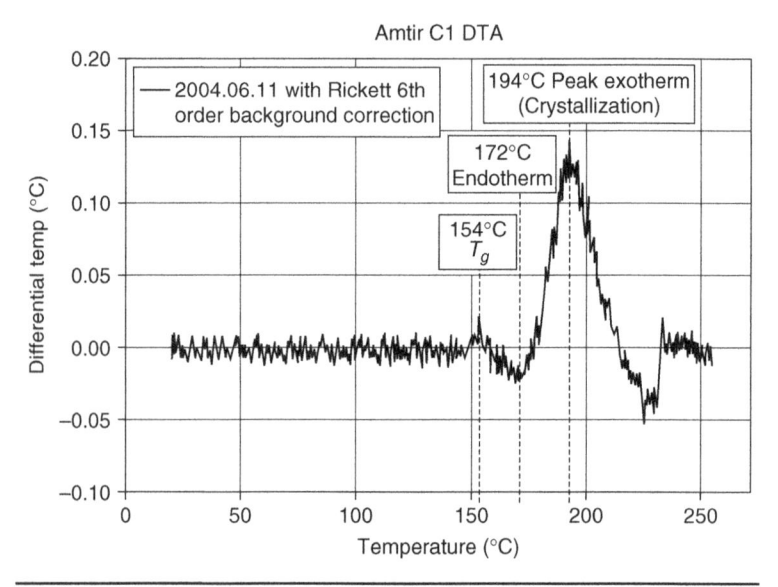

FIGURE 2.3 Differential thermal analysis (DTA) plot for AMI C1 glass.

attributed to nucleation of very small crystallites. The second region, well above the reference line, indicates rapid growth of the crystallites, an exothermic process. After the peak is past, the curve begins to fall, indicating the crystallites are now dissolving in the surrounding low-viscosity glass.

AMI C1 is an arsenic-selenium-tellurium glass which has a tendency to crystallize. Figure 2.4 shows microscope photographs of crystals forming on the surface of C1 glass after heating for an extended period. Figure 2.4*a* shows nucleation at 215°C for 8 h. Figure 2.4*b* shows much larger crystals grown at 238°C for 15 h. When producing the glass, one must quench (rapidly cool) the glass in a timely manner beginning at a temperature above the crystallization range down to a temperature below the nucleation range. These factors must be taken

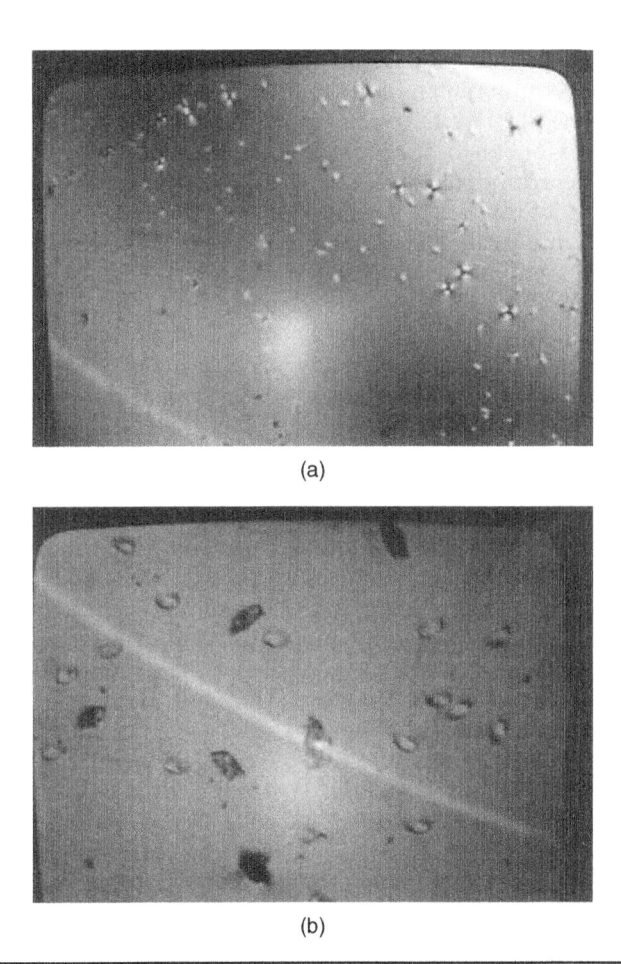

(a)

(b)

FIGURE 2.4 Crystallite formation in heated AMI C1 glass. (*a*) C1 glass heated 8 h @ 215°C; (*b*) C1 glass heated 15 h @ 238°C.

into account when casting, molding, slumping, extruding, or drawing a glass into fiber. The rules of Zachariasen[30] first pointed out that to form a glass it was important to select substances with a low coordination number (4 or less) for both the cation and the anion. Coordination number is not the only factor.

There must be rigid directional requirements for the chemical bonding which is found in covalent solids such as chalcogenide glasses. Position in the periodic table for element A and element B will determine the coordination number for their binary compound AB. According to Mooser and Pearson,[31] the average principal quantum number N_x of a crystalline binary AB compound is a measure of the metallic character of the bonds present. The difference in Pauling electronegativity Δ is a measure of the ionic character of the bonds. A plot of these two factors by Mooser and Pearson[31] showed that compounds of the same crystalline structures fell in the same general area of the diagram (Fig. 2.5). In fact, similar crystal structures with the same coordination number (3, 4, 5, 6, or 8) fall in rather specific areas of the plot. We have already dealt with the electronegativity differences for bonds. The principal quantum number of the valence electrons corresponds to the row of the periodic table on which the element is located.

The same treatment may be applied to chalcogenide glasses. However, most compositions will not be binary, only two elements. Most will contain three or more elements. So the average principal quantum numbers and average electronegativity numbers must be calculated based on composition percentages for each element. Consider a ternary system with a composition represented by $Ax + By + Cz$. You calculate an average N_x by multiplying the N_x of each element by its composition fraction and adding for a total. The same procedure is followed for the electronegativity differences. The point for each composition could then be plotted as shown in Fig. 2.5. Multicomponent glasses based on silicon, sulfur, selenium, and tellurium were treated in this manner.[32] Note all the chalcogenide glasses lie in the coordination number 4 area. The silicate glasses start in the 4 region but fall in the coordination number 6 region as more metal oxides are added. Note that pure SiO_2 glass has a network structure that is open. Pure silica when heated is slightly permeable to helium and to a lesser extent hydrogen. The glass optical properties are altered by adding metal oxides to the composition which fill in the voids in the network structure. The heavy metal oxides are from higher row numbers of the periodic table. Bonding to silicon which already has a coordination number of 4 leads to areas of coordination number 6. Silicate glasses containing metal oxides are not permeable to helium or hydrogen. For chalcogenide glasses formed from a melt, coordination number and bond type as determined by chemical composition are the two most important factors in glass formation.

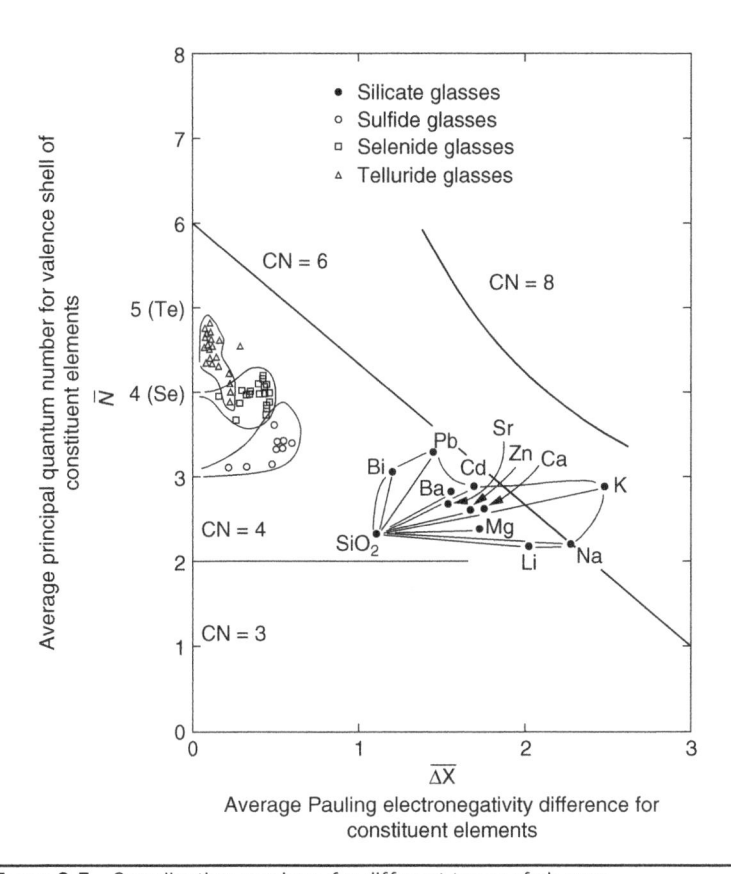

FIGURE 2.5 Coordination numbers for different types of glasses.

2.3 Evaluating Possible Glass Forming Systems

The discussion to follow describes the methods used and the results of the first exploratory program carried out at Texas Instruments (TI) over the period from 1962 to 1967. After the first program was finished, what was learned led to other programs which are described later. At this point, only one system had just been investigated,[13] the Si-As-Te glass forming system. Development of methods and selection of systems to be investigated were just starting.

In selecting elements A and B to use in a potential glass forming system, the first question to ask is, do they form a chemical bond with each other? The answer may be found in the library by using a binary compound book such as the one by Hansen.[33] If a ternary three-element system is considered, the diagrams for compounds AB, AC, and BC should be found. The diagrams have temperature on both sides as the y axis while the x axis is composition going from

100 percent A on the left to 100 percent B on the right. Percentages may be either weight percent or atomic percent with the latter preferred. The results presented represent equilibrium results, i.e., the reactants at a specific concentration held at a specific temperature for periods of time reaching days or even weeks. The purpose is to identify the crystalline compounds formed under the conditions of the experiment. If there is no AB compound formed, the diagram may be essentially two lines from the melting point of A to the melting point of B describing solid solutions of A and B. An example is the diagram for Ge-Si.[33] A diagram might show different solid compounds in equilibrium with a liquid phase. Above all the areas outlined will be a line representing a temperature above which A and B mixtures are liquids. The *liquidus curve* for an AB mixture identifies a melt temperature at each indicated composition point. The same description fits the AC and the BC diagrams. If a compound is found, a straight line will be drawn down from the temperature point to the composition point. Later, we will see that the size of the glass forming area in a diagram is affected by the number of compounds that may form from the specific combination of elements. From the diagrams one may find the temperature required to produce a homogeneous melt to test the ability of the composition to form a single-phase glass when quenched.

After all these years, most of the IVA-VA-VIA elements have been extensively investigated in various combinations. The investigator may wish to find a unique combination of elements with a chalcogen or even more than one chalcogen. There may be a desire to revisit previous investigations of a particular combination of elements. This discussion is only meant to serve as a guide for the start of the investigation. Obviously, the first step is to go to the literature to find any previous work that may be pertinent.

The qualitative investigation of a glass forming combination of elements requires making many 25- to 50-g small samples. Quartz vials are usually used because of their purity and high-temperature softening points. High-purity reactants are used and weighed accurately with an analytical balance.

The tube is evacuated to remove the atmosphere and sealed with a hydrogen-oxygen torch. Samples are then placed in some type of a rocking furnace in a hood and heated to a temperature about the same as the boiling point of the most volatile constituent element. The rocking action is begun to start the reaction of the elements. After some time, the furnace is raised in temperature to produce a homogeneous melt of the reactants. The estimated temperature is based on the binary compound diagrams. Mixing continues for some time (hours) above the melt temperature. Rocking is stopped with the vial in a vertical position. The melt is allowed to cool at the melt temperature. The vial is removed from the furnace, held in a vertical position, and quenched in air. Caution must be exercised. The vial is held by metal tongs. A face mask and protective gloves are worn. The danger of an explosion for some combinations should not be ignored. As mentioned above, the rate of cooling affects the extent of the composition range which is judged to be glass forming.

After cooling, the vial is opened and the sample is removed for examination. Usually, a sample can be identified as glass or crystalline by visual examination. Sometimes use of a microscope sensitive to some degree in the near infrared is useful. A glass sample can have two phases, glass and crystal or even two immiscible glasses. If x-ray diffraction is used, a completely blank film demonstrates a single-phase amorphous glass. The glass samples are cylinders with round bottoms. Slices are carefully sawed out, some with different thickness, 2 to 5 mm. The disks are polished on both sides using an optical abrasive such as aluminum oxide. The last sample with the round bottom is saved for reflection measurements. The one flat side is polished in the same manner as the disks.

The infrared transmission T is measured using a standard spectrophotometer covering the wavelength range of interest, generally 1 to 20 µm. A reflection attachment, if available, is used to measure the infrared reflectivity R from the flat side of the bottom sample. A qualitative measure of the refractive index n and the bulk absorption coefficient α can be obtained from solving the simple equation

$$T = (1 - R)^2 e^{-\alpha x}$$

where x is thickness in centimeters and $R \sim (n - 1)^2 / (n + 1)^2$. In the absence of a direct measurement of R, n and α can be calculated from the measured T for the two samples of different thickness.

Another number very important to the glass evaluation is an approximate glass softening point. An ASTM method or a dilatometric measurement is impractical for small survey samples. A simple apparatus much like the one shown in Fig. 2.6 may be devised and easily built. One of the polished disk samples is placed in the bottom of the chamber. A thermocouple is placed inside next to the sample. Inert gas may be slowly circulated if needed. As the disk sample is heated and begins to soften, the glass rod in the center drops down, moving the indicator arm and signifying the softening point has been reached. The size of the weight, the weight of the rod, and the thickness of the sample are important variables. Calibration with known glasses should be carried out to support the results.

Determining the glass forming compositions for a binary system is of course the easiest since you have only two elements. The results are then plotted on a composition percent straight line with glass or crystalline marked at each composition point on the line.

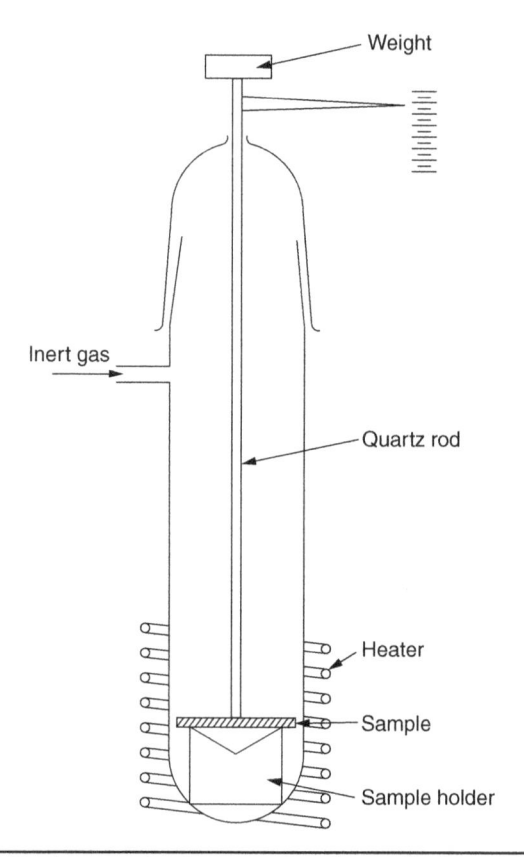

FIGURE 2.6 Simple apparatus for measuring glass softening point.

For a three-component ternary system, many samples are required. Table 2.1 lists over 50 samples[14] prepared in the evaluation of the Si-P-Te system, a system evaluated early in the exploratory program at TI. The 39 samples evaluated, identified by number, are plotted in their respective composition points in the compositional triangular diagram shown in Fig. 2.7. The glass forming composition region is designated to lie within the dashed lines. Again, note that the extent of this region is dependent on the quenching procedure as well as the composition. For example, quenching in a cool liquid would enlarge the area. The evaluation results indicated a system that was not promising.

The glass forming region is smaller than that of the first system evaluated at TI, the Si-As-Te system.[13] Comparison of the sizes of glass forming regions indicates differences in the glass forming tendencies of different element combinations. The conclusions reached

Chalcogenide Glasses

Sample Number	Si (Atom %)	P (Atom %)	Te (Atom %)	Softening Point (°C)	Remarks
3	20	20	60	190	Stable glass
12	25	25	50		Crystalline
15	40	20	40		Crystalline
16	30	20	50		Unstable glass
19	15	20	65	150	Stable glass, poor IR transmission
20	15	25	60	200	Stable glass, ~10% IR transmission
21	20	25	55	185	Stable glass, ~10% IR transmission
22	25	20	55	325	Stable glass, poor IR transmission
23	30	25	45		Unstable glass
24	35	20	45	385	Unstable glass
25	35	15	50		Unstable glass
26	30	15	55		Unstable glass
27	25	15	60		Unstable glass
28	20	15	65	250	Stable glass
29	15	15	70	180	Very stable glass, 10% IR transmission
30	10	20	70		Reacted violently with the atmosphere
31	10	25	65		Reacted violently with the atmosphere
32	10	30	60		Reacted violently with the atmosphere
33	15	30	55		Reacted violently with the atmosphere
34	35	25	40		Crystalline, unstable
35	40	15	45		Crystalline, unstable
36	40	10	50		Crystalline, unstable
37	35	10	55		Crystalline, unstable
38	30	10	60	400	Unstable glass

TABLE 2.1 Composition of Samples Used in the Evaluation of the Si-P-Te System[14] *(Continued)*

Sample Number	Atom % Si	Atom % P	Atom % Te	Softening Point (°C)	Remarks
39	25	10	65	300	Stable glass, 10% IR transmission
40	20	10	70	200	Stable glass, 10% IR transmission
41	15	10	75	175	Very good glass, 35% IR transmission
42	15	5	80	160	Very good glass, 35% IR transmission
43	20	5	75	200	Very good glass, 20% IR transmission
47	25	5	70	250	Stable glass
48	30	5	65		Glassy
49	35	5	60		Glassy
50	40	5	55		Crystalline
51	30	0	70		Crystalline
52	25	0	75		Glassy
53	20	0	80		Glassy
54	15	0	85	375	Glassy
55	10	5	85		Crystalline
56	10	10	80		Crystalline

TABLE 2.1 Composition of Samples Used in the Evaluation of the Si-P-Te System[14] (*Continued*)

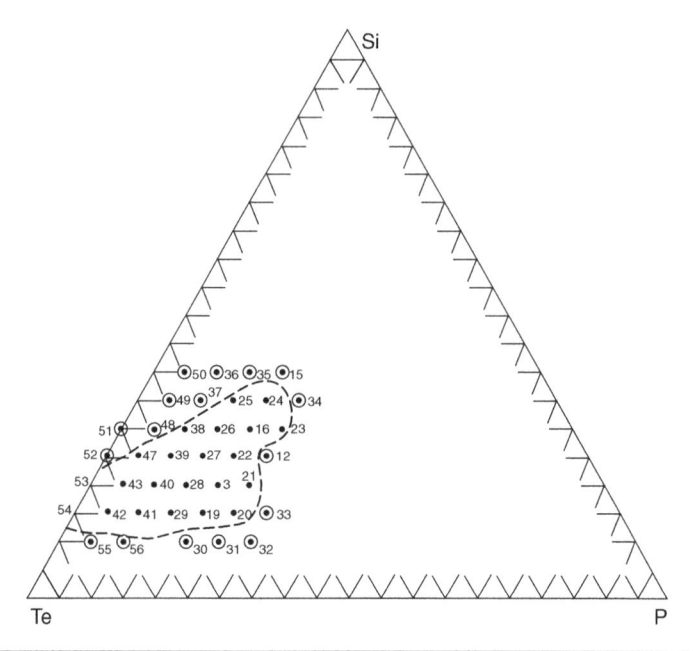

FIGURE 2.7 The glass forming composition diagram for the Si-P-Te ternary system.[14]

regarding glass forming tendency in the IVA-VA-VIA ternaries were

$$S > Se > Te$$
$$As > P > Sb$$
$$Si > Ge > Sn$$

The reversal between As and P may be due to the fact that As forms bonds with Si and Ge while P does not.[33] Efforts to form glasses with Sn were unsuccessful except in very low Sn concentrations.

2.4 Qualitative Evaluation of Compositions for Development

For the Si-P-Te system, only one-third to one-half produced glasses of good quality, and those had softening points below 200°C. Disks were cut from the glass sample, and polished and evaluated relative to infrared transmission as a function of wavelength, and a qualitative estimate of the refractive index of each was determined. Figure 2.8 shows the results of this effort. Note that at this time, measurements were crude in comparison to today and could only be considered qualitative.

Following this procedure developed in the exploratory program, seven ternary IVA-VA-VIA systems were evaluated during the three-year period from 1962 to 1965. Results of the evaluation are given in Table 2.2. Each system was evaluated with respect to softening point, approximate refractive index, and infrared absorption in the two

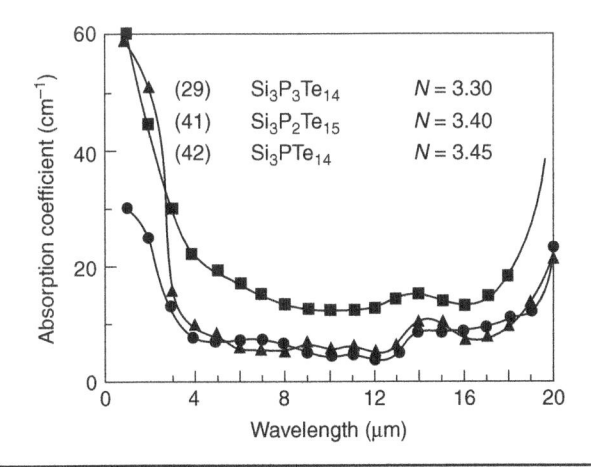

Figure 2.8 Measured infrared transmission of some Si-P-Te glasses.

System	Max. Softening Point	Refractive Index	Absorption 3 to 5μ	Absorption 8 to 14μ
Si-P-Te	180°C	3.4	No	Slight
Si-Sb-Se	270°C	3.3	Yes	Yes
Si-Sb-S	280°C	—	Yes	Yes
Ge-P-Se	420°C	2.4–2.6	Slight	Yes
Ge-P-S	520°C	2.0–2.3	Very slight	Yes
Si-As-Te	475°C	2.9–3.1	No	Slight
Ge-As-Te	270°C	~3.5	No	Very slight
Ge-P-Te	380°C	~3.5	No	Very slight

TABLE 2.2 General Properties of the Best Infrared Transmitting Glasses from Each of the Ternary IVA-VA-VIA Systems

atmospheric windows of 3 to 5 μm and 8 to 14 μm. The Si-As-Te, Ge-As-Te, Si-P-Te, and Ge-As-Te systems were rated useful in both bands relative to transmission. The Si-Sb-S and the Si-Sb-Se systems produced glasses unstable and reactive with the atmosphere. Glasses with the highest softening points were Ge-P-S, G-P-Se, and Si-As-Te. Sulfur-based glasses begin to absorb strongly at wavelengths beyond 8 μm and are not useful in the 8- to 12-μm thermal window. The conclusion was that none of these systems promised to produce glasses meeting our original goal, with physical properties comparable to those of silicate-based optical glasses. Efforts were made to improve the glasses by blending two systems together over the full composition range. In this case, the system contained four elements rather than three. An example is the blending of the Si-As-Te ternary with the Ge-As-Te ternary forming Si-Ge-As-Te glasses. The only way to evaluate the usefulness of these glasses was to prepare them in high-quality batches up to 1 to 2 kg so that they could be characterized physically and optically in a more quantitative manner. Table 2.3 lists the glasses characterized and their individual results. Note the sulfur glasses are somewhat better physically than those based on Se and Te. However, they are not useful for the thermal window, the goal of the program. Figure 2.9 is a photograph of a large cast Si-Ge-As-Te glass plate. Also shown are glass prisms to be used to measure the infrared refractive index quantitatively as a function of wavelength. Figure 2.10 shows an attachment built at TI for the Perkin Elmer 13 U spectrophotometer for performing the prism minimum deviation measurement in the infrared. The results are precise index numbers good to three to four decimal points. We will discuss this method in detail when the instrument used presently at AMI is described in Chapter 4. Figure 2.11 shows results obtained for two

Chalcogenide Glasses 37

Composition	Refractive Index	Softening Point (°C)	Deformation Point (°C)	Thermal Coefficient of Expansion [in/(in·°C) × 10^6]	Hardness (Knoop Scale)
$SiAsTe_2$	2.93	317	250	13	167
$Si_3Ge_2As_5Te_{10}$	3.06	320	284	10	179
$GeAs_2Te_7$	3.55	178	140	18	111
Ge_3PS_6	2.15	520	375	15	185
Ge_7PS_{12}	2.20	480	360	13	179
Ge_2S_3	2.30	420	360	14	179
$Si_6As_4Te_9Sb$	2.95	475	350	9	168

TABLE 2.3 Physical Constants of Chalcogenide Glasses in This Program Prepared in Amounts Sufficient for Detailed Evaluation

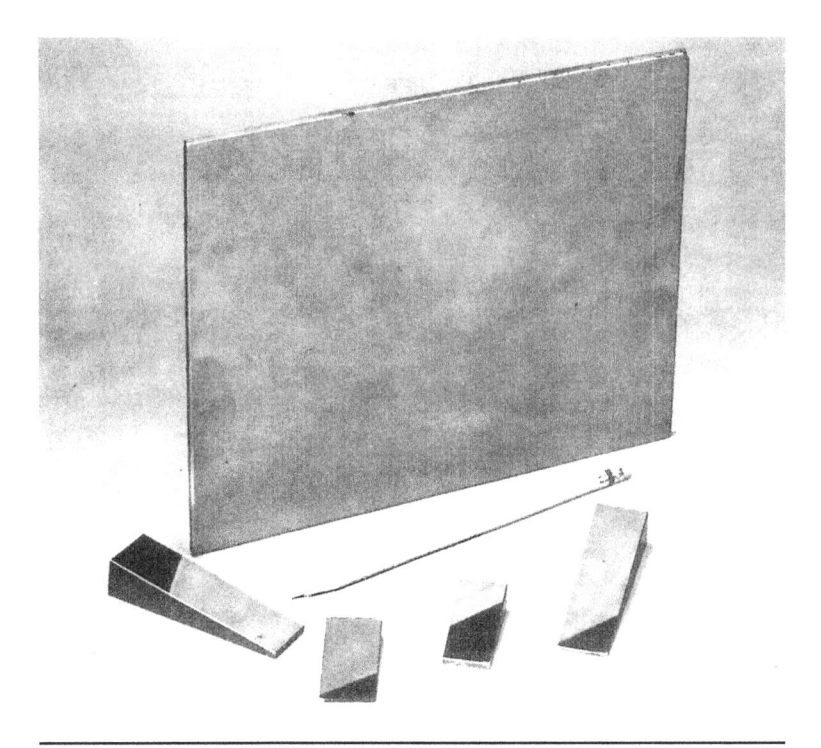

FIGURE 2.9 Photograph of a large cast chalcogenide glass plate and four glass prisms fabricated for IR index measurements.

38 Chapter Two

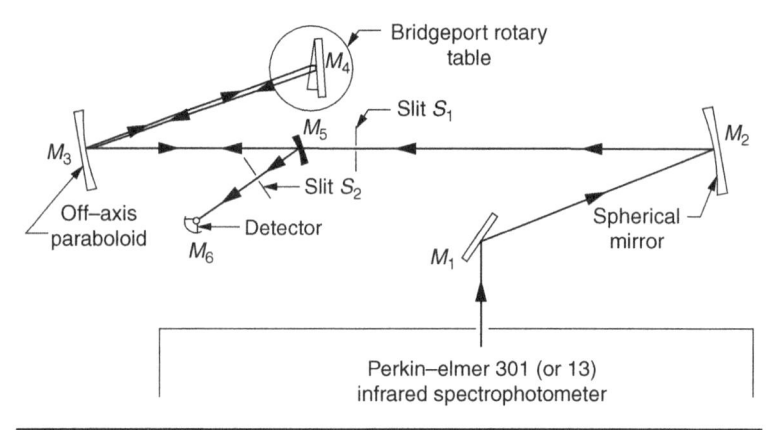

FIGURE 2.10 Infrared refractometer attachment for the Perkin Elmer 13 Spectrophotometer.

prisms of a Si-Ge-As-Te glass composition made from separate batches. The precise index measurements were used to calculate the bulk absorption coefficient as a function of wavelength. The absorption coefficient α is shown in the bottom curve and given in units of cm^{-1}. The index differences between the two prisms were 0.0026 at 3 μm dropping to 0.001 at 10 μm. These are not bad results for the first time producing the glasses in large amounts, fabricating the prisms, and

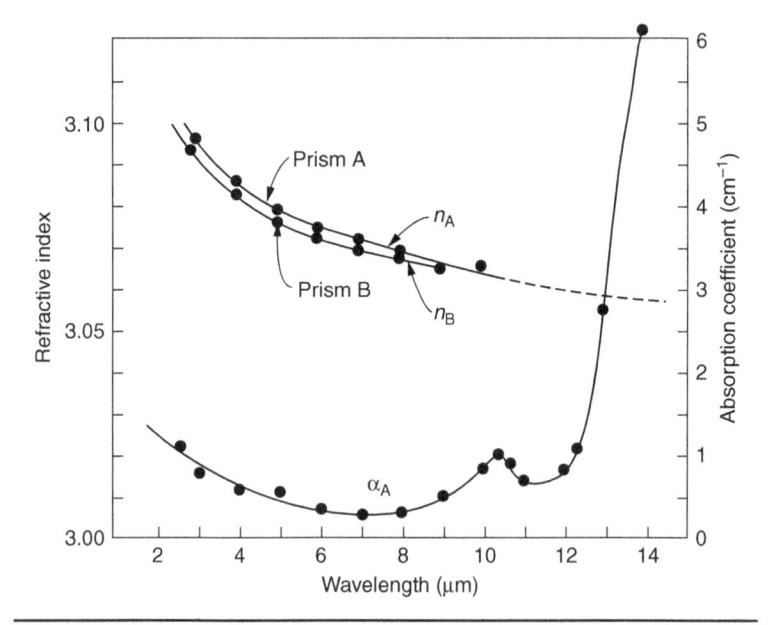

FIGURE 2.11 Precise infrared index results for two prisms of a Si-Ge-As-Te glass composition made from different melts.

Chalcogenide Glasses

making the measurements. With a criterion, at this early stage, of an absorption coefficient of less than 1 cm^{-1}, the Si-Ge-As-Te glass was judged good from 2.5 to 12.5 μm. The Si-As-Te glasses were judged good to only 9 μm, and the Ge-As-Te glasses good from 2.5 to 20 μm. Blending of two ternary systems was carried out for six combinations. The one that produced the most useful results was the Si-As-Te with the Ge-As-Te system. Table 2.4 lists the compositions prepared

Sample No.	Composition	Softening Point (°C)	Hardness (Knoop)
239	Si$_6$As$_8$Te$_{26}$	196	108.4
242	Si$_5$GeAs$_8$Te$_{26}$	190	126.5
245	Si$_4$Ge$_2$As$_8$Te$_{26}$	124	126.5
248	Si$_3$Ge$_3$As$_8$Te$_{26}$	200	136.8
251	Si$_2$Ge$_4$As$_8$Te$_{26}$	190	127.0
253	SiGe$_5$As$_8$Te$_{26}$	180	126.5
255	Ge$_6$As$_8$Te$_{26}$	Crystalline	—
258	Si$_6$As$_9$Te$_{45}$	160	108.4
260	Si$_5$GeAs$_9$Te$_{45}$	136	105.8
261	Si$_4$Ge$_2$As$_9$Te$_{45}$	148	110.9
262	Si$_3$Ge$_3$As$_9$Te$_{45}$	146	108.7
263	Si$_2$Ge$_4$As$_9$Te$_{45}$	148	109.0
264	SiGe$_5$As$_9$Te$_{45}$	150	113.4
265	Ge$_6$As$_9$Te$_{45}$	162	113.7
240	Si$_5$As$_5$Te$_{10}$	310	166.9
266	Si$_4$GeAs$_5$Te$_{10}$	290	156.5
267	Si$_3$Ge$_2$As$_5$Te$_{10}$	293	179.0
268	Si$_2$Ge$_3$As$_5$Te$_{10}$	256	151.2
269	SiGe$_4$As$_5$Te$_{10}$	Crystalline	—
241	Si$_7$As$_5$Te$_8$	434	207.8
244	Si$_6$GeAs$_5$Te$_8$	380	195.6
247	Si$_5$Ge$_2$As$_5$Te$_8$	394	198.6
250	Si$_4$Ge$_3$As$_5$Te$_8$	379	195.0
251	Si$_3$Ge$_4$As$_5$Te$_8$	Crystalline	—

TABLE 2.4 Blended Glasses Formed from the Ternary Systems Si-As-Te and Ge-As-Te

40 Chapter Two

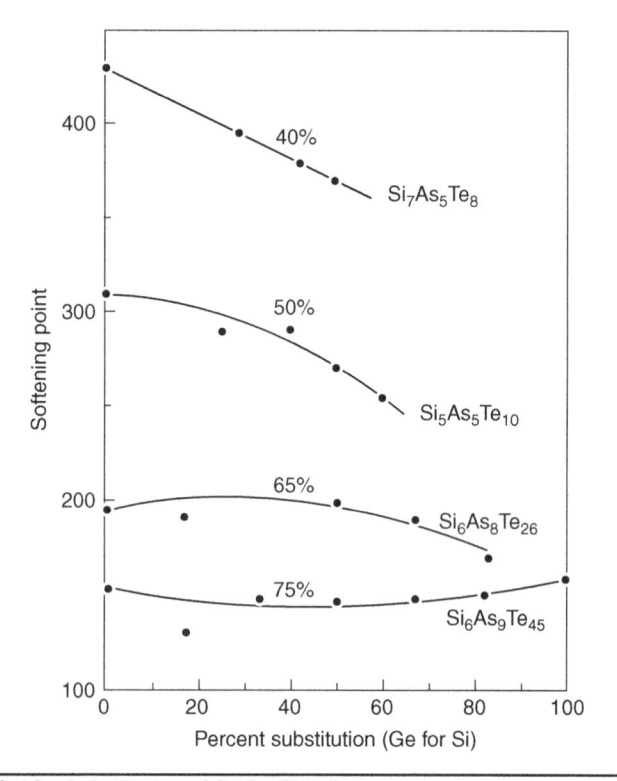

FIGURE 2.12 Substitution of Ge for Si in Si-As-Te glasses change in softening point.

with the measured softening point and Knoop hardness for each composition. Figure 2.12 shows the change in softening point of the individual Si-As-Te glass compositions as germanium is substituted for silicon. Note the softening points all decline when germanium is introduced. The chalcogenide glass plate in Fig. 2.9 shows the first glass plate cast at TI made from a blended Si-Ge-As-Te glass. The plate was polished and an antireflection coating was later applied.

2.5 General Physical Properties of Chalcogenide Glasses

2.5.1 Softening Points and Hardness

The higher the softening points, the harder the glass. Figure 2.13 shows the plot of measured hardness for about 100 compositions plotted against their measured softening points. Some of the glasses contained

Chalcogenide Glasses 41

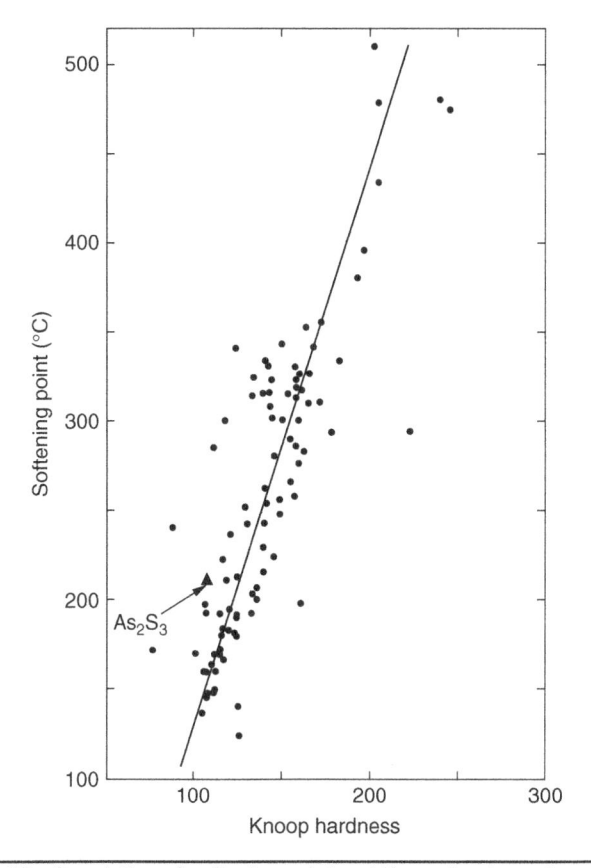

FIGURE 2.13 Correlation of softening point and Knoop hardness for chalcogenide glasses.

four elements. Even with a softening point of 500°C, Knoop hardness was less than 250.

2.5.2 Thermal Coefficients of Expansion
The higher the softening point, the smaller the thermal expansion coefficient. Results obtained from 30 samples are plotted in Fig. 2.14. The correlation is very general in part because the glasses measured are from different systems some with four elements. Many factors may affect results.

2.5.3 Density
The densities of selenium and tellurium glasses containing silicon, germanium, arsenic, and phosphorus are almost a linear function of the calculated average molecular weight of the glass composition.

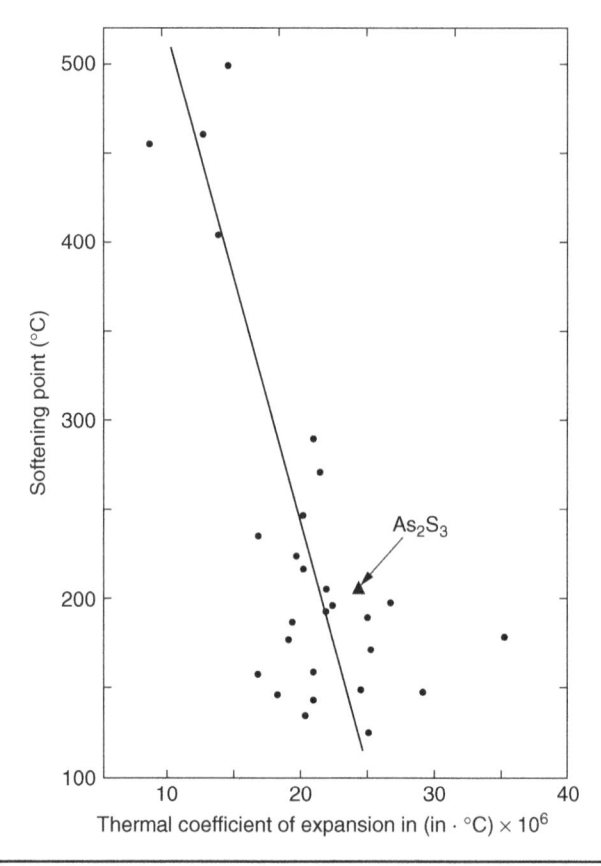

FIGURE 2.14 Correlation of softening point and thermal expansion for chalcogenide glasses.

Figure 2.15 plots the results for 28 glass compositions. The densities for Te, Se, S, Si, and Ge are added for reference. Some of the samples were small in mass. The values obtained from large samples, the prisms and the cast plate, were given greater consideration in drawing the straight line. To check the validity of the straight line, densities for 15 samples of Ge-As-Se glasses reported in the literature[34] were calculated and compared to reported values and found an average error in agreement of only –3 percent.

2.5.4 Molar Refraction

The apparent linear relationship between density and molecular weight suggests that other properties are additive and can be predicted. One such property is the refractive index. The refractive index is related to the molar refraction and molecular volume of a

Chalcogenide Glasses

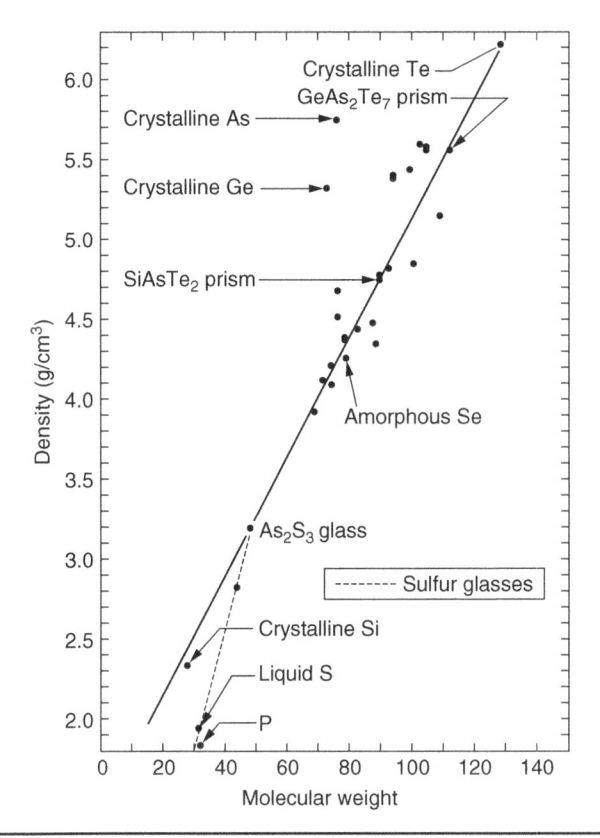

FIGURE 2.15 Density versus molecular weight for chalcogenide glasses.

substance. From the Lorentz-Lorenz equation,[35] molar refraction is given by

$$R = \frac{N^2-1}{N^2+2}V = \frac{N^2-1}{N^2+2}\frac{MW}{d}$$

where R is the molar refraction, N is the refractive index at some non-dispersive wavelength, and V is the molar volume equal to average molecular weight divided by the density d. For a nonpolar amorphous glass, molar refraction is almost equal to the molar polarization. Molar refraction has the units of volume and can be thought of as the additive sum of the volumes of each atom (or ion) in the molecule. Molar refraction is related to the radius of the individual molecule by

$$R = \frac{4}{3}\pi A\alpha = \frac{4}{3}\pi A r^3$$

where A is Avogadro's number, α is the polarizability of the atom or ion, and r is the radius of the conducting sphere formed by the molecule.

This equation has been applied to the study of bonding in organic and inorganic compounds including oxide glasses.[36] For a molecular compound of the form $A_xB_yC_z$, where x, y, and z are the atomic fractions of the constituents A, B, and C, the molar refraction becomes

$$R = xR_A + yR_B + zR_C$$

where R_A, R_B, and R_C are the atomic (or ionic) refraction values resulting from their presence in the molecule. The approach applies well to the covalent bonded chalcogenide glasses so that single values for each element can be determined and used in many different glass compositions. The atomic refraction values should be close to the cube of their accepted covalent radii. Calculating directly from accepted covalent radii would yield low values because the atomic spheres are loosely packed. Amorphous selenium was chosen as the starting point in calculating atomic refraction values for use with chalcogenide glasses. Using available experimental data. The atomic refraction for selenium was calculated and used as a reference. The index wavelength chosen was 5 µm. The atomic refractions for silicon, germanium, phosphorus, arsenic, sulfur, and tellurium were calculated from the cube of their covalent radii and normalized to selenium. From these atomic refraction literature-derived values, the molar refractions for the 28 glass compositions used in the density plot of Fig. 2.15 were calculated and compared to the measured values. Agreement was ±4.1 percent. The results are given in Table 2.5. Another approach that yielded better agreement was to treat the glass formulas of atomic refraction values for glasses with different concentrations of the same elements as simultaneous equations and solve directly for the experimental atomic refraction values of each constituent element. When the 28 glass compositions were recalculated, the agreement with experimental values was ±1.1 percent. Table 2.6 lists the atomic refraction values determined from the literature and from solving the simultaneous equations. Values from glasses based on S, Se, and Te are given for comparison.

An illustration of the worth of the method follows: The refractive index for chalcogenide glasses at 5 µm can be calculated within a few percent by using the density vs. molecular weight plot in Fig. 2.15 and the atomic refraction values in Table 2.6 to calculate molar refraction:

$$R = xR_A + yR_B + zR_C$$

Then solve for N from $R = (N^2 - 1)/(N^2 + 2) \times$ molecular weight/density.

This procedure was applied to 20 As-Se-Te glasses reported by Jerger and Billian at Servo Corporation.[5] The accurate values they reported and the values estimated agreed within +3 percent. The results are shown in Table 2.7.

Chalcogenide Glasses 45

Composition	R Measured	R Calculated Literature Value	% Error	R Calculated Average Value	% Error
PS$_4$	7.36	8.49	+15.4	7.36	0
Ge$_3$PS$_6$	8.26	9.23	+11.7	8.27	+0.1
Ge$_2$S$_3$	8.90	9.30	+6.5	8.90	0
As$_2$S$_3$	9.44	9.42	−0.2	9.44	0
Se	11.55	11.51	−0.4	11.55	0
P-Se$_9$	11.17	11.44	+2.4	11.31	+1.3
PSe$_4$	11.35	11.35	0	11.06	−1.7
AsSe$_9$	11.85	11.52	−2.9	11.62	−1.9
AsSe$_4$	11.55	11.54	−0.1	11.68	+1.1
Si-Se$_9$	11.65	11.31	−2.9	11.65	0
Ge$_{16}$As$_{47.3}$Se$_{36.7}$	11.33	11.56	+2.0	11.36	+0.3
Ge$_{15}$As$_{45}$Se$_{40}$	11.45	11.54	−0.3	11.35	−0.8
Ge$_3$P$_3$Se$_{14}$	10.70	11.36	+6.2	10.70	0
Ge-Se$_9$	11.23	11.49	+2.3	11.23	0
SiAsTe$_2$	13.42	14.55	+8.4	13.53	+0.8
Si$_2$PTe$_7$	16.05	15.96	+0.4	15.21	−5.2
Si$_3$PTe$_{16}$	16.55	16.76	+1.3	15.77	+1.3
Si$_3$As$_2$Te$_5$	13.70	14.43	+5.3	13.70	0
Si$_3$As$_3$Te$_4$	12.95	13.74	+6.1	13.23	+2.2
Si$_2$As$_3$Te$_5$	13.90	14.66	+5.5	13.90	0
GeAs$_4$Te$_5$	14.07	15.09	−7.3	14.63	+4.0
GeAs$_2$Te$_7$	15.95	16.47	+3.2	15.87	−0.5
Ge$_3$P$_4$Te$_{13}$	15.83	15.90	+0.4	15.81	−0.1
Ge$_3$As$_{10}$Te$_7$	13.40	14.03	+4.7	13.58	+1.3
GeAs$_{10}$Te$_9$	14.30	14.75	+3.1	14.52	+1.5
GeAs$_{12}$Te$_7$	13.65	14.05	+2.8	13.96	+2.3
Ge$_3$As$_2$Te$_{15}$	16.25	16.77	+3.2	15.83	−2.6
Ge$_2$As$_3$Te$_{15}$	16.15	16.79	+4.0	16.02	−0.8
		Average	±4.1		±1.1

TABLE 2.5 Molar Refraction Values of Chalcogenide Glasses

Element	R Calculated from Literature Values	R Calculated from Covalent Radii	R Calculated for S Glasses	R Calculated for Se Glasses	R Calculated for Si-Te Glasses	R Calculated for Ge-Te Glasses
Si	9.45	10.1		12.5	9.6	
Ge	11.35	13.4	9.7	8.3		9.9
P	10.6*	8.8	3.3	9.1	13.8	15.0
As	11.6[†]	12.5	11.0	12.2	11.6	12.2
S	7.95	7.85	8.4			
Se	11.51	11.55		11.55		
Te	18.55	18.21			17.05	17.8

*From P-Se4 glass.
[†]From As-Se4 glass.

TABLE 2.6 Atomic Refraction Values for IVA, VA, and VIA Elements When Used in Chalcogenide Glasses

Servo No.	Composition	N at 5 μm	N Calculated	% Error
1	$As_{38.7}Se_{61.3}$	2.79	2.62	−6.0
2	$As_{27.5}Se_{72.5}$	2.65	2.59	−2.3
3	$As_{40}Se_{35}Te_{25}$	2.88	2.90	+0.7
4	$As_{40}Se_{25}Te_{35}$	3.07	3.06	−0.3
5	$As_{30}Se_{30}Te_{40}$	3.08	3.11	+1.0
6	$As_{20}Se_{60}Te_{20}$	2.74	2.80	+2.2
7	$As_{35}Se_{45}Te_{20}$	2.90	2.71	−6.5
9	$As_{45}Se_{45}Te_{10}$	2.77	2.75	−0.7
10	$As_{25}Se_{45}Te_{30}$	2.91	2.93	+0.7
11	$As_{30}Se_{60}Te_{10}$	2.76	2.71	−1.8
12	$As_{10}Se_{60}Te_{30}$	2.73	2.89	+5.9
13	$As_{20}Se_{50}Te_{30}$	2.84	2.94	+3.5
14	$As_{20}Se_{70}Te_{10}$	2.65	2.68	+1.1
16	$As_{35}Se_{55}Te_{10}$	2.83	2.72	−3.9
17	$As_{30}Se_{55}Te_{15}$	2.82	2.76	−2.0
18	$As_{25}Se_{55}Te_{20}$	2.80	2.81	+0.4
19	$As_{20}Se_{55}Te_{25}$	2.64	2.84	+7.5
20	$As_{15}Se_{55}Te_{30}$	2.78	2.92	+5.0
			Average ±2.9%	

TABLE 2.7 Measured and Calculated Refractive Index Values for Servo As-Se-Te Glasses[5]

Glass	Dielectric Constant	Frequency (cps)	Resistivity ($\Omega \cdot cm$)(300K)
$Ge_{15}As_{15}Se_{70}$			5×10^{10}
$Si_{15}Sb_{15}S_{70}$			9.6×10^{7}
$Ge_{15}P_{15}Se_{70}$			9.3×10^{10}
$Si_{15}Sb_{35}S_{50}$	14	100	2×10^{9}
$Si_{6}As_{9}Te_{45}$			5×10^{5}
$Ge_{2}As_{3}Te_{15}$			2×10^{7}
$Si_{3}Ge_{2}As_{5}Te_{10}$	24	1 kc	1×10^{8}
$Ge_{3}P\,S_{6}$			9×10^{9}
$GeAs_{4}Te_{5}$			5×10^{5}
$Si_{4}As_{3}Te_{3}$			5×10^{9}
$GeAs_{2}Te_{7}$			2.8×10^{4}

TABLE 2.8 Electronic Conduction of Some Chalcogenide Glasses

2.5.5 Electrical Properties

It was mentioned earlier in Chap. 1 that chalcogenide glasses are electronic conductors and their properties have been the subject of intense study, but not in this program. Early on at TI, it was found that some glasses containing Te and Sb could become good conductors with low resistivity but poor infrared transmission. Some electrical measurements carried out on 11 compositions are listed in Table 2.8. From the values, it is clear that the glasses used optically may be called high-resistivity semiconductors ranging in values from 10^{4} to 10^{10} ohm·cm at room temperature. The two measured dielectric constants were high. What is not shown in the table is the fact that these glasses are electronic conductors with very low mobility for their electrons or holes involved in the conduction process. The poor mobility is the result of the nature of glass. All glasses are disordered solids.

2.5.6 Physical Strength

At this point in time, the large samples with flawless physical quality required for meaningful measurements were not available. Some attempts to measure the tensile strength breaking fibers were made but with poor results. Most all the fibers had surface flaws. Ultimate tensile strength ranged from 500 to 1000 psi for three samples judged flawless. Later, we will discuss physical strength of glasses based on rupture modulus tests and the determination of Young's modulus and the shear modulus from acoustical measurements.

2.5.7 Softening Points

Softening point decreases with increasing molecular weight.

$$S > Se > Te$$
$$P > As > Sb$$
$$Si > Ge > Sn$$

2.6 Chemical Bonding in Chalcogenide Glasses

2.6.1 Composition Location in the Glass Forming Diagram

For ternary systems, glass forming compositions are experimentally found to exist within an area designated within a triangular composition diagram. We have not mentioned a factor very important in interpreting the chemical bonding in the different regions—stoichiometry. Figure 2.16 presents the results[16] determined for the glass forming region in the Ge-Sb-S system. Note the stoichiometric compound GeS_2 is designated as well as the stoichiometric compound Sb_2S_3. The compounds are stoichiometric because their atomic ratios are correct for Ge with a valence of +4 and S with a valence of −2, while Sb has a valence of +3 with again S with a valence of −2. A line is drawn in the diagram connecting the two stoichiometric compounds. Any glass composition along that line is stoichiometric because the compounds'

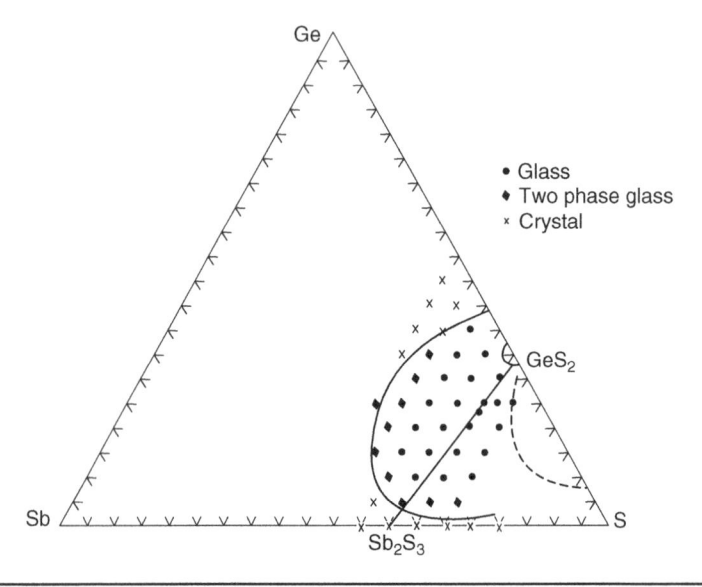

FIGURE 2.16 The glass forming composition diagram for the Ge-Sb-S system.

atomic ratios are correct. Now, chemical bonds Ge-S and Sb-S are thermodynamically favored, from a free energy of formation standpoint, over S-S bonds. On the sulfur-rich side of that line, there is more than enough sulfur to satisfy the bonding requirements of both Ge and Sb. The remaining sulfur is bonded to other sulfur atoms in chains or rings. However, at the stoichiometric line, all the sulfur is bonded to either Ge or Sb. There are no longer S-S primary bonds. Across the line in the metallic rich areas, there is not enough S to go around. The free energy of formation for Ge-S is greater than that for Sb-S. Ge will use up its share of S first. But well away from the line we may expect to find Ge-Sb bonds, Ge-Ge bonds, and Sb-Sb bonds. The Ge-S bond has a high enough energy level that in binary form it can transmit visible light. The Ge-Ge, the Ge-Sb, and the Sb-Sb are all metallic and do not transmit visible light. Figure 2.17 is a diagram[16] depicting transmission in the visible band for Ge-Sb-S glasses as a function of sulfur content. When the composition contains less than 55 percent sulfur, visible transmission is lost because metallic bonding becomes appreciable. This type of discussion can be applied to all the IVA-VA-VIA ternary systems regarding the bonds formed. In the chalcogen-rich area, the metallic elements bond to their share of the chalcogen. Chalcogen-chalcogen bonds exist. Across the stoichiometry line, after all the chalcogen is used up, metal-metal bonds will have to form. The change in the bonding accounts for the variation in physical and optical properties for glasses formed within the system.

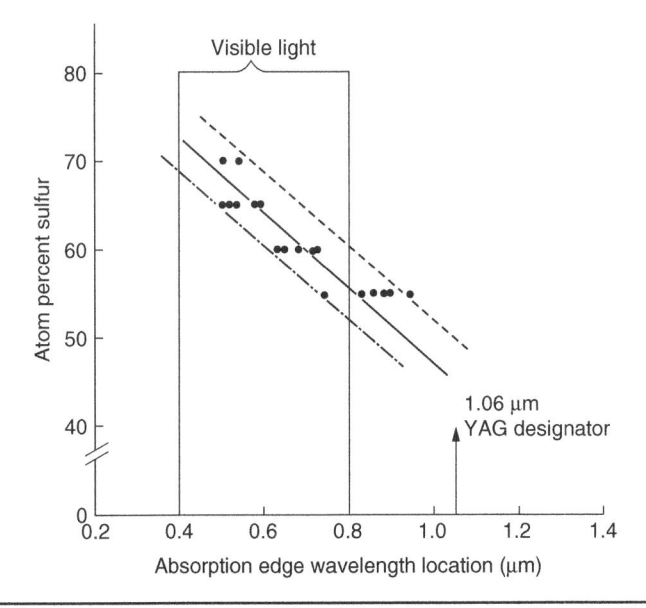

FIGURE 2.17 Absorption edge wavelength location as a function of sulfur content location in Ge-Sb-S glasses.

2.6.2 Molecular Vibrations of Constituent Atoms

Some insight into the molecular nature of the chalcogenide glasses may be gained by the use of some standard methods. One method already mentioned in Chap. 1 is far infrared reflection spectroscopy to observe the strong Restrahlen like bands due to the constituent atom pairs. The word *like* is added because the term is normally used to describe infrared reflection for crystalline materials, not glasses. The greater the ionic character of the bond formed between the atom pair, the more intense the absorption which in turn increases the magnitude of the reflection. From inspection of the shape of the curve, one can deduce the frequency[37] of the vibration between the two atoms to a fair degree of accuracy. Of course, if the sample can be ground thin enough and polished again, the absorption frequency may be directly measured through transmission. Absorption results may also be obtained by powdering the material and pressing into a pellet using KBr or TlBr and measuring IR transmission. Another way, and the most accurate, is to use a Raman spectrophotometer which directly measures the frequency of all the intense vibrations. Keep in mind, such instruments were not readily available at the time of the results of the program being described.

Figure 2.18, also seen in Chap. 1 as Fig. 1.7, shows the measured reflectivity for several chalcogenide glasses. The curves for the binary glasses As_2S_3 and Ge_2S_3 will yield the harmonic oscillator frequencies for the As-S bond and the Ge-S bond. The other two glasses are three

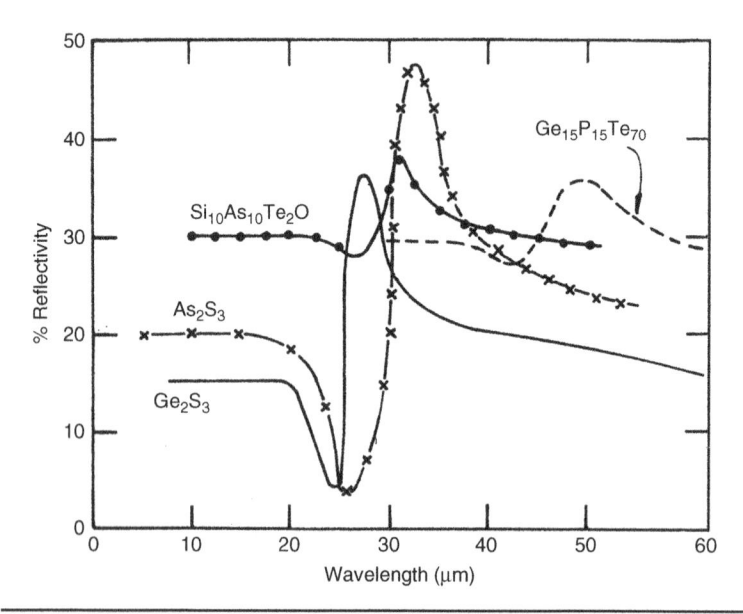

Figure 2.18 Far infrared Restrahlen-like reflection bands of some chalcogenide glasses.

component glasses and will yield the harmonic oscillator frequency for the dominant pair which in this case comprises Si-Te and Ge-Te. The other bond pairs absorb but not with the intensity required to produce a reflection band. Also, the intensity is affected by the concentration of the element in the glass composition.

The optical constants are interdependent. That is, the refractive index is really a complex number $N = n - ik$, where n is the real part of the refractive index and k is the imaginary part of the refractive index, sometimes called the extinction coefficient. The bulk absorption coefficient α can be calculated from $\alpha = 4\pi k/\lambda$, where λ = wavelength in centimeters. Reflectivity R is calculated from

$$R = \frac{(n-1)^2 + k^2}{(n+1)^2 + k^2}$$

In the transparent region, k is very small and is omitted in the calculation. However, in the region of the Restrahlen band, it becomes large, even the dominant term. The curves in Fig. 2.18 show the expected shape. The peak of reflectivity, the maximum absorption wavelength, and the wavelength for the harmonic oscillator do not coincide because of the interrelationship of the optical constants. The wavelength of the harmonic oscillator for each bond pair and the maximum absorption were determined, when possible, by using the inspection method described by Moss.[37] The values needed for the calculation are the magnitude of maximum reflectivity, the wavelength of maximum and minimum reflection, and the short wavelength refractive index. The calculated results found from the curves in Fig. 2.18 and for a number of other samples are shown in Table 2.9. In some cases only absorption results were available. From the results, we see that binary glasses are straightforward, yielding their oscillator frequency, As-S, Ge-S, Ge-Se, As-Se. A small change is observed when a third element is present. The absorption of the third element is not intense enough to produce a second band whether due to the concentration or the ionic character of the bond. At the very least, the different mass of the atom when coupled to the primary structure produces a frequency change. As an example, compare the oscillator frequency for the Ge-S binary and in the ternary Ge-S-Te.

Tellurium is an atom much heavier than sulfur which lowers the frequency of vibration. The force constant for each atom pair may be calculated from

$$\nu_o = \frac{1}{2\pi C} \times \left(\frac{\kappa}{\mu}\right)^{1/2}$$

where ν_o = wave number of harmonic oscillator frequency
C = speed of light
κ = force constant
μ = reduced mass

52 Chapter Two

System	Constituent Atoms Involved	Wavelength of Reflection Max. (μm)	Wave Number of Calculated Harmonic Oscillator Frequency v_o (cm^{-1})	Wave Number of Calculated Maximum Absorption Frequency v_{max} (cm^{-1})	
As-S	As-S	32	291	307	[313] T1Br
Ge-S	Ge-S	27.5	349	360	[370] T1Br
Ge-S-Te	Ge-S	28	342	355	—
Ge-P-S	Ge-S	27	358	366	—
Si-As-Te	Si-Te	31	307	322	[323] T1Br
Ge-P-Te	Ge-Te	50	196	205	[212] T1Br
Si-Se	Si-Se	—	382	—	[392] T1Br
Ge-Se	Ge-Se	40	234	250	—
As-Se	As-Se	44	217	—	[226] KBr
P-Se	P-Se	—	350	—	[363] KBr
P-S	P-S	—	525	—	[535] KBr
Ge-As-Te	Ge-Te	50	196	205	—
Ge-P-Se	Ge-Se	39	244	255	—
Ge-As-Se	Ge-Se	41	233	247	—
P-S	P=S	14.7	675	—	[710] KBr

TABLE 2.9 Wave Number of the Harmonic Oscillator for Glass Bond Pairs and Wave Numbers for Maximum Absorption Calculated from Far IR Reflection and Absorption Measurements

If it is assumed that the vibration between the atoms is a simple diatomic vibration, an estimate of the equilibrium interatomic distance can be calculated from the force constant by using Gordy's rule.[38]

$$Gordy's\ rule: \kappa = 1.6N\left(\frac{X_A X_B}{d_{AB}^2}\right)^{3/4} + 0.30$$

where κ = force constant (units dyne cm^{-1} × 10^{-5})
N = bond order = 1 in this case
X_A, X_B = Pauling electronegativities for A and B
d_{AB} = equilibrium distance between A and B

The interatomic distances for nine atom pair vibrations identified in the manner described were calculated and compared to the sum of

Bond	v_o cm^{-1}	dÅ Calculated (Gordy's Rule)	dÅ from Addition Covalent Radii	Δ
Ge-S	349	2.29	2.24	+0.05
Ge-Se	234	2.56	2.38	+0.18
Ge-Te	196	2.61	2.57	+0.04
As-S	291	2.87	2.21	+0.66
As-Se	217	2.89	2.35	+0.54
Si-Te	307	2.35	2.46	−0.11
Si-Se	382	2.15	2.27	−0.12
P-S	525	2.02	2.08	−0.06
P-Se	350	2.42	2.22	+0.20

TABLE 2.10 Interatomic Bond Distances Calculated Using Gordy's Rule Compared to Sum of Covalent Radii

covalent radii for each of the atoms in the bond pair. The results are shown in Table 2.10.

Both Ge and Si are probably in tetrahedron structures where bonds are symmetric and equal. The sum of covalent radii agrees quite well with the calculated value when a bond order of 1 is assumed and used in Gordy's rule. However, the agreement between the calculated and the addition of covalent radii is not as good for As-S and the As-Se. Their structures are probably pyramidal, which does not fit the simple diatomic model. This fact suggests that a more detailed analysis of the infrared vibrations may yield information concerning the molecular arrangements of the constituent atoms.

The molecular units may be thought of as individual molecules free to absorb and vibrate independent of their nearest neighbors and surroundings. In the close association of the solid environment, the vibrations will decrease in frequency generally. Since there is no uniform orientation from molecule to molecule, symmetry considerations used in analyzing the vibrational spectra of crystalline materials do not apply. In free molecules all vibrational modes are infrared active if a change in the electric dipole occurs due to the vibration. Some normal vibrations not infrared active can be observed by the Raman effect. The simplest approach is to assume that molecular units may involve three atoms and the structure may be X-Y$_2$ linear or X-Y$_2$ nonlinear. For four atoms, the structure may be X-Y$_3$ pyramidal. For five atoms, the X-Y$_4$ structure would likely be tetrahedral. The equations for the normal mode vibrations of these molecules are found in

Herzberg.[39] The expressions involve atomic masses, bond lengths, bond angles, and two force constants k and $k\delta$. The $k\delta$ is a measure of the restoring force opposing a change in the angle between the two valence bonds. The magnitude is about 10 percent of the k value and was assumed to be such in all the calculations. Interatomic distances were taken to be the sum of the covalent radii. With the equations and using Gordy's rule, the vibrational frequencies for four molecular gases typical of the four molecular configurations considered were calculated and compared with observed frequencies. The four gases were CO_2, SO_2, $AsCl_3$ and $SiCl_4$. Agreement between observed and calculated was poor except for the $AsCl_3$, X-Y_3 pyramidal case. A method better suited for polyatomic force constant prediction was developed by Somayajulu.[40] This method utilizes the elemental covalent force constants and electronegativity to predict force constants. The expression used is

$$K_{AB} = (K_{AA} K_{BB})^{1/2} + \Delta$$

where $\Delta = (X_A - X_B)^{1/2}$

Tables for elemental force constants are given by Somayajulu.[40] Values for single, double, and triple bonds are given including constants for hybridized orbitals such as sp^3, the tetrahedral structure. For silicon and germanium two force constants are given, single and sp^3 tetrahedral. Both the X-Y_2 linear and nonlinear molecules have three vibrational modes. In both cases the v_1 wave number corresponds to the symmetric stretching vibration, v_3 corresponds to the unsymmetric stretching mode, while the v_2 frequency is the bending mode. The v_1 modes are not infrared active because of the molecule symmetry but can be seen in Raman spectra. All four of the pyramidal molecule modes are infrared active. The calculated frequencies for pyramidal molecules in close agreement with observed has been mentioned before.

The observed frequencies for Si-Te, Si-Se, Ge-Te, and Ge-S in each case agree very well with the v_1 mode frequency calculated for the nonlinear X-Y_2 symmetric molecule. The vibrations for the As-S and As-Se molecules are quite different. They both match the v_1 mode of the pyramidal structure. The absorptions were found in binary glasses but not observed when a group IVA element was present. Surprisingly, the P-S and P-Se vibrations fit the X-Y_2 nonlinear configuration rather than the pyramidal configuration.

The differences may be related to the chemical differences between arsenic and the other VA elements phosphorus and antimony. The differences were pointed out while explaining glass forming tendencies. Also, we should keep in mind the change in bonding that will occur when compositions move from the chalcogen-rich region through the stoichiometric line into chalcogen-deficient compositions. No doubt, in this investigation or one similar, Raman spectra would have been beneficial in trying to untangle observed frequencies as related to structures. Later investigations using Raman results have been reported.[41]

2.6.3 Mass Spectrometric Investigation of Bonding in the Glasses

Chalcogenide glasses are made from volatile elements. An investigation using a mass spectrometer equipped with a Knudsen cell will yield information from emitted species concerning the types of bonds present and their relative stability. The Knudsen cell has a small hole in the top which allows vapors, under equilibrium conditions with the heated sample, to flow into the Bendix time of flight (TOF) mass spectrometer for analysis.

Ideally, the partial pressure of a constituent follows Raoult's law[35] which states the partial pressure of species A (P_A) is equal to the atomic fraction of A (X_A) times the pressure of pure A (P_A^0) at that temperature. Deviation from this value indicates bonding of A in the solid or liquid. Measuring the pressure of a species as a function of temperature yields thermodynamic information that can be related to the binding energy of the species in the solid or liquid phase. The slope of a plot of ln P_A versus $1/T$ yields a differential heat of solution for species A in the solid or liquid. A change in the slope over a temperature range will indicate

	Vapor Species	Appearance Temperature (°C)	ΔH	Softening Temperature (°C)
$Si_{15}Te_{85}$	Te	386	24 kcal	173
$Si_{15}As_{15}Te_{70}$	As	278	35	207
	Te	377	18	
$Si_{15}As_{45}Te_{40}$	As	300	—	292
$Si_{30}As_{15}Te_{55}$	—	—	—	359
$Ge_{10}As_{20}Te_{70}$	As	262	28	178
	Te	386	34	—
$Ge_{15}As_{45}Te_{40}$	As	233	36	300

TABLE 2.11 Detected Vapor Species from Heated Chalcogenide Glasses

a change in bonding within the solid or liquid. So the instrument identifies the chemical nature of the vapor species, the temperature at which it appears, and the amount with temperature change. To ensure the Knudsen cell would function as expected, it was calibrated using pure arsenic. The value of ΔH_V obtained for sublimation of arsenic was 31.2 kcal which matches the accepted value of 31 kcal.[42] The temperature range of operation was 25 to 500°C, low enough to ensure equilibrium conditions inside the Knudsen cell were maintained. The appearance temperature of each species was noted and a heat of vaporization determined when possible. A summary of results is presented in Table 2.11.

The heats of vaporization given are based on initial slopes. Note that in the glass $Si_{30}As_{15}Te_{55}$ there was no vapor species detected up to 500°C, well above its softening point, which is very unusual. The low-silicon high-tellurium glass showed only tellurium at 386°C, well above the softening point. The low-As, low-silicon glass gave off As and Te vapors. The high-As glass emitted As vapor in large amounts. The high softening Si-As-Te glass gave off no vapors. The high-tellurium Si-As glass showed As and Te emission. The Ge-As-Te glass emitted only As vapors. Figures 2.19 and 2.20 illustrate the differences between two Ge-As-Te glasses, one high Te and one high As.

High heats of vaporization, greater than for pure As, indicate strong bonds being formed in the glass. Only one glass was so stable in its bonding that it emitted no vapors. All the others emitted Te, As, or

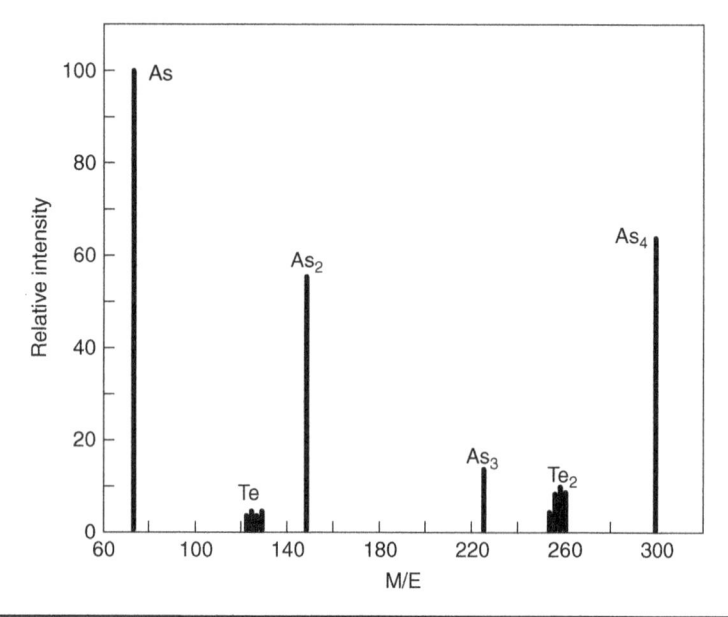

FIGURE 2.19 Mass spectrum of glass $Ge_{10}As_{20}Te_{70}$.

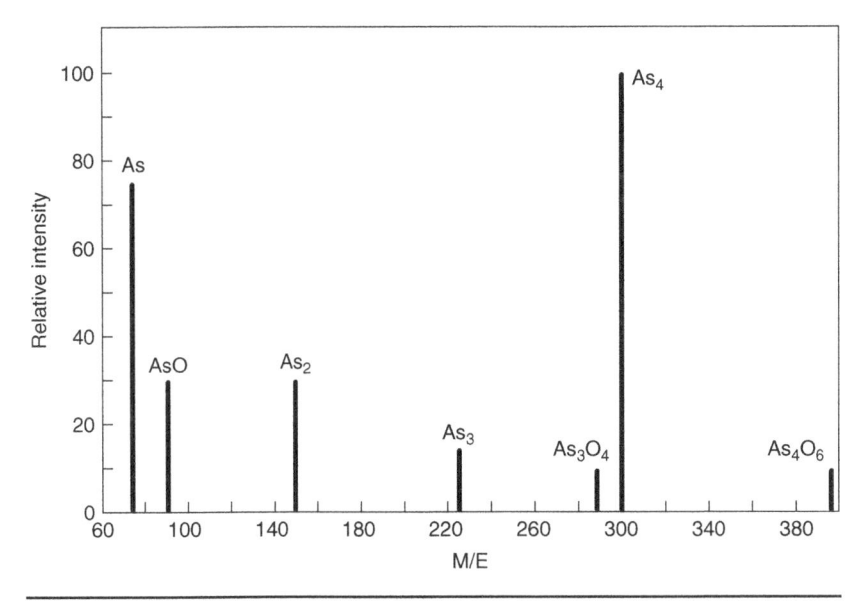

[Figure 2.20: Mass spectrum with peaks labeled AsO, As, As₂, As₃, As₃O₄, As₄, As₄O₆ plotted as Relative intensity vs M/E]

FIGURE 2.20 Mass spectrum of glass $Ge_{15}As_{45}Te_{40}$.

both. There were no three-component molecules detected in the vapor of any of the glasses. There were no vapors containing Si or Ge. Four forms of As vapor were detected: As, As_2, As_3, and As_4. The equilibrium of As vapor above high-arsenic-containing glasses becomes complicated. It appears that in compositions containing high concentrations of arsenic in the low-chalcogen regions, the arsenic is not bonded in the glass network, only captured. Similar results were observed in Ge-P-S glasses with the evolution of phosphorus molecules on quenching the melt, resulting in explosion of the quartz vials.

In studies carried out on Ge-Sb-Se glasses, the major vapor phase species was GeSe between 450 and 550°C. The heat of vaporization measure was 44.5 kcal. At 575°C, GeSe disappeared and Sb appeared at the melting point of Sb_2Se_3. Also note that the appearance of Te in Si-As-Te glasses and Ge-As-Te glasses corresponds roughly to the melting point of As_2Te_3. Such data are vital when heating and casting glasses in an open system.

2.6.4 X-ray Radial Distribution Analysis of Chalcogenide Glasses

It is well known that the molecular structure of crystalline compounds may be determined by X-ray diffraction analysis. What is not well known is that application of the method will yield information concerning atomic nearest neighbors sometimes even second-nearest neighbors of amorphous solid materials. Debeye[43] pointed out that

	$f_1 f_1$
Si–Si	196
Si-As	462
Si-Te	730
As-As	1040
As-Te	1664
Te-Te	2700
Ge-Ge	1020
Ge-As	1060
Ge-Te	1760

Radial Distribution Areas for Si-As-Te and Ge-As-Te Glasses			
	R_I (Å)	R_{II} (Å)	$R_I : R_{II}$
Si Te$_4$	2.62	4.14	1:8
Si$_{15}$As$_{15}$Te$_{70}$	2.58	4.12	1:12
Si$_{15}$As$_{45}$Te$_{40}$	2.52	3.95	1:5
Si$_{30}$As$_{15}$Te$_{55}$	2.50	4.12	1:4
Ge$_{15}$As$_{45}$Te$_{40}$	2.50	4.02	1:4

Note: R_I and R_{II} are distances for nearest and second-nearest neighbor interactions from the radial distribution function. $R_I : R_{II}$ is the area ratio between nearest neighbor peak and second-nearest neighbor peak.

TABLE 2.12 Relative Scattering Power between Various Atomic Interactions

one or two broad, diffuse diffraction bands were produced by liquids, glasses, resins, and unoriented polymers. The method was applied to Si-As-Te and Ge-As-Te glasses. X-ray scattered intensity measurements were taken using a standard Norelco wide-range goniometer. The relative scattering power for possible atomic pairs was calculated and is shown in Table 2.12. The radial distribution functions were calculated and plotted for each of glass systems SiTe$_4$, Si$_{15}$As$_{15}$Te$_{70}$, Si$_{15}$As$_{45}$Te$_4$, Si$_{30}$As$_{15}$Te$_{55}$, and Ge$_{15}$As$_{45}$Te$_{40}$. All the radial distribution function curves show maxima at R values less than 2 Å which are necessarily false. The curve for SiTe$_4$ is shown in Fig. 2.21. The area ratios R_I:R_{II} yield the most useful information. The results for the glasses studied are found in the lower portion of Table 2.12. In the case of SiTe$_4$ glass, the 1:8 ratio can only be explained if R_I consists of as many Si-Te bonds as possible with excess Te forming Te-Te bonds with R_{II} consisting of Te,Te interactions. The Si-Te interactions have only one-fourth the scattering power of the Te-Te. The structural interpretation is that Si-Te bonding in Si-Te glasses is nonlinear consistent with the infrared assignment of X-Y$_2$ molecules for Si-Te and

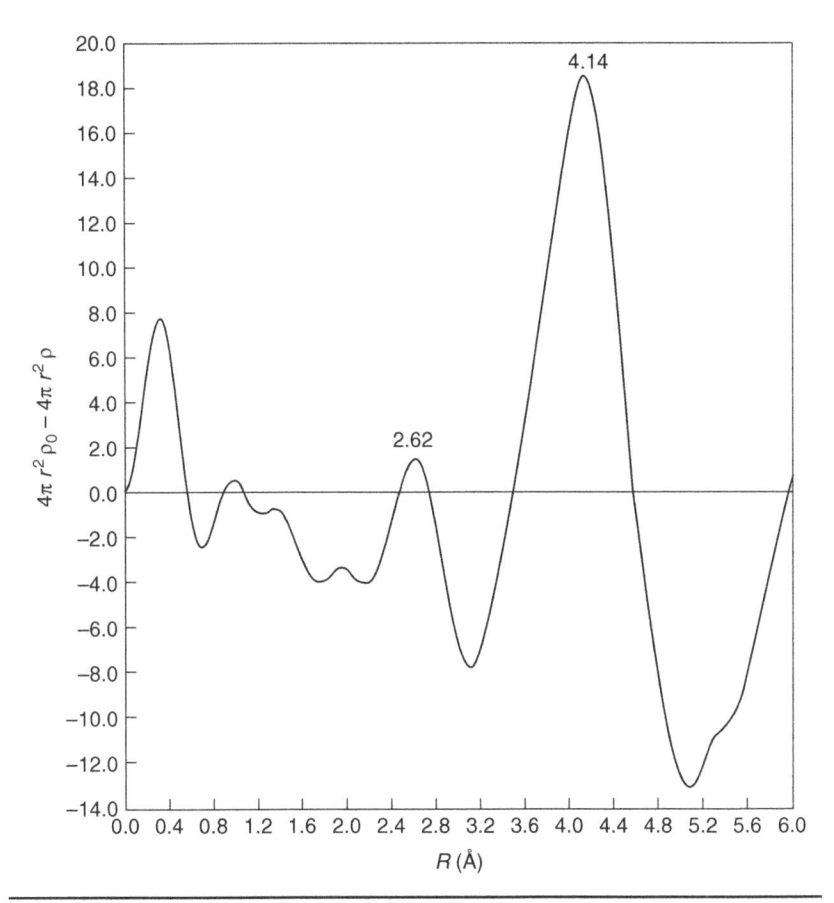

FIGURE 2.21 Radial distribution function of SiTe$_4$ glass.

Ge-Te glasses. The X-ray results did not support the existence of Si-Te$_4$ tetrahedral molecules which one would expect to be present.

2.6.5 Conclusions from the TI Exploratory Programs of 1962 to 1965

1. Seven ternary IVA-VA-VIA glass forming systems were investigated in this study. The Si-As-Te, the Ge-As-Te, and the Ge-P-Te glass systems showed the best infrared transmission in the 8- to 14-µm region. Attempts to form glasses using Sn or containing B failed. Silicon formed an unstable bond with selenium with a tendency to react with water to form hydrogen selenide. Glasses containing four elements were formed by combining the Si-As-Te glasses with the Ge-As-Te glasses.

2. The chemical bonding and thus the properties change with compositions within the glass forming region. The extent of

the glass forming region is dependent upon which elements are selected and the number of binary compounds they form with one another.

3. The chalcogenide glasses are characterized by covalent bonding. The molar refraction approach can be used to predict the refractive index of a glass using the covalent radius of each constituent atom, its atomic fraction in the glass, and the predicted density.

4. Glasses containing IVA elements Si and Ge are harder and stronger than those based more on VA elements P, As, and Sb. In some areas of the glass forming regions, excess P or As exists as free molecules which are emitted as such when the glass is heated. In ternary systems, no vapor molecules containing all three elements were observed.

5. The goal of finding chalcogenide glass compositions with physical properties comparable to those of oxide optical glasses was not reached and appears unlikely to be attained using the IVA-VA-VIA elements.

2.7 Chalcogenide Glasses Containing Transition Elements[44]

At this time, it was concluded that the chalcogenide glasses evaluated were never going to meet the window requirements of an airborne infrared optical system. Elements other than those in IVA and VA groups would have to be used in the glass composition. These must be elements with lower electronegativities and which form stronger chemical bonds with chalcogens. These elements can form multibonds, valences of +3 or +4, and form more than one stoichiometric compound with chalcogens. Titanium (Ti) and zirconium (Zr) from group IVB and vanadium (V) from group VB were selected. The crystalline chalcogenide compounds of these elements have melting points in excess of 1200°C. A high melting point ensures a low-viscosity melt and thus a homogeneous mixture. The methods previously used never exceeded 1000°C so a new approach must be developed. Basically, either an open or a closed system may be used for high-temperature compounding of materials that contain volatile constituents. In the open system the reactants are melted so rapidly that a homogeneous melt is obtained before an appreciable portion of the volatile constituent is evolved. The open method is simple and rapid. In the closed system, the reactants are sealed in a vial and slowly raised to a compounding temperature. With this system, the beginning composition is maintained. However, it is difficult to find materials that can withstand the high temperatures and resulting vapor pressure.

The open system was tried, first concentrating on placing the reactants in recesses in the water-cooled copper plate and using the

Chalcogenide Glasses 61

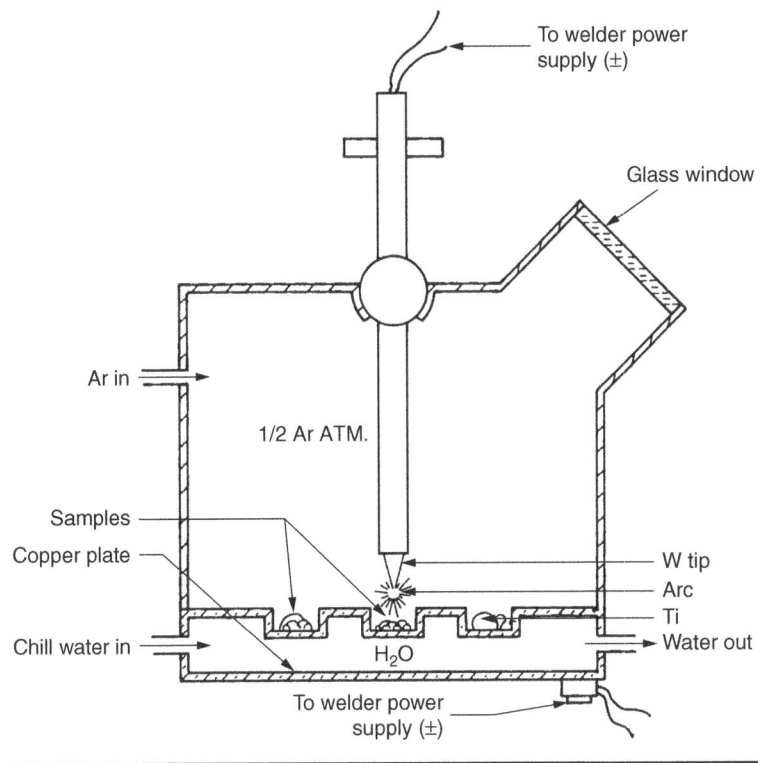

FIGURE 2.22 Arc-melter reactor.

arc directly to form a melt. Figure 2.22 shows a diagram of the arc-melter reactor. Notice in the diagram an argon atmosphere was used to prevent oxidation of the reactants. Reactants were badly scattered. Pressing reactants into pellets using organic binders did not help. Next, graphite containers were used to contain the reactants by using graphite lids and then some made of tungsten. Only partial melting occurred. The graphite containers were placed on small graphite rods to thermally isolate the chambers. Complete melting was obtained with long arcing but with great loss of Te as vapors which changed the composition considerably. Mixtures tested were Ti-Si-Se, Ti-Ge-Se, Ti-Ge-Te, Ti-Te-Se, and Ti-Se-S. Formation of $Ti-Si_2$ or $Ti-Ge_2$ from the elements using the arc was easy to accomplish. However, when they were brought in contact with molten mixtures containing S, Se, or Te, a violent evolution of the chalcogen occurred. The extreme stability of Ti-Si and Ti-Ge bonds is believed to be the cause of the excessive loss of the chalcogens. For this reason attention was turned to Ti-V based compositions that had lower melting points than the Ti-Si or Ti-Ge systems. However, attempts with the Ti-V-Te compositions produced similar results. No attempts were made using Zr in place of Ti. The chemistries of the two elements were very similar. The open system was abandoned after 67 attempts with no real success. The last attempt

62 Chapter Two

failed even though reactants used were known to form a stable glass. The method failed. Attention turned to the closed system. A diagram of the system used is shown in Fig. 2.23. After a few attempts with unsupported quartz vials failed, all samples were placed in evacuated, sealed quartz vials that were supported in chambers made from graphite or BN. The sample was lowered into the bottom chamber to a distance where the thermocouple fit into a recess in the bottom of the BN/graphite support chamber all inside the hollow SiC heating element. After everything is in place, the entire system, upper and lower chambers, is evacuated. When the desired reaction temperature is reached and maintained for a period of time, the sample is ready for quenching. The sample is pulled up into the upper chamber. The flap valve between the upper and lower chambers is closed. The pump valve is closed to maintain the vacuum in the lower chamber to avoid heat shocking the SiC heating element. Air is then let into the upper chamber to aid in the quench of the sample. The quench rate is very rapid, estimated to be 5 to 10°C/s. After cooling the sample is removed through the top of the upper chamber. Using this procedure, melt temperatures of up to 1700°C were reached and held. First attempts concentrated on making a glass from the Ti-V-Te system.

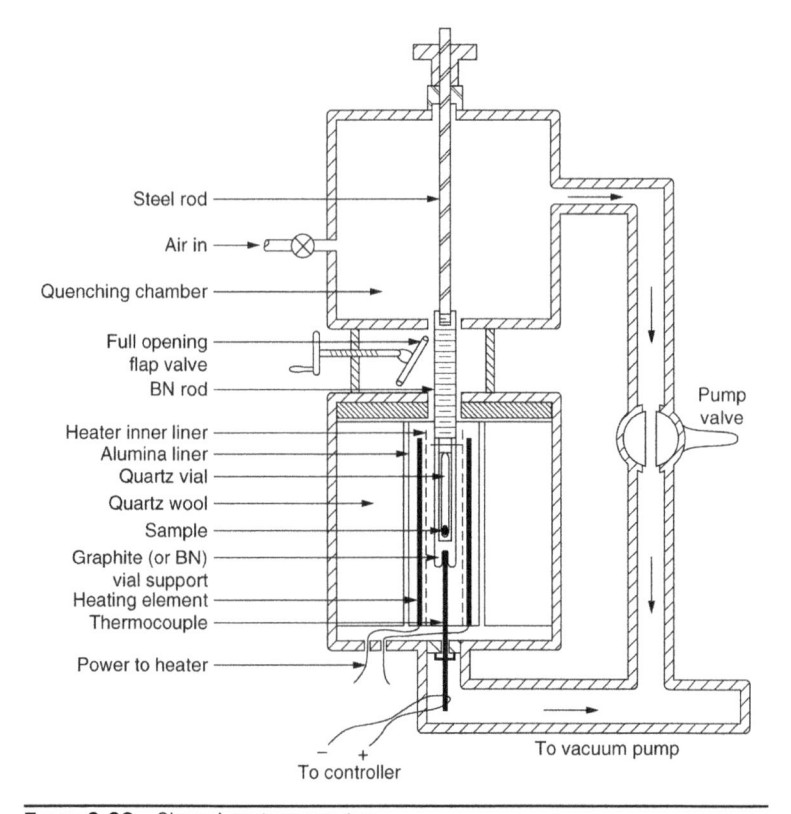

FIGURE 2.23 Closed system reactor.

Altogether there were 30 attempts. The composition of $Ti_{15}V_{15}Te_{70}$ was chosen as the most likely to succeed.

Figure 2.24 shows photographs of the results with that composition heated to 1300 to 1600°C. Notice that only at 1600°C were there signs that a melt had formed at the bottom of the vial. Obviously a higher temperature would be required. Raising the temperature to 1700°C and holding it longer produced homogeneous melts that

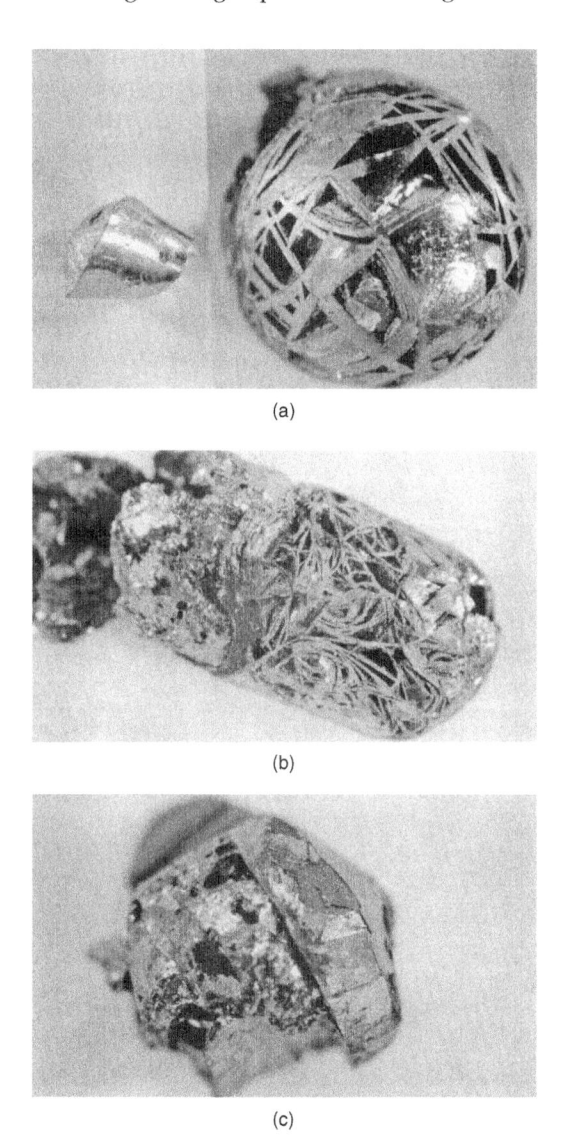

FIGURE 2.24 Photographs of the results from attempts to form glass from the composition $Ti_{15}V_{15}Te_{70}$. (*a*) Reactor temperature 1300°C, (*b*) reactor temperature 1500°C, and (*c*) reactor temperature 1600°C.

64 Chapter Two

cooled to crystalline solids. Photographs of the results are shown in Fig. 2.25. The solid was different in composition because Te vapor condensed on the walls of the vial. The true composition could be calculated by weighing the lost Te. For three runs the final compositions were calculated to be $Ti_{30}V_{30}Te_{40}$, $Ti_{23}V_{23}Te_{54}$, and $Ti_{11}V_{11}Te_{78}$.

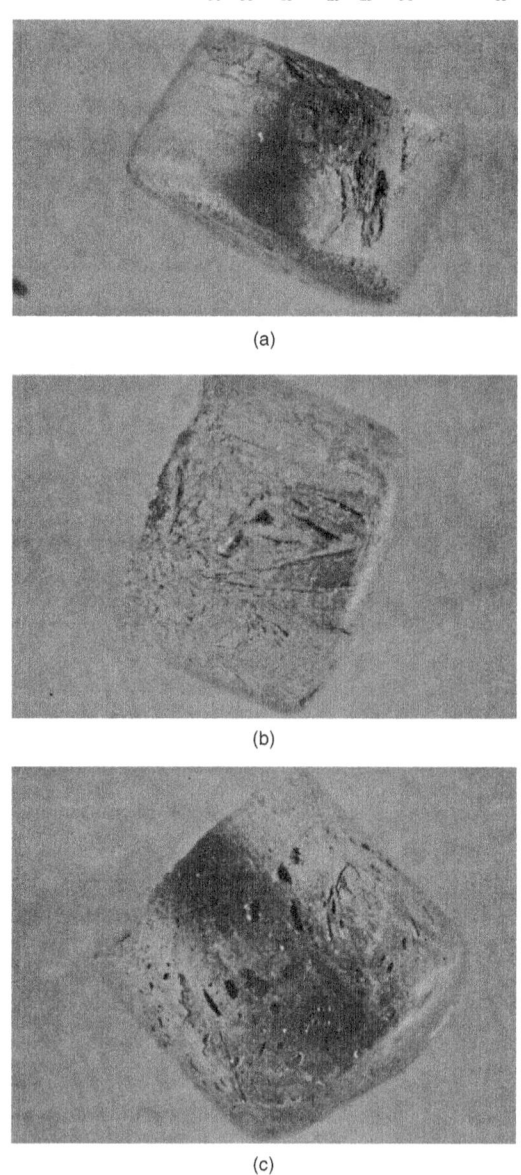

(a)

(b)

(c)

FIGURE 2.25 $Ti_5V_{15}Te_{70}$. (a) Sample no. 95 reactor temperature 1700°C, (b) sample no. 96 reactor temperature 1600°C for 2 h, and (c) sample no. 97 reactor temperature 1600°C for 1 h.

Chalcogenide Glasses 65

It was decided to use elements other than V with Ti to reduce the melt temperature. Ni and Ge were used along with Se for Te. The incidence of explosions increased. And still no glasses formed from homogeneous melts. Compositions containing Ti were abandoned. Perhaps Ti glasses did not form because the Ti-Te formation broke down any Te-Te amorphous chain-based structure. The coordination number for Ti is perhaps 6 or 8 Te atoms at the high melt temperatures. Similar considerations exist for the use of V or Zr. The net result is that crystallization occurs because of the decrease in viscosity of the melt. The need for a coordination number of 4 to form glasses is again supported by the results.

Attempts to form chalcogenide glasses using Ni and Ge with Se and S were carried out with melt temperatures of 1000 to 1200°C. First appearance of the samples was that of a homogeneous, single-phase glass. A photograph of the results using a Ni-Ge-Se composition is shown in Fig. 2.26. Unfortunately, this was not the case when the

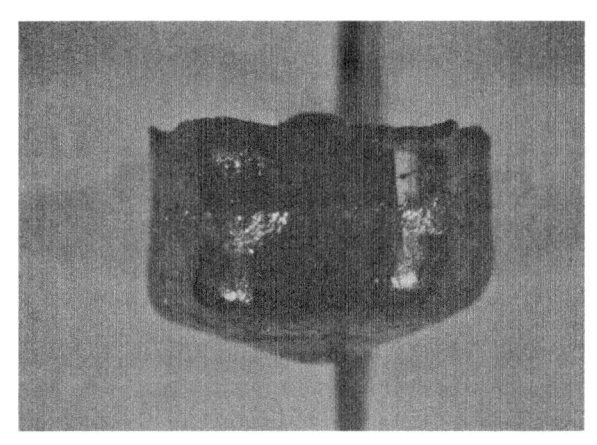

FIGURE **2.26** Results from a Ni-Ge-Se composition.

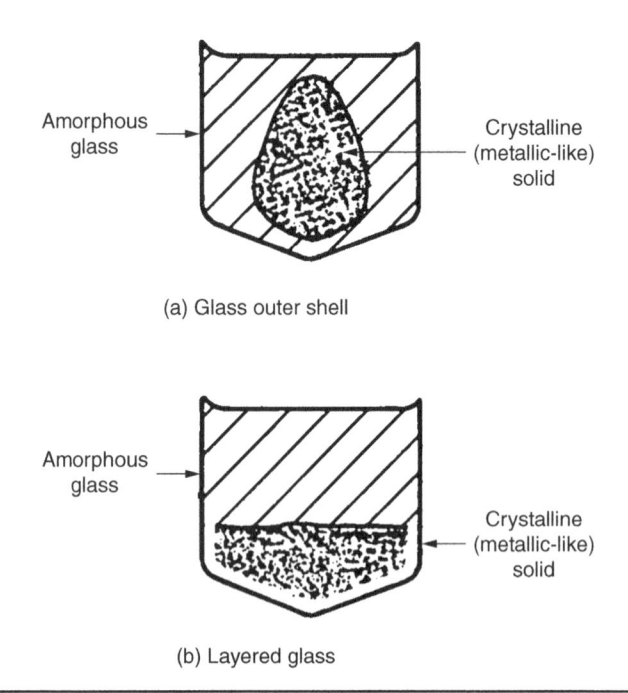

FIGURE 2.27 Configuration of two phase samples.

sample was sliced for evaluation. As shown in Fig. 2.27a, the outer shell was an amorphous glass (GeSe) while the inside was a second-phase crystalline (NiSe) solid. Other samples represented in Fig. 2.27b were layered where the amorphous lower-density glass was on top of a denser crystalline phase solid. Similar results occurred with Ni-Ge-Te, Ni-Ge-S, and Ni-Ge-Se.

The use of Zn and Mn with Ge in Zn-Ge-Se and Mn-Ge-Se did produce glasses. Compounding temperatures were 1200 to 1300°C. Softening points of the glasses were around 300°C. Their appearance and infrared transmission were similar to those of other Ge-Se glasses. The Te-based compositions Mn-Ge-Te, Ni-Zn-Te, Ni-Ge-Te, and Zn-Ge-Te were prepared and compounded at 1300°C. All had the same metallic luster and were very crystalline inside. Photographs of samples from three of the systems are shown in Fig. 2.28. Table 2.13 summarizes the results obtained for all compositions attempted using the closed system.

2.8 Discussion of Results

This effort was based on the assumption that a chalcogenide glass could be formed using elements known to form high melting compounds if the reactants could be heated hot enough to form a melt

Chalcogenide Glasses 67

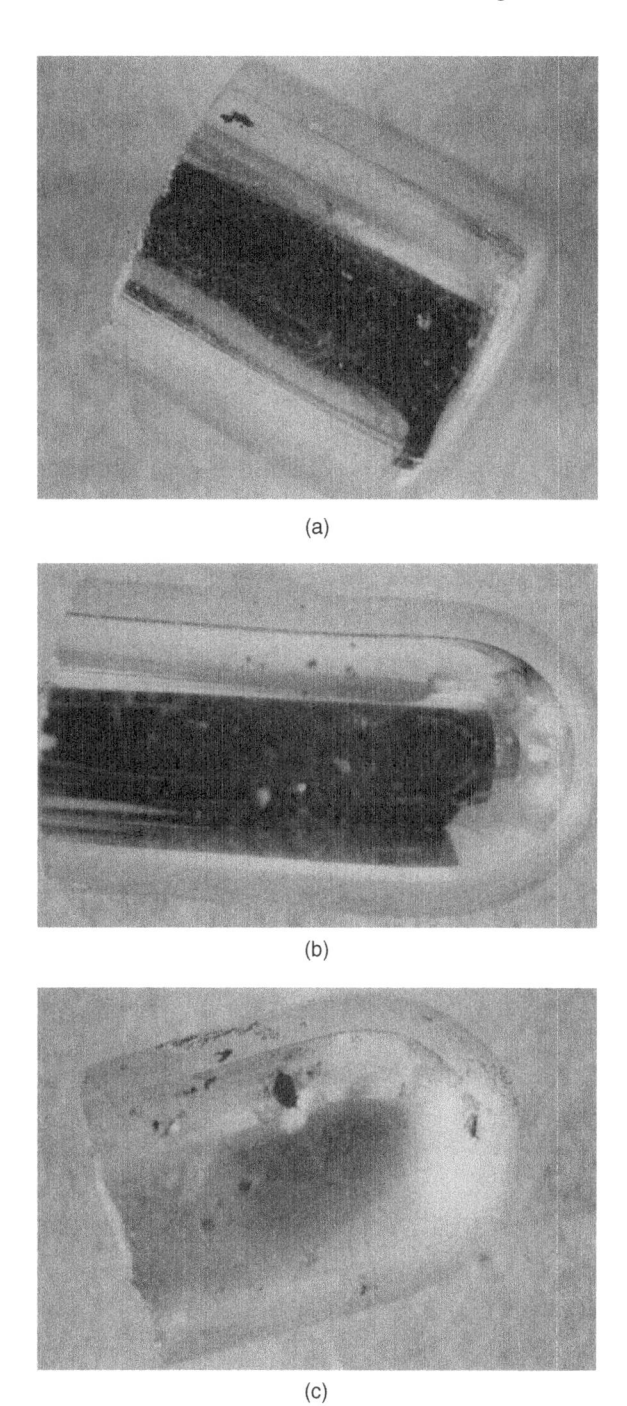

FIGURE 2.28 Photographs of tellurium samples. (*a*) Ni-Zn–Te, (*b*) Zn-Ge–Te, and (*c*) Mn-Ge-Te.

System	Homogeneous Melt Obtained	Glass	Remarks
Tellurium Compositions			
Ti-V-Te	Yes at 1600–1700°C	No	Dense, hard, crystalline solid
Mn-Ge-Te	Yes at 1300°C	No	Crystalline solid, metallic appearance
Ni-Zn-Te	Yes at 1325°C	No	Crystalline solid, metallic appearance
Ni-Ge-Te	Yes at 1300°C	No	Crystalline solid, metallic appearance
Zn-Ge-Te	Yes at 1300°C	No	Crystalline solid, metallic appearance
Ti-Ni-Te	No at 1500–1600°C	—	Porous, crystalline
Ti-Si-Te	No at 1350°C	—	Porous, crystalline
Selenium Compositions			
Ni-Zn-Se	Yes at 1225°C	Yes	Some infrared transmission, low softening point
Mn-Ge-Se	Yes at 1300°C	Yes	Low infrared transmission, contains crystallites, softening point ~290°C
Zn-Ge-Se	Yes at 1200°C	Yes	Low infrared transmission, contains crystallites, softening point ~310°C
Ni-Ge-Se	Yes at 1200°C	Yes	Two-phase, Ge-rich glass, Ni-rich crystalline phase
Ti-V-Se	No at 1400°C	—	Porous, crystalline
Ti-Ge-Se	No at 1400°C	—	Porous, crystalline
Sulfur Compositions			
Ni-Ge-S	Yes at 1000°C	Yes	Ge-rich glass on top of Ni-rich crystalline phase

TABLE 2.13 Results of Attempting to Compound High-Temperature Chalcogenide Glasses Using the Closed System Method

before quenching. The transition elements titanium and vanadium would be paired with silicon and germanium in sulfur-, selenium-, and tellurium-based compositions. It was found that an open system based on an arc welder would not work because the chalcogens were immediately evolved. A closed system was devised in which the ingredients were sealed in quartz vials supported by cylinders made from BN or graphite. Melt temperatures of 1200 to 1700°C were reached. It was found that Ti-Te, Ti-Se, and Ti-S bonds that were formed were so strong that an amorphous chalcogen chain-ring structure could not form. Homogeneous melts at high temperatures did form and were quenched, but the results were all crystalline. In place of Ti and V, the other elements Zn, Mn, and Ni were used with Ge. Some glasses with low softening points that transmitted infrared did form. In other cases, two-phase glass and crystalline materials formed. No tellurium-based glass was formed. No glass comparable to oxide glasses resulted.

References

1. R. Frerichs, *Phys. Rev.* 78, 643 (1950).
2. Rudolph Frerichs, *Opt. Soc. Am.* 43, 1153 (1953).
3. C. Schultz-Sellack, *Ann. Physik* 139, 162 (1870).
4. W. A. Fraser and J. Jerger, *J. Opt. Soc. Am.* 43, 332 (1953).
5. C. J. Billian and J. Jerger, Contract No. Naval Office of Research-3647(00), January 1963.
6. J. Jerger, Contract No. Air Force 33(657) 8560, June 1963.
7. J. Jerger and R. Sherwood, Contract No. Naval Office of Research-4212(00), August 1964.
8. *The Structure of Glass*, vol. 2, *Proceedings of the Third All Union Conference on the Glass State*, 1959, Consultants Bureau, New York, 1969.
9. S. Nielsen, *Infrared Phys.* 2, 117 (1962).
10. J. A. Savage and S. Nielsen, *Phys. Chem. Glasses* 5, 82 (1964).
11. J. A. Savage and S. Nielsen, *Phys. Chem. Glasses* 7, 56 (1966).
12. J. A. Savage, *Infrared Optical Materials and Their Antireflection Coatings*, Adam Hilger Ltd., Bristol and Boston, 1985.
13. A. R. Hilton and Maurice Brau, *Infrared Phys.* 3, 69 (1963).
14. A. R. Hilton, Naval Office of Research 3810(00), September 1965.
15. A. R. Hilton, N00014-66-C0085, July 1966.
16. A. R. Hilton, Defense Advanced Research Projects Agency (DARPA) Contract No. N00014-73-C-0367, June 1974.
17. *Physics and Chemistry of Glasses*, vol. 7, 105-126 (1966).
    Part 1, A. R. Hilton, C .E. Jones, and M. Brau
    Part 2, A. R. Hilton and C. E. Jones
    Part 3, A. R. Hilton, C. E. Jones, R. D. Dobrott, H. M. Klein, A. M. Bryant, and T. D. George
18. Valentina Kokorina, *Glasses for Infrared Optics*, CRC Press, Boca Raton, Fla., 1996.
19. A. David Pearson, Electrochemical Society Meeting, Los Angeles, Calif., 1962.
20. *Semiconductor Effects in Amorphous Solids*, W. Doremus, ed., North Holland Publishers, Amsterdam, 1969.
21. *Amorphous and Liquid Semiconductors*, M. H. Cohen and G. Lucovsky, eds., North Holland Publishers, Amsterdam, 1971.
22. R. J. Patterson, 15th National Infrared Information Symposium (IRIS) at Ft Monmouth N.J. (1966), 1967; U.S. Patent 3,360,649.

23. Infrared Transmitting Materials, Materials Advisory Board, MAB-243.
24. J. Pappis and B. A. di Benedetto, *Chemical Deposition of Multispectral Domes*, Air Force Materials Laboratory (AFML), AFML-TR-75-27, 1975.
25. Charlie Jones and Harold Hafner, Final Technical Report, Contract No. AF 33 (615)-3963, 1968.
26. Harold C. Hafner, Final Technical Report, AFML-TR-72-54, 1972.
27. Linus Pauling, *The Nature of the Chemical Bond*, Cornell University Press, Ithaca, N.Y., 1948.
28. R. L. Myuller, *Solid State Chemistry*, Consultants Bureau, New York, 1966.
29. Thomas J. Loretz, Computer Engineering Service, Personal communication, 2004.
30. W. H. Zachariasen, *J. Am. Chem. Soc.* 54, 3841 (1932).
31. E. Mooser and W. B. Pearson, *Acta Crystallog.* 12, 1015 (1959).
32. A. R. Hilton, *Phys. Chem. Glasses* 9, 148 (1968).
33. Max Hansen, *Constitution of Binary Alloys*, McGraw-Hill, New York, 1958.
34. R. L. Myuller, L. A. Bardakov, and Z. U. Borisovaz, *J. Univ. Leningrad* 10, 94–101 (1962).
35. Samuel Glasstone, *Text Book of Physical Chemistry*, D. Van Nostrand, Princeton, N.J., 1958, p. 529.
36. W. A. Weyl and E. C. Marboe, *The Constitution of Glasses: A Dynamic Interpetation*, Interscience, New York, 1962.
37. T. S. Moss, *Optical Properties of Semiconductors*, Academic, New York, 1959.
38. Walter Gordy, *J. Chem. Phys.* 14, 305 (1946).
39. Gerhard Herzberg, *Molecular Spectra and Molecular Structure: II, Infrared and Raman Spectra of Polyatomic Molecules*, D Van Nostrand, Princeton, N.J., 1956.
40. G. R. Somayajulu, *J. Chem. Phys.* 28, 814 (1958).
41. G. Lucovsky, J. P. De Neufville, and F. L. Galeener, "Study of Optic Modes of $Ge_{30}S_{70}$ Glass by Infrared and Raman Spectroscopy," *Phys Rev.* 9, 1591 (1974).
42. A. Glassner, "The Thermochemical Properties of Oxides, Fluorides and Chlorides to 2500 ^0K," Argonne National laboratories, ANL-5750, U.S. Government Printing Office, Washington.
43. P. Debye, *Ann. Physik* 46, 809 (1915).
44. A. Ray Hilton, "Titanium Chalcogenide Infrared Transmitting Glasses," Contract No. 14-66-COOO85, Naval Office of Naval Research, 1967.

CHAPTER 3
Glass Production

3.1 Reactants

After a glass composition has been selected for use, a method of preparing the glass in quantity with high quality must be developed. Of first importance is identifying a reliable source of the required elemental reactants in high-purity form. The transport properties, electrical conductance, of crystalline materials may be dominated by small concentrations of impurities in their reactants. Chalcogenide glasses have already been described as poor electronic conductors, so the metallic impurity effect on conductance is of minor consequence. However, their optical properties may be adversely affected by small concentrations of impurities.

Because of the great importance of crystalline semiconductor materials, tremendous effort has been spent producing important reactants in high-purity form and developing precise methods to verify the purity of the product. Reactants used in chalcogenide glasses have benefited greatly from these efforts. Most of the elements used to produce what we might call electronic materials are by-products of primary metals production. The most important are copper, lead, zinc, silver, and aluminum. A good example is the production of pure copper at the Asarco Plant in Amarillo, Texas. Plates of 99 percent copper are electroplated on titanium plates in almost 0.5-mi-long electroplating facility. A photograph[1] of the tank house is shown in Fig. 3.1. The design goal of the unit is over 400,000 tons/yr of refined copper. The resulting copper is 99.90 percent pure. A slime of waste, less than 0.1 percent of the beginning copper, falls to the bottom of the cell. In that waste are low concentrations of silver, gold, platinum, palladium, antimony, arsenic, cadmium, indium, selenium, tellurium, and thallium. These elements are chemically separated and then purified by chemical means and sold to industry. In similar processes, germanium is a by-product in the production of zinc and gallium, used to make the crystalline semiconductor gallium arsenide, which is a by-product of the production of aluminum.

The costs of reactants used in chalcogenide glasses are not stable and may vary greatly. The supply of each depends upon the rate of

FIGURE 3.1 The Asarco copper refining electroplating tank house at Amarillo.

production of the primary metals. When primary metal production is down and demand for the by-products exceeds the available supply, the price may rise dramatically, as for any other commodity. Another factor is purity. To be useful in electronic materials production, the purity standards are five 9s at a minimum sometimes reaching purity levels of seven 9s. There has been a tremendous improvement in purification and verification of purity techniques since the start of the semiconductor industry. The production of high-purity chalcogenide glasses has benefited as well from this development. For example, from one supplier the listed impurities in a high-purity six 9s arsenic in parts per million (ppm) are C, 0.09; Na, 0.03; and Cl, 0.08.

For five 9s sulfur from a supplier, the impurities listed in parts per million were Ca < 0.2, Cu < 0.5, Fe < 0.5, Mg < 0.2, Si < 1.0, and Ag < 0.2 with the values for the common impurities C and H_2O not even mentioned. Residual impurities vary among the sources of the ore used in the production of the primary metal and the production means used.

Major impurities affecting optical performance of chalcogenide glasses are gas molecules of water, hydrogen sulfide, hydrogen selenide, and hydrogen telluride. Carbon, silica, and other oxides produce unwanted absorption in wavelength regions of interest. Metals such as iron produce absorption at the short wavelengths near the beginning of transmission. Sulfur glasses and selenium glasses are considered insulators or semi-insulators in their electronic conduction. The impurity iron can dope the glasses, just as if they were crystalline, forming a deep energy level in their forbidden gaps which results in increased absorption in visible or near-infrared light. Unfortunately, the analytical results furnished with the high-purity element may have been developed for other larger, more important applications than infrared glass. The impurities of consequence may not have been measured. The metallic impurities are often covered by routine emission spectrographic analysis. Negative impurities are more difficult and are often neglected. Arc-source mass spectroscopy will measure negative impurities such as chlorine or sulfur. Often, the only way to find out for sure if a reactant meets the need of the glass is to make a batch, perhaps a kilogram of the glass, and evaluate the outcome, usually by measuring infrared transmission through a sample 1 to 2 cm thick. The test is expensive but often necessary.

3.2 Compounding Methods

The simple method of compounding a glass is illustrated in Fig. 3.2. Reactants are accurately weighed and placed in a glass tube, preferably a pure silica tube. Semiconductor-grade quartz is extremely pure and can be used without fear up to temperature of 1000°C under vacuum. A cap is provided for the end of the tube so that it can be evacuated after the reactants are in place. The loaded tube is placed in a rocking furnace and heated while pumping so that any moisture can be removed from the chamber. The temperature is raised enough to melt the chalcogen being used to remove dissolved gases. One must keep in mind sulfur melts at 119°C and boils at 445°C, selenium melts at 217°C and boils at 685°C, while tellurium melts at 450°C and boils at 990°C.

After a period of time, the pump tube is sealed and removed so that the rocking can begin to form the glass into a homogeneous melt. Temperatures are kept below the boiling point of the volatile chalcogen, allowing time for the metallic elements to react which lowers the volatility of the melt. Sulfur-based glasses require special care because of sulfur's low boiling point and because reaction with arsenic, as an example, is very exothermic. The heat generated may increase the internal pressure due to the presence of sulfur that has not reacted. Some germanium-containing glasses are reacted for hours to ensure

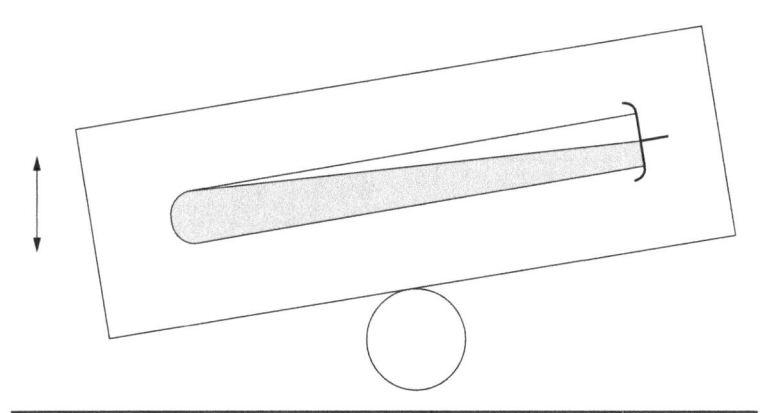

FIGURE 3.2 Reactants sealed in a tube placed in a rocking furnace.

a complete reaction of all constituents. With the quartz chamber under vacuum, reaching the boiling point of the chalcogen offsets the atmospheric pressure from outside. Each glass composition is different relative to internal pressure and heating. Care must be taken to avoid excessive internal pressure and quartz failure with the resulting reaction of the molten glass with the atmospheric oxygen. Compounding should be carried out in a closed room or hood with provisions made for exhausting any fumes in case of quartz failure.

After the reaction period is over, while still rocking, the temperature is lowered to a point well above the softening point of the glass where the melt is still fluid. At this point, the glass is ready to be quenched. The furnace is placed in a near-vertical position. Prior arrangements have been made to prevent the tube of glass from falling from the furnace. Room air or blown air comes in contact with the surface of the tube. The temperature is monitored to ensure the glass reaches and stays well below T_g. After the furnace has cooled, the glass may be left to cool in room air. After cooling is complete, the quartz tube is vented to air. The quartz tube is carefully fractured and removed from the glass. The result of the operation is a 1- to 2-kg cylinder, sometimes called a *boule*, of the specified glass composition ready for evaluation.

3.3 Compounding with Reactant Purification

In the early period of producing chalcogenide glasses at TI, the rocking furnace method described was used. In the early 1970s, the U.S. Air Force launched a great effort dedicated to developing large windows for high-energy CO_2 lasers. TI became involved because in its active program large glass windows had already been cast. Reduction of absorption at 10.6 μm was a top priority. Examination of the glass using an infrared microscope identified the existence of particulate matter in the glass that turned out to be carbon from the selenium. An effort[2,3] was made to remove all impurities and oxides by distillation using filters to establish the intrinsic transmission for TI 1173 and TI 20 glasses. The apparatus designed to purify the compounded glasses is shown in Fig. 3.3.

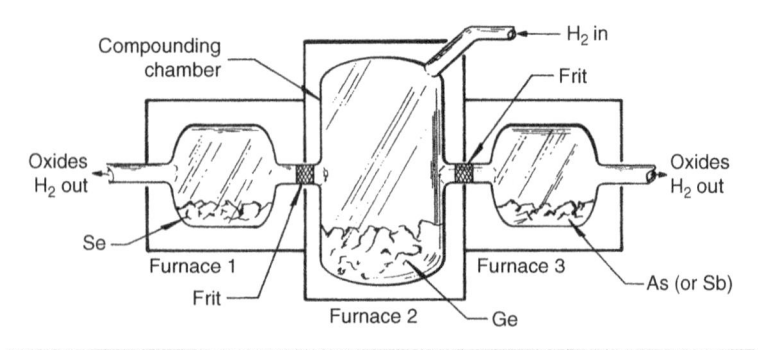

FIGURE 3.3 Preparation of high-purity TI 1173 and TI 20 glasses.

Germanium was considered to be the reactant of highest purity with no particulate matter, so it was loaded in the center chamber, with selenium on one side and arsenic or antimony on the other. All the materials were heated slightly, and hydrogen flowed through the chambers with the hope of reducing surface oxides. The next step was to seal the two side tubes, evacuate the chambers, and seal off the top tube. Then selenium on one side and arsenic or antimony on the other were heated and distilled through the porous quartz filter (frit) to remove particulate matter. It is worth mentioning that the distillation of antimony was extremely difficult and time-consuming. Antimony has only 10 mm of vapor pressure at 1000°C. Particulate matter (C) was not present in the resulting glass. The absorption in both glasses decreased relative to glass produced using the production method current at the time. A definite correlation for both glasses was found between increased silicon levels and absorption at 10.6 and 9.4 µm. The silica contamination, 2 to 5 ppm, probably occurred during the quartz fabrication using the hydrogen oxygen torch. The resulting oxygen concentration for both glasses, 5 ppm, was due to silica and the 8 ppm in the reactant germanium. Addition of 5 ppm Al as an oxide getter reduced absorption at 9.4, 10.6, and 13 µm. Table 3.1 lists the impurity levels found in the glasses. In the table, P refers to glass prepared using the existing production process. Most of the effort was concentrated on the TI 1173 glass which was currently used in existing systems. The reductions for TI 1173 are shown in Fig. 3.4 and compared

Glass	[Si] ppm	[O] ppm ±2	[Al] ppm	β 10.6 µm cm^{-1}	β ~ 13 µm cm^{-1}	β ~ 9.4 µm cm^{-1}
1173-P	2.5	4	0.3†	0.06	0.65†	—
1173-92	1.4	5.5	6.1	0.012	0.19	0.0075
1173-107	2.6	5.5	6.3	0.019	0.25	0.022
1173-109	2.2	4.5	6.1	0.024	0.25	0.025
1173-122	1.1	—	5.7	0.013	0.22	0.015
20-P	2.2	6	0.4†	0.056	0.65†	—
20-98	4.4	5	5.3	0.053	0.34	0.045
20-102	3.3	—	5.6	—	—	—
Ge	N.D.	8	N.D.	—	—	—
Sb	N.D.	—	N.D.	—	—	—
Se	N.D.	—	N.D.	—	—	—

†No aluminium added. N.D. = not detected.

TABLE 3.1 Impurities and Absorption in TI 1173 and TI 20 Glass Samples Used in Reactant Purification Evaluation. P Designates Standard Production

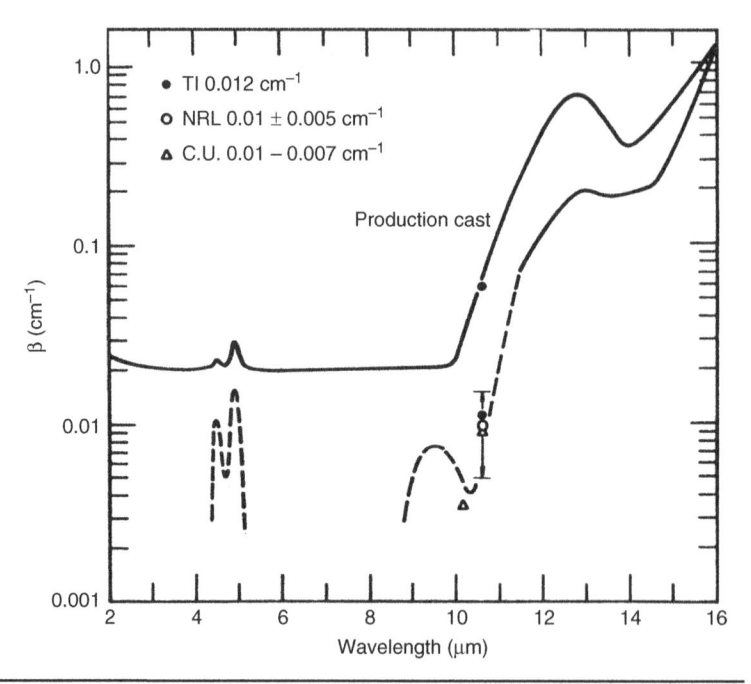

FIGURE 3.4 TI 1173 absorption as a function of wavelength comparing standard production glass to glass made with reactant purification.

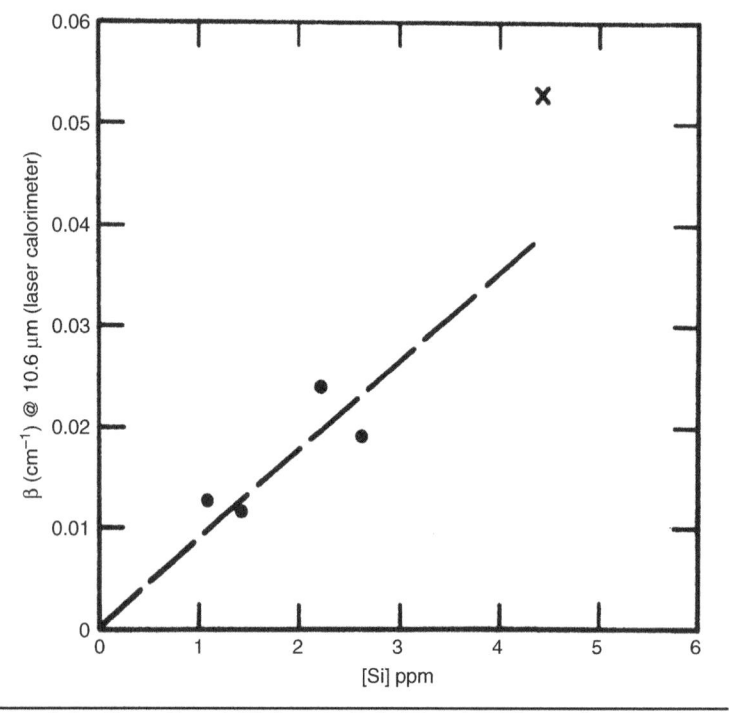

FIGURE 3.5 Absorption at 10.6 μm in TI 1173 and TI 20 as a function of silicon content.

to results from the standard production. Also shown are results of absorption at 10.6 µm, determined by measuring the heat rise in a sample while transmitting a CO_2 laser beam of known intensity. The measurement, termed *laser calorimetry*, was applied to the same glass by TI, the Naval Research Laboratory, and the Catholic University. Results were from 0.012 to 0.007 cm^{-1}. A correlation between Si content and absorption at 10.6 and 9.4 µm was found for both TI 1173 and TI 20. The 10.6-µm results are shown in Fig. 3.5. The X in the figure is for the one TI 20 glass tested. Obviously, the reactant purification step would substantially improve quality and the transmission if adopted into the production of these two infrared-transmitting chalcogenide glasses.

3.4 Open Casting Methods

Early preparations of glass did not involve casting. The first casting method employed at TI was termed a *pour caster*. A diagram of the unit is shown in Fig. 3.6. All action takes place in a sealed chamber equipped with outside controls and an observation window. The atmosphere inside is controlled flow of inert gas. The glass to be cast is placed in the crucible and moved into the melting furnace where it can be melted and stirred. After mixing and reaching of the proper temperature, the stirring is stopped, allowing any bubbles to

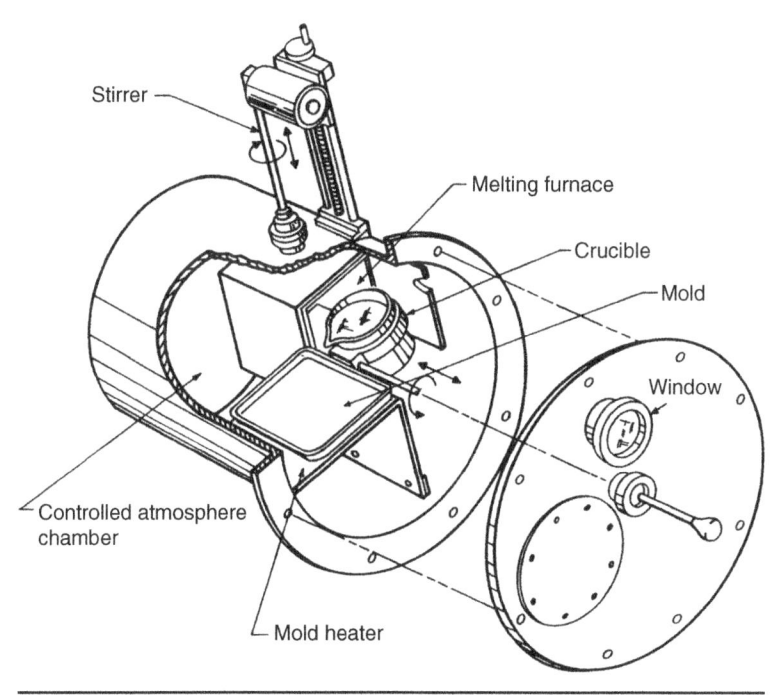

FIGURE 3.6 Diagram of the TI pour glass casting unit.

Chapter Three

rise to the surface. The crucible is moved from the furnace and tilted, and the glass is poured into the mold. The glass is cooled and then annealed. Plates 5 × 5 in up to 5 × 7 in were produced. The homogeneity of glass thus produced was not good because of excess striae.

A second casting method was developed in which the melted glass in the crucible was allowed to flow through a bottom hole tube into a mold directly below. This method was referred to as the *bottom hole* caster. Initially, the bottom hole was plugged by the first glass to melt. When casting time arrived, a heater around the bottom tube was turned on, and the glass plug melted so the glass flowed freely into the mold below. Figure 3.7 shows two photographs comparing the resulting homogeneity of glass cast using the two methods. The photographs are made from a striae scope. Collimated near-infrared light is passed through the glass plate, and the image is photographed. Light phase cancellation occurs when the light passes through striations in the glass, producing an image.

The bottom hole casting was a big improvement. Optical homogeneity had become an important performance specification for high-resolution infrared optical systems. The positive results from this casting change allowed an upgrade of the required **MTF** (modulation transfer function) image spoiling test score used to test infrared glass. The measurement required at 10 lines per millimeter an MTF score of

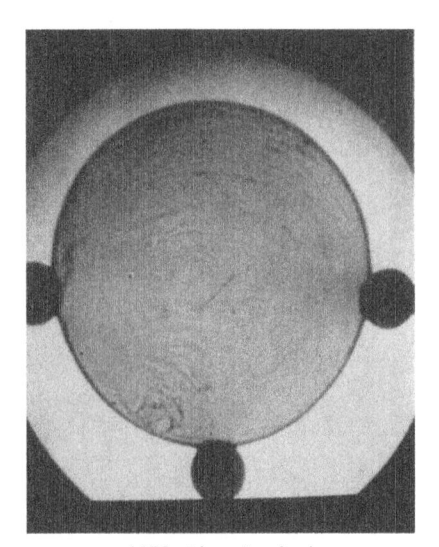

1173 striae standard
Cast: April 1971
(old casting method)

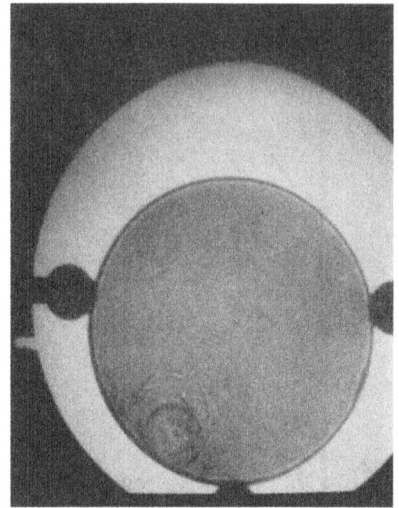

Blank: 34173
Cast: 6/11/73
(bottom cast)

FIGURE 3.7 Striae comparison of two 5.5-in-diameter TI 1173 glass blanks cast by different methods.

85 percent. With the new casting method, the MTF score was raised to 94 percent, equal to that of germanium. However, for TI 1173, the bottom glass plug was found to form a large amount of crystallites grown during the glass mixing process prior to casting. When casting occurred, the crystallites flowed out and ruined a sizable portion of the cast plate. The yield of striae free large diameter lens blanks was very low.

The author was asked to assume control of the TI glass production in the fall of 1974. The first change made was to implement the element purification process for compounding all the TI 1173 in the production process. It was not necessary to do the same for TI 20 glass since it was no longer in production. In working with the glass blowers, a process was devised so that double-chamber quartz tubes could be fabricated on the glass lathe. The chambers were separated by a porous quartz filter called a frit. One chamber end, the one to hold the finished glass, was left rounded and closed. The other had a wide mouth so the reactants could be easily loaded. The charge in the tube was enough to make 7 kg of glass. About 10 ppm of pure aluminum wire was added to the reactants to serve as a getter for the 5 ppm of oxides anticipated. After loading the cap was sealed, the tube evacuated, heated to melt the chalcogen, and finally sealed off for compounding. After the reactions were believed complete, the temperature was raised and the entire 7 kg distilled through the frit into the glass chamber, where it was rocked and mixed prior to quenching using blown air. The goal was high-purity glass free of particles and oxides.

The second major change made was to eliminate the particles caused by the glass plug used to control the flow of glass into the mold. Again, in working with the glass blowers, a procedure was developed by which a ground glass joint was placed in the bottom of the crucible to control the flow of glass into the mold. Thus, the entire cast plate would be useful material. The ball joint was controlled by a long glass tube that reached outside the top of the chamber. After much effort, the standard production became striae-free, high–purity, 12 in × 12 in, 7-kg plates. Part of the TI success was due to vapor pressure and viscosity information supplied by Larry Swink. Figure 3.8 is a diagram of a sealed pressure measuring device produced at that time by TI. In the diagram, TI 1173 or TI 20 glass was sealed in the bottom chamber. When the glass was heated, the resultant vapor pressure produced a twist of the Bourdon tube that deflected the attached mirror. The vapor pressure was measured by the amount of nitrogen pressure applied in the top chamber to return the mirror to its original position. Vapor pressure results of the measurements are shown in Fig. 3.9 covering the elements S, Se, As, and Sb. Notice Sb is only showing 10 mm vapor pressure at 1000°C. Pressures for both TI 1173 and TI 20 are about 20 to 50 mm at 600°C near their casting temperatures.

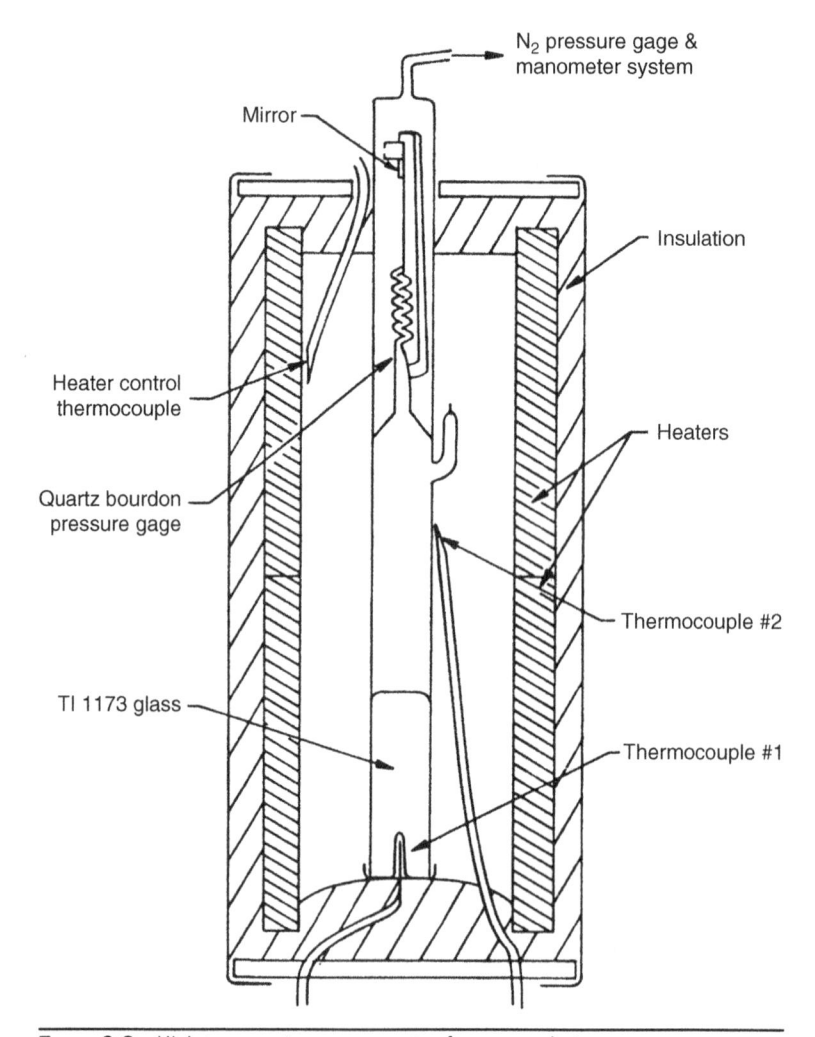

FIGURE 3.8 High-temperature manometer for measuring vapor pressure above chalcogenide glass melts.

The device constructed for direct measurement of viscosity is shown in Fig. 3.10. A double quartz chamber is separated by a tube of known inside diameter and length. An amount of glass is placed in the chamber; then the chamber is evacuated, sealed, and placed in the furnace that has a hole at the center such that a HeNe laser beam can pass through the quartz flow tube. The furnace is mounted so it can rotate about the center hole. In a vertical position, the glass is melted in the bottom chamber and the temperature noted. The furnace is inverted, and the glass flows from the upper chamber through the tube down to the bottom chamber. The time required is measured by noting the time from when the laser light is blocked by the glass to

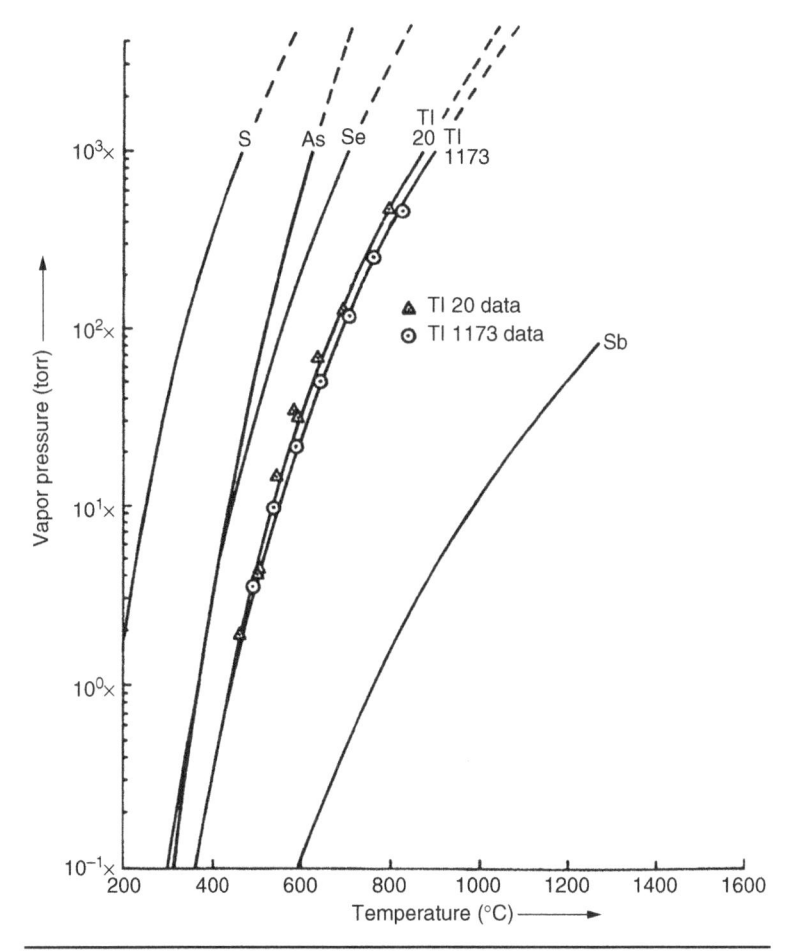

FIGURE 3.9 Measured vapor pressure of elements and glasses.

the time the light reappears. Viscosity then may be calculated from viscous flow fundamentals involving the diameter and length of the tube and the time required for the glass to flow through the tube. Results of the measurements for TI 1173 are presented in Fig. 3.11 and designated as experimental data. Prior extrapolated data are also shown. Later we will show how measured T_g and the softening point are used to form a viscosity versus $1/T$ plot for glasses. The problem in that method lies in assigning a viscosity number to T_g and the softening point. The casting temperature used in this plot is indicated by an arrow that corresponds to a viscosity value at casting of 500 to 1000 p.

One disadvantage of an open casting system is that vapors are constantly escaping from the melt of the heated glass during the mixing process. As has already been pointed out, the vapors are different in

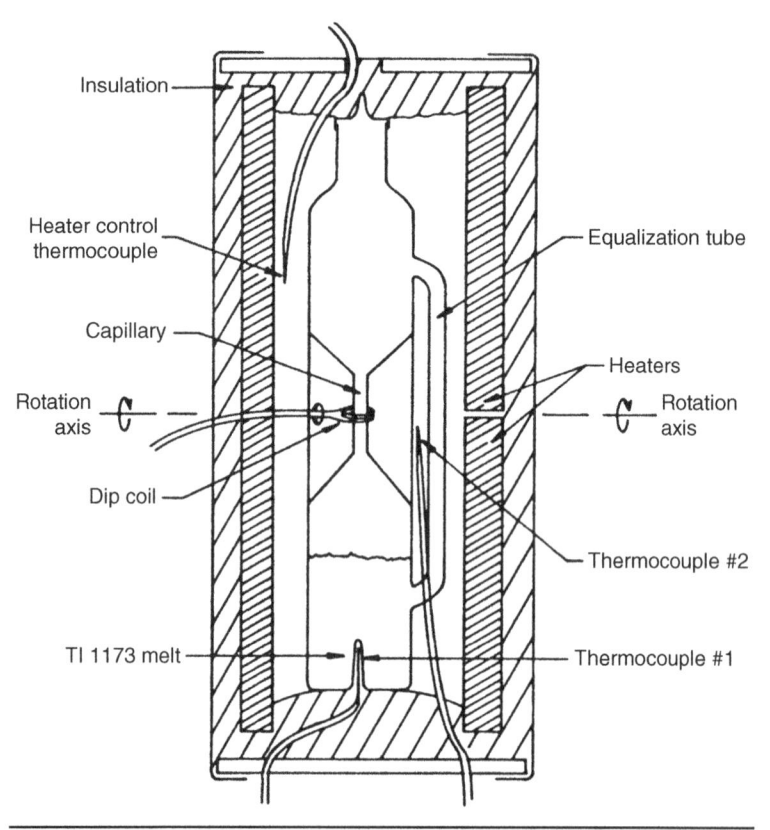

FIGURE 3.10 High-temperature chalcogenide glass viscometer.

composition from the glass, which means the refractive index changes slightly. The question becomes, How slightly? Now as the glass is processed into the form of blanks, one is left with a certain amount of scrap that is valuable because it contains germanium. The desire is to clean up the scrap and reprocess it. When a boule of glass is compounded, scrap may be mixed with fresh elements as a cost-reducing means. A large producer such as TI desires to reduce cost but without degrading the system performance due to index change. The decision was made to measure the change when glass is recast. Three prisms were fabricated from standard glass production. Using the TI infrared refractometer, the index of all three was measured over the range of 2.5 to 14 μm. The values for all three were averaged and found to agree with one another over the wavelength range by ±0.0002. The process was repeated with three prisms made from glass recast once. An average value for the recast glass prisms was calculated and did not agree as well as the standard glass production ±0.0010. When the

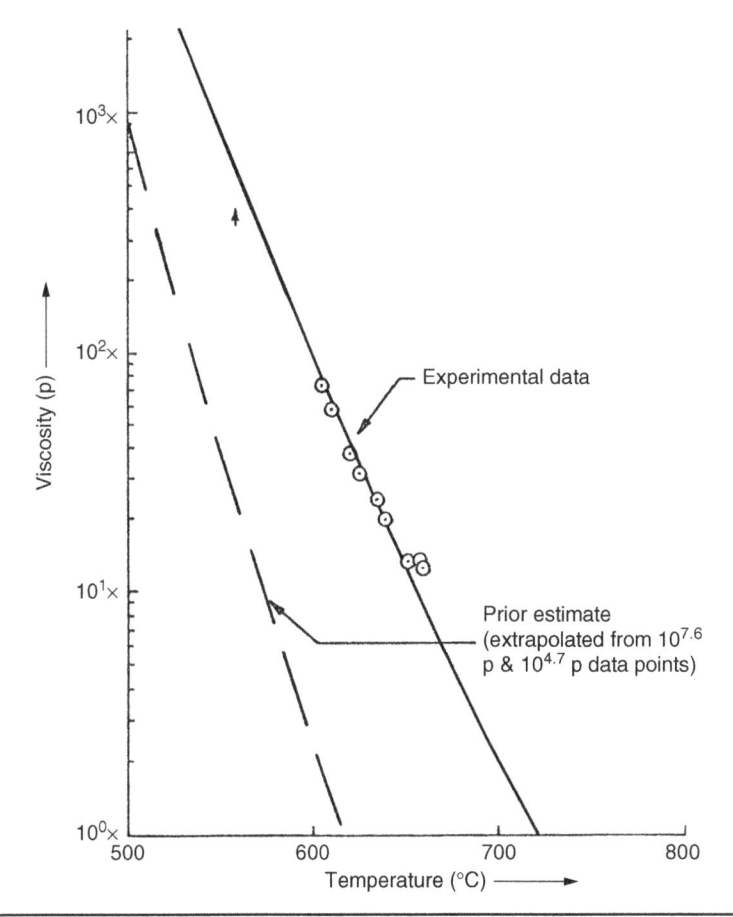

FIGURE 3.11 Viscosity of TI 1173 glass measured in liquid range.

two average results were compared, the recast index increased by about +0.0040. The increase was thought to be due to loss of Se and Ge-Se vapors. The optical designer could then decide if glass could be recast once, twice, or even 3 times and used based on the index change and system design parameters such as focus. The dispersion parameters for the glass were left relatively unchanged.

Figure 3.12 shows the final configuration of the casting furnace used at TI for TI 1173 glass. The chamber enclosure is round and large with a single door in front. Inside is a large quartz crucible that holds all the glass to be cast. Heaters are provided to melt and control the temperature of the glass. Provisions are made for a powered stirring device controlled from the top. A rod from the top is connected to the ball joint in the bottom of the crucible that allows glass to flow into the mold in the bottom chamber. Provisions are made for heating the

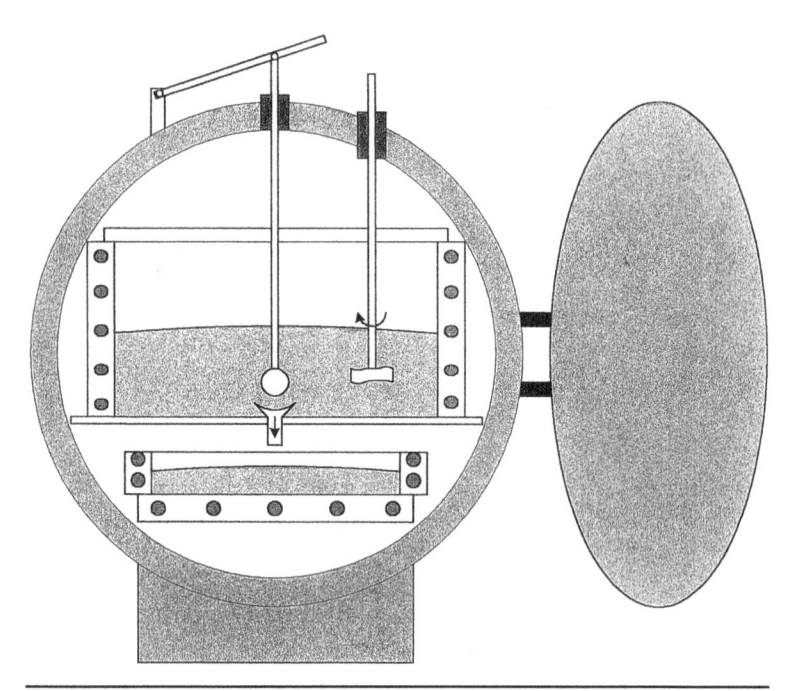

FIGURE 3.12 Diagram of an open down hole glass casting unit.

mold and controlling the glass temperature after it has been cast. Using a system like this, TI, and later Raytheon, cast many tons of TI 1173 glass each year.

3.5 Purification, Compounding, Casting—One Closed Operation

The changes made by the author in the TI glass casting process before leaving were not major. For one thing, the glass was in production and supplying orders for existing systems. Permission to make more drastic changes proposed by the author was refused on the valid grounds that changes could cause delays in deliveries.

When Amorphous Materials (AMI) was started in May 1977, such restraints no longer existed. The fabrication of compounding chambers with porous filters on a glass lathe had already demonstrated improved purity. The drawback of the open casting system where the escape of glass vapors changed the refractive index had been investigated and the effect measured. The decision was made to combine element purification, compounding the glass, and casting in a closed sealed system into one operation. The first requirement was to recruit the most skilled master glass blower available. Glen Whaley was such a person who had been thoroughly trained in the Central Research Laboratories at TI. The author had worked with him many times. The most difficult job they worked on together was the building of a far infrared (95.8-μm) pure He laser. The author felt Whaley was the best person for the job and recruited him as the first employee of Amorphous Materials. He joined the company as a stockholder and a vice president. Whaley developed quartz fabrication techniques that all his former colleagues from TI told him were impossible.

The task at AMI was different from the one at TI. It has been previously pointed out that AMI was not allowed to produce TI 1173 and was forced to revert to the Ge-As-Se glass, called TI 20. The glass was renamed Amtir 1. The fact that the glass contained arsenic—not antimony—made it better suited for distillation. Besides, the composition had been carefully selected so that it had no tendency to form crystallites as did the troubled TI 1173. Over a period of months, the system shown in Fig. 3.13 was developed and the glass processing steps worked out that produced high-purity striae-free 8-in-diameter plates weighing 9 kg. In the diagram, the cap covering the mouth of the compounding chamber has been sealed on and then closed after the reactant purification step. The reactants contain about 10 ppm aluminum wire to getter the oxides. The unit is placed in a two-zone furnace that moves in a horizontal plane. The temperature is raised in the compound side, and horizontal motion mixes the reactants compounding the glass. After a period of time, motion is stopped, and the compound tube side is raised in temperature to distill the glass through the filter into the round chamber to form the 8-in plate. After distillation is complete, motion is started again to mix the glass in the

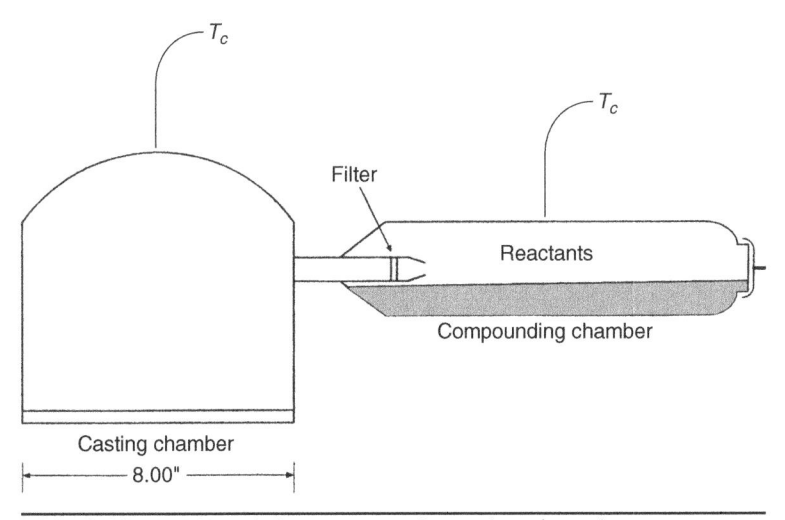

FIGURE 3.13 AMI closed glass compounding and casting unit.

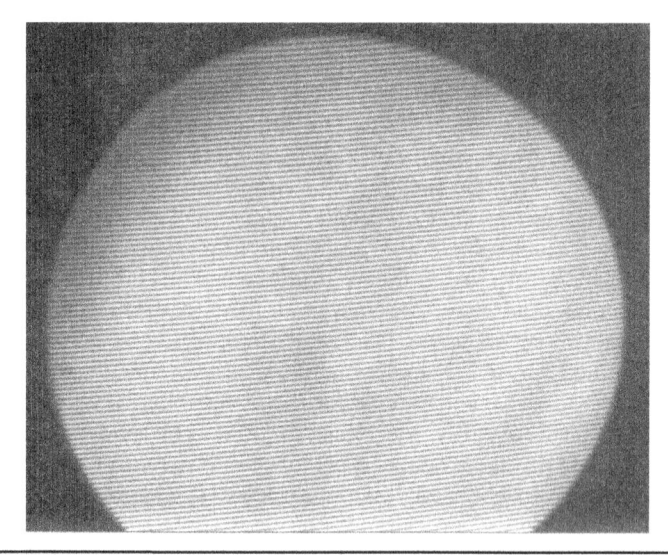

FIGURE 3.14 Striae scope photograph looking through an 8-in-diameter striae-free Amtir glass plate.

round chamber thoroughly. The glass is cooled to slightly above T_g, the furnace is opened, and air is blown on the chamber to quench the glass. After cooling, the quartz chamber is broken and the glass plate is removed. The plate undergoes a preliminary evaluation and then is placed in an oven to go through its anneal cycle. After annealing, the glass is ready for processing by sawing, core drilling, or slumping into shapes required for blanks to be fabricated into optical elements used in imaging systems.

The AMI closed compounding casting glass process has produced from 1978 to 2007 over 35 tons of Amtir 1 glass in 9-kg plates. Figure 3.14 shows a striae scope photograph taken looking through an 8-in-diameter striae-free plate of Amtir glass.

3.6 Summary

1. High-purity elements are required to produce low-absorption infrared chalcogenide glasses. The elements used are by-products from the production of primary metals.
2. Glass may be made in kilogram quantities in simple tubes placed in a rocking furnace. Optical quality will not be good.
3. Early TI casting units were open with melted glass poured from a crucible into a mold. Again, optical quality was not good.
4. Quality improved at TI when the glass was allowed to flow through a tube in the bottom of the crucible into a mold below.

5. Improvement in transmission at TI was obtained when glass was distilled through porous filters to remove particulate material and aluminum wire was added to getter oxide impurities.
6. An all-quartz closed system was developed at AMI that combined reactant purification, compounding glass, and casting all into one operation under vacuum. Optical quality was excellent.

References
1. Asarco in Texas, Eng. Mining J., September 1981.
2. A. R. Hilton, D. J. Hayes and M. D. Rechtin "Chalcogenide Glasses for High Energy Laser Applications," Contract N00014-73-C-0367, June 1974. Sponsored by Advanced Research Projects Agency (ARPA), ARPA N0 2443.
3. A. R. Hilton, D. J. Hayes, and M. D. Rechtin, "Infrared Absorption of Some High Purity Chalcogenide Glasses," J. Non. Cryst. Solids 17, 319 (1975).

CHAPTER 4
Characterization of Glass Properties

A glass composition has been selected for development based on a qualitative evaluation. Now the glass has been prepared in quantity and in quality so that quantitative measurements may be made to fully evaluate the potential usefulness of the glass. Considerable effort is required. AMI started with one glass and presently produces seven. AMI thus has gone through this process a number of times. The AMI established method for preparing a glass for sale in the marketplace is a major part of the company quality control plan. Others may develop a different approach while emphasizing other properties and other methods.

4.1 Thermal Expansion, Glass Transition Temperature, and Softening Point

Earlier, it was pointed out that the thermal expansion measurement defines the difference between a crystalline solid and a glass. Now it is nice if you are in possession of a sophisticated piece of equipment capable of carrying out the measurement automatically and then producing the data plotted in a nice color slide similar to some reproduced in this book. But if you are not so fortunate, you may build and use a simple device such as the AMI system shown in Fig. 4.1. A quartz glass tube inside a circulating air furnace has a quartz glass rod inside which rests on a vertical 2-in sample of the chalcogenide glass to be measured. The glass sample is in a chamber that has two sides open so that air flows freely through and around the sample. A thermocouple is placed in the chamber next to the sample, and the temperature is read from the meter on top of the furnace. Also on top, the rod is pushing against a linear in length, mildly spring loaded transducer which is monitored by an indicator. The display is set to show numerical readings in this case ±20,000. Before it is used, the transducer is calibrated in linear sensitivity to determine the numerical change per 0.001-in movement. The transducer is calibrated using a good micrometer.

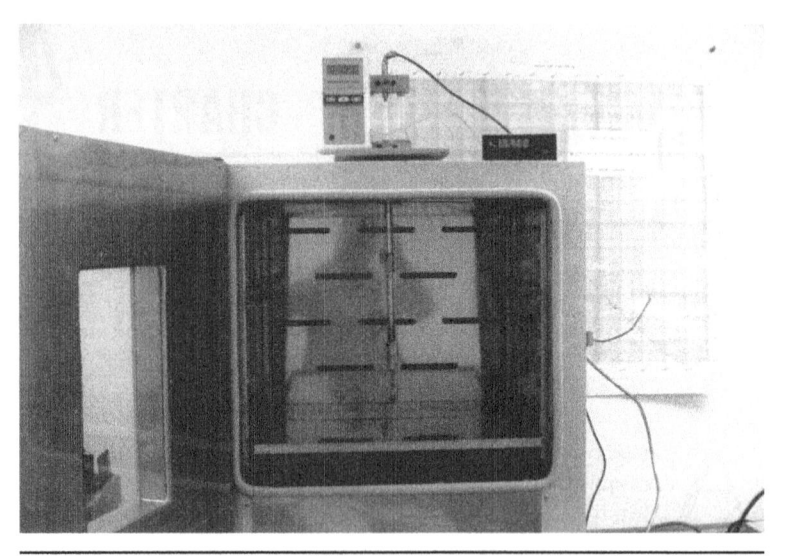

FIGURE 4.1 Photograph of the AMI glass thermal expansion apparatus.

The door is shut and the temperature started up in a slow, programmed manner. Numerical data are taken from the display at indicated temperatures. The sample expands linearly as a solid does, as the temperature increases to a point where it begins to expand much more rapidly. The slope changes to a greater expansion typical of a liquid. When the data are plotted, where the two slopes intercept is the glass transition temperature T_g. As the temperature continues to increase, the transducer slows its numerical increase, stops, and then decreases. That indicates the glass is being compressed by the force of the transducer. This temperature point is taken as the softening point T_d (dilatometer soft point) for the glass. The data for Amtir 3 are shown plotted by hand on linear graph paper in Fig. 4.2. Such data are a good starting point. However, for more accurate data, the material should be measured again using an instrument specifically designed for this purpose as in the results shown earlier in Fig. 2.2 obtained from Tom Loretz[1] of CES. Also shown earlier, CES is equipped to carry out differential thermal analysis (DTA) on powdered samples. The crystallization range may be accurately determined in this manner, also shown in Fig. 2.3. Both T_g and T_d are information vital to the production of the glass composition and its use in molding, extrusion, and drawing fibers. However, the dilatometer measurement does not determine the absolute viscosity for the glass at these two points. A general value is assumed. The viscosity of a glass increases exponentially with temperature. Thus a plot of log to base 10 of viscosity versus 1 over absolute temperature will yield a straight line.

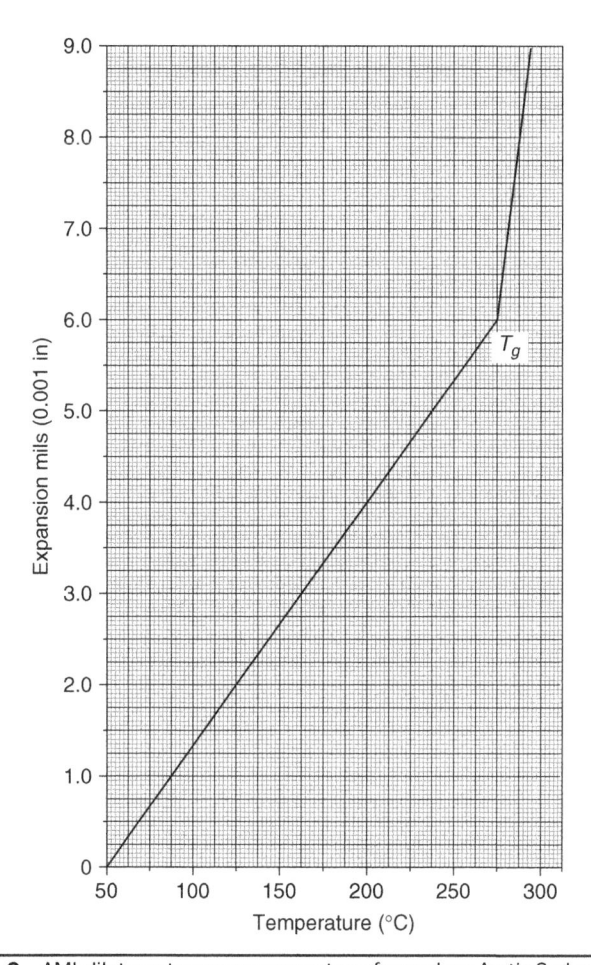

Figure 4.2 AMI dilatometer measurement performed on Amtir 3 glass.

Earlier, it was mentioned that Robert Patterson[2] in the TI Air Force Development Program developed the Ge-Sb-Se glass designated as TI 1173. Patterson decided to develop a device for use with chalcogenide glasses that would measure the softening point, strain point, and glass transition temperature in the absolute terms of viscosity. A diagram of this device is shown in Fig. 4.3. The device used a pointed spring-loaded (70-g) quartz rod against polished glass samples of standard size and thickness placed in a heated enclosure. Sample temperature was measured by a thermocouple placed underneath. When the penetration depth in the heated sample reached 0.05 mm, the temperature was recorded as the softening point by method A. Method B, with identical procedures, was performed except the penetration depth was increased to 0.45 mm. Softening by method B was recorded. A third ASTM softening point method sometimes used involves

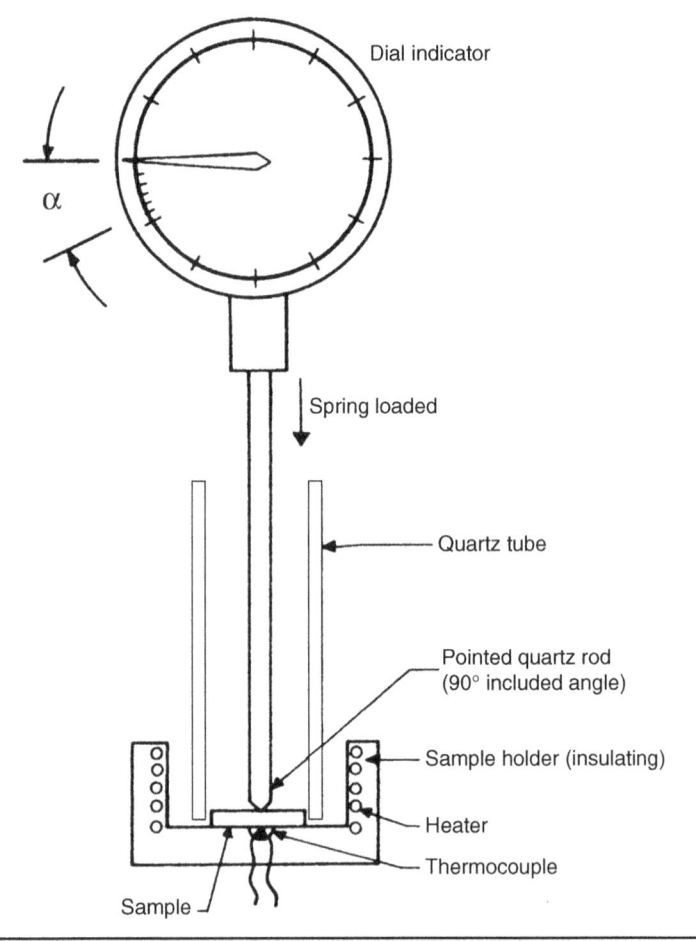

FIGURE 4.3 Device used to standardize softening point measurements of chalcogenide glasses used by Harold Hafner at TI.

a glass fiber of specified length and dimensions heated while in a vertical position until its length begins to increase. The viscosity for this method of measuring the softening point was set at $10^{7.6}$ poise (p) by Littleton.[3] Three viscosity-temperature points certified by the National Bureau of Standards (NBS) for glass 712 are plotted by Patterson[2] on the $1/T$ straight line shown in Fig. 4.4. The measured points using method A and method B performed on glass NBS 712 have been added to the plot thus calibrating these methods in absolute viscosity terms. The methods were then used to specify the strain point viscosity as $10^{14.6}$ p. The strain point is thought of as the upper use temperature for a glass. The anneal point corresponds to T_g and is specified as $10^{13.4}$ p. The softening point using the ASTM method occurs at $10^{7.6}$ p. Two of the TI glasses, TI 1173 (Amtir 3) and TI 20

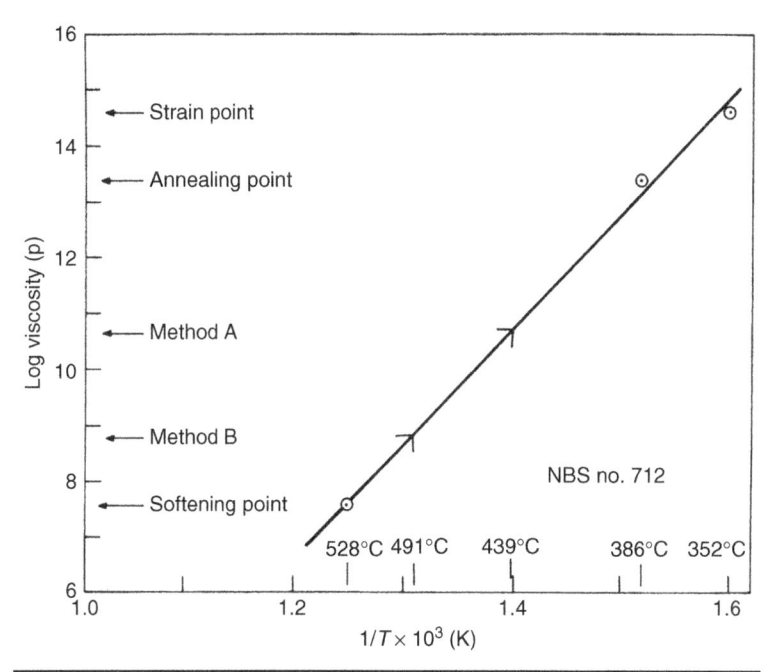

FIGURE 4.4 Log viscosity versus reciprocal absolute temperature for NBS glass no. 712.

(Amtir 1) were measured using method A and method B and their strain points, their anneal points T_g certified long ago. Now AMI makes both of these glasses and has dilatometer results for only Amtir 3, shown in Fig. 4.2. Combining the TI results with AMI results allows us to be able to convert the AMI data to a viscosity scale.

Figure 4.5 shows a plot of several AMI glasses in a viscosity versus $1/T$ plot. Notice Amtir 1 and Amtir 3 have three points determined early by TI: strain point, T_g, and ASTM softening point. However, Amtir 3 has an extra point, an AMI T_d softening point from a dilatometer measurement corresponding in viscosity to a value of $10^{11.6}$. The data for Amtir 6 and C1 glass were dilatometer only. Now armed with the T_d viscosity number, we can make a valid viscosity versus $1/T$ plot for any of our glasses that can be used as a guide in using the glass. Only the two dilatometer points are needed to determine the $1/T$ straight-line plot for the glass. In Fig. 4.5, the temperature required to draw C1 fiber places a mark on the line corresponding to 15,850 p. Notice that the mark for Amtir 6 drawing fiber corresponds to the same viscosity. An extrapolation, not shown, for the $1/T$ line for Amtir 4 for fiber drawing agreed exactly with the viscosity value and temperature of C1 core glass. The bottom hole casting temperature for striae-free Amtir 3 glass determined at TI was 575°C. The intersection on the Amtir 3 line corresponds to a low viscosity value of around 10 p.

94 Chapter Four

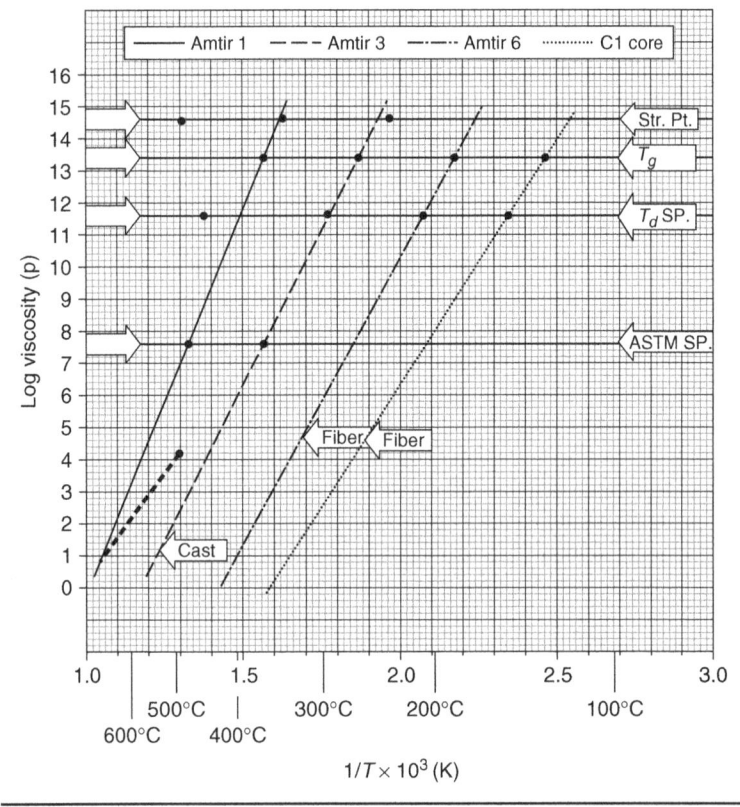

Figure 4.5 Log viscosity versus reciprocal absolute temperature for some AMI chalcogenide glasses.

The experimental direct measurement of the viscosity of Amtir 3 described in Chap. 3 is shown as a dashed line close to the $1/T$ line for Amtir 3. The viscosity data of the dashed line at 575°C indicate a value of 150 p for casting. Treatment of dilatometer data in this manner provides a useful guide when using a glass for new applications or considering the use of a new glass.

4.2 Transmission, Precise Refractive Index, and Thermal Change in Refractive Index

Transmission and absorption are measures of quality and are therefore evaluated on each glass plate and recorded. In order for a glass composition to be used by designers in production of systems, much more information concerning the optical and related physical properties must be carefully measured. Complete characterization is a long and difficult task.

Early infrared spectrophotometers were designed to be double-beam with one used for reference I_0 and the other for the sample I. Optical paths were equal in length and intensity so that the detected output signals were always I/I_0. Some instruments used the ratio of the infrared detected electric signals while others, termed *optical null instruments*, had a wedge or comb placed in the reference beam to balance the two signals during the scan while recording the I/I_0 ratio continuously based on the position of the wedge or comb.

The transmission accuracy of commercial instruments under ideal conditions was usually considered 1 to 2 percent. For absorbing materials this may be fine. But for low-absorbing, transparent materials, it means thicker samples 2 cm or more are required for better accuracy. Unfortunately, a thick sample with a large refractive index increases the optical path in the sample beam, leading to loss in accuracy. The instruments in this generation were designed to work with organic compounds that were thin and had low refractive indexes.

The appearance of Fourier transform instruments was a great step forward. In this instrument, there is only one beam and it is polychromatic. The beam transmitted through the sample is made to interfere with itself optically by using a scan mirror producing a pattern that when analyzed mathematically (Fourier transforms) reveals the variation in transmitted energy as a function of wavelength. Multiple scans are used to increase signal-to-noise results. The outcome is compared to a previously recorded, no-sample same-number-of-scans reference outcome. The results are displayed in transmission or absorption terms and printed out after desired additions to the display. Each scan takes only a few seconds. For poorly transmitting samples, increased signal-to-noise accuracy may require 50 to 100 scans and the results averaged, eliminating the influence of noise. The instruments are very versatile and useful, a real advance in the state of the art.

Fourier transform infrared (FTIR) instruments used at AMI are a Perkin Elmer Paragon 1000 and a Nicolet AVATAR 320 utilized in the production area. The wavelength range generally used is 2 to 14 µm although the scan range may be changed for slightly shorter wavelengths than 2 µm or longer than 14 µm, out to 20 µm. Figure 4.6 shows a Perkin Elmer FTIR transmission scan for an Amtir 1, 8-in-diameter 9-kg plate 6 cm thick. The scan is used with the standard QC documentation preserved for each plate produced. Notice the two narrow absorptions at 4.9 and 4.5 µm due to dissolved H_2Se molecules in the glass that couple to the Ge atom (4.9) and the As atom (4.5). The magnitude of absorption for the gas is low, 0.04 to 0.05 cm^{-1}, of little consequence for lenses less than 1 cm thick. The glass tested is considered low-absorbing and has flat parallel faces so that the full expression given below for transmission should be used to calculate absorption, assuming multiple reflections.

$$T = \frac{I}{I_0} = \frac{(1-R)^2 e^{-\alpha x}}{1 - R^2 e^{-2\alpha x}}$$

Perkin Elmer

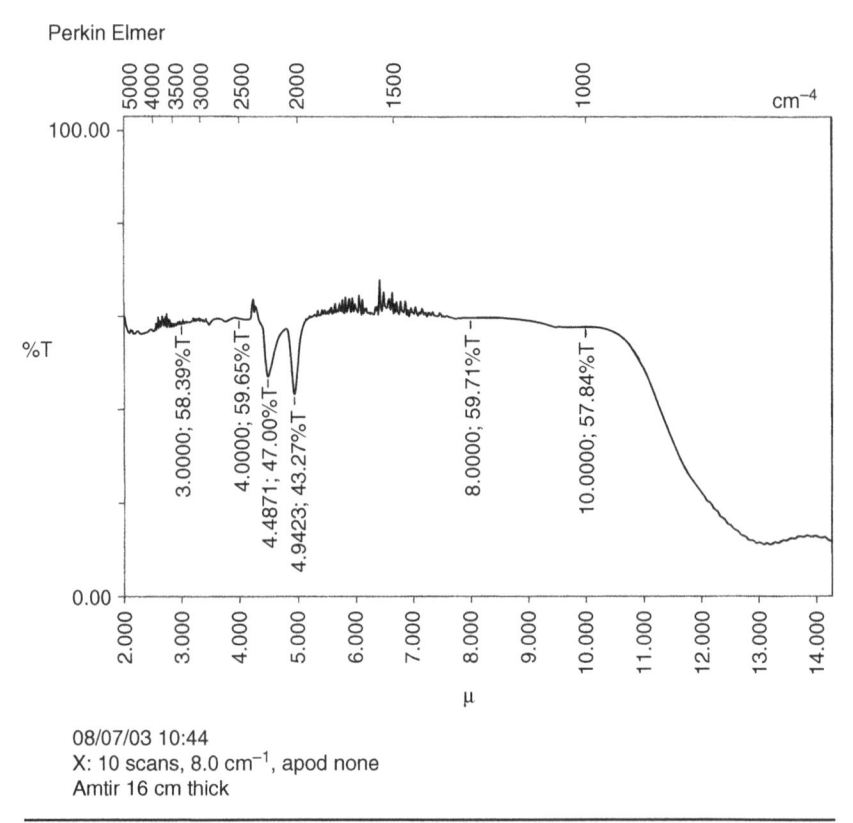

08/07/03 10:44
X: 10 scans, 8.0 cm^{-1}, apod none
Amtir 16 cm thick

FIGURE 4.6 Perkin Elmer FTIR transmission scan for Amtir 1 plate 08-28.

where T = transmittance
R = Fresnal reflection coefficient calculated from $R = (N-1)^2/(N+1)^2$
α = absorption coefficient, cm^{-1}
x = thickness, cm

Computer programs can be written to solve for α as a function of wavelength from transmission values of plates of known thickness. For QC, computer programs are written setting minimum transmission values required for plates of known thickness as a function of wavelength. Sometimes the absorption values are set as a specification by the user. An example of how this procedure has been used is shown in Table 4.1. The specified values for Amtir 1 were set by the Army. Required transmission values for different thicknesses and wavelengths are given in the table. Conversely, a program can be written to calculate transmission as a function of thickness based on the precise infrared refractive index as a function of wavelength,

Wavelength (μm)	R	$(1-R)^2$	R^2	$T_{x=0}$	$T_{x=0.25"}$	$T_{x=0.37"}$	$T_{x=0.5"}$	0.75"	7.0"
3	0.186	0.6626	0.0347	0.6864					
4	0.186	0.6626	0.0344	0.6862					
5	0.185	0.6642	0.0343	0.6878					
6	0.185	0.6642	0.0342	0.6877					
7	0.184	0.6659	0.0340	0.6893					
8 (.02 cm^{-1})	0.184	0.6659	0.0339	0.6893	0.6800	0.6756	0.6708	0.6617	0.6529
9 (.02 cm^{-1})	0.184	0.6659	0.0338	0.6892	0.6807	0.6763	0.6715	0.6624	0.6536
10 (.02 cm^{-1})	0.183	0.6675	0.0336	0.6907	0.6814	0.6770	0.6722	0.6631	0.6543
11	0.183	0.6675	0.0335	0.6907					
12 (.2 cm^{-1})	0.182	0.6691	0.0332	0.6921	0.6049	0.5674	0.5296	0.4638	0.4075
12.8 (.4 cm^{-1})	0.181	0.6708	0.0328	0.6935	0.5308	0.4680	0.4085	0.3147	0.2439

TABLE 4.1 Calculated Transmission Requirements as a Function of Thickness for Amtir 1 to Meet Specifications

Chapter Four

Wavelength (μm)	Refractive IR Index	Absorption Coefficient (cm⁻¹)	Thickness (cm)	Reflectivity	Transmission
\multicolumn{6}{c}{Calculated IR Transmission for Amtir 4}

Wavelength (μm)	Refractive IR Index	Absorption Coefficient (cm⁻¹)	Thickness (cm)	Reflectivity	Transmission
		Fiber Results			
1.00	2.7788	0.04	2.6	0.221588	0.568757
1.50	2.6995	0.005	2.6	0.211036	0.642296
2.00	2.6778	0.003	2.6	0.208115	0.649923
2.5	2.6686	0.003	2.6	0.206873	0.651618
3.00	2.6637	0.002	2.6	0.20621	0.654372
4.00	2.6588	0.002	2.6	0.205547	0.655281
5.00	2.6561	0.002	2.6	0.205181	0.655783
6.00	2.6541	0.001	2.6	0.20491	0.658012
7.00	2.6524	0.001	2.6	0.204679	0.658329
8.00	2.6508	0.001	2.6	0.204462	0.658628
9.00	2.6491	0.009	2.6	0.204231	0.644245
10.00	2.6472	0.012	2.6	0.203973	0.639178
11.00	2.6452	0.014	2.6	0.203702	0.635961
12.00	2.643	0.024	2.6	0.203403	0.618777
13.00	2.6407	0.35	2.6	0.20309	0.257349
14.00	2.638	0.47	2.6	0.202723	0.187959

TABLE 4.2 AMI Program to Calculate Transmission of Amtir 4 as a Function of Thickness and Wavelength

incorporating absorption coefficient results calculated from the plate scan. Table 4.2 illustrates this procedure when applied to Amtir 4.

Specifications for wavelengths less than 2 μm (NIR) require the use of a different kind of an instrument. In this case, the AMI Beckman UV 5240 spectrophotometer covers the wavelength range of 0.5 to 2.5 μm. AMI uses a reflection attachment specifically designed for the instrument to evaluate antireflection coating results. The attachment mounts inside the sample area that may be closed when it is necessary to keep light out of the instrument. In double-beam operation, the instrument is an electronic null instrument not designed for high-index solid materials. Care must be exercised in taking measurements, and the results must be carefully evaluated. The fundamental absorption edge, where the glasses begin to transmit, cannot be measured using either of the FTIR instruments. The Beckman instrument must be used. As an example, a plot of the transmission edge for an 8-in-diameter Amtir 4 glass plate is shown in Fig. 4.7. Notice the plate is 2.6 cm thick.

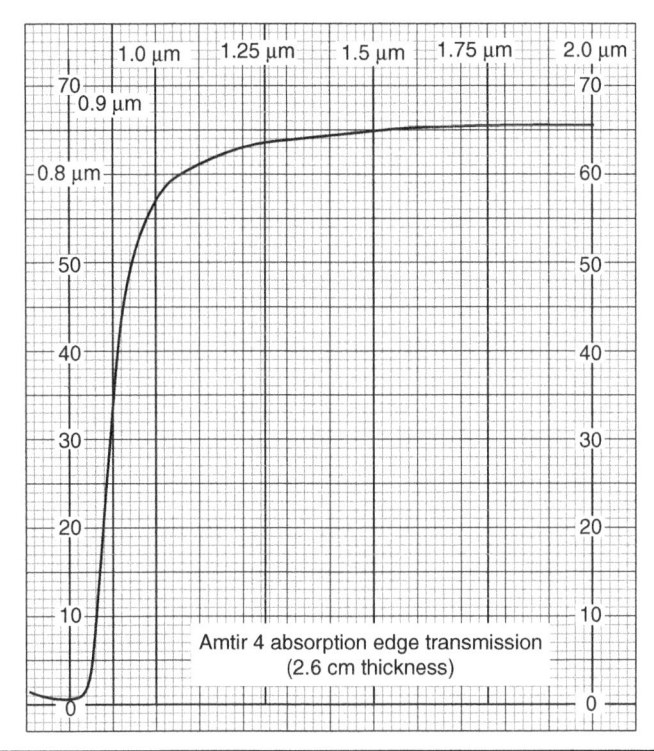

FIGURE 4.7 Measured NIR transmission of an Amtir 4 plate at the absorption edge.

For optical designers, the most important information in designing a lens is accurate, precise refractive index numbers covering the wavelength range of operation along with the sign and magnitude of the thermal change in index. The attachment to the Perkin Elmer 13 spectrophotometer used to measure the infrared refractive index at TI is shown in Fig. 2.10. The Perkin Elmer spectrophotometer serves as a monochromatic source of light. Later at TI, the attachment was changed so that the thermal change in refractive index $\Delta N/\Delta T$ for infrared optical materials could be measured. A diagram of the equipment is shown in Fig. 4.8. The Bridgeport Rotary Table with 5 seconds of arc accuracy turns a hollow copper chamber with a vertical flat mirror side where the glass prism is placed. The chamber may be filled from outside with coolant. The chamber is surrounded by a heat shield and an outer wall sealed to the base. Flat NaCl windows are provided to transmit the monochromatic infrared light needed to perform the minimum deviation measurement. The enclosure is evacuated, the coolant is added, and the heat shield helps maintain the reduced temperature. The temperature of the prism is measured by a thermocouple. Measurements are made at room temperature and at lower temperatures brought about by using dry ice or liquid

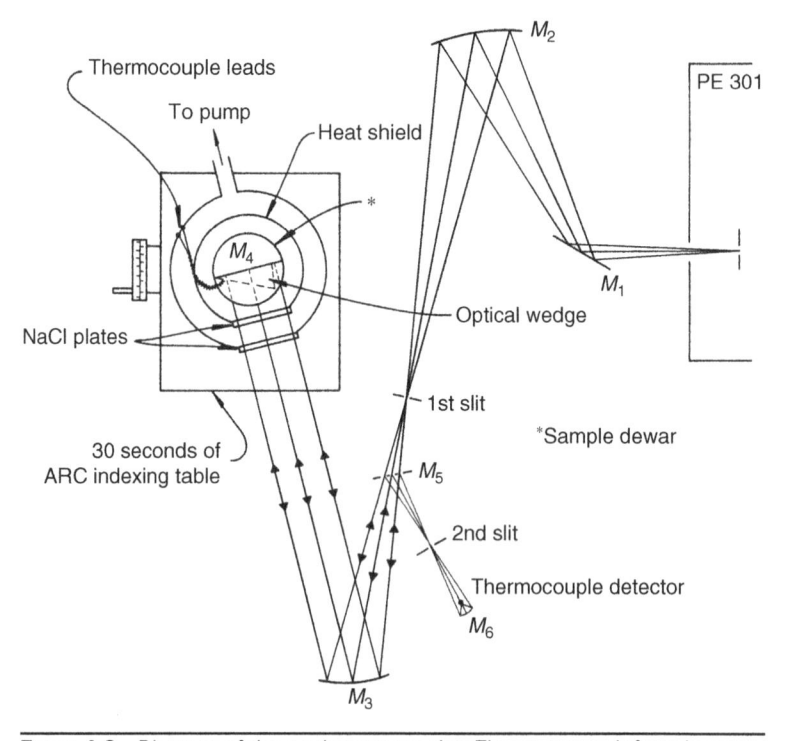

FIGURE 4.8 Diagram of the equipment used at TI to measure infrared refractive index and the changes with temperature.

nitrogen as coolant. The $\Delta N/\Delta T$ is calculated from the results for the two temperatures. The value calculated is from room temperature down. However, in systems the materials are usually used from room temperature up. Howard Kennedy, an optical designer at TI, told the author the value listed in the literature for germanium was much too low. Those reported in the literature were also measured from room temperature down. The designer said he could demonstrate he was correct by taking a germanium lens and measuring change in focal length as the temperature increased. The author decided to test his opinion by remeasuring several materials from room temperature up, using the same equipment. All that was necessary was to pour oil in the chamber instead of liquid nitrogen. Heating elements were inserted, and the results for several materials are shown in Table 4.3. Earlier values for room temperature down are included for comparison. The data show Kennedy was right.

AMI performs the minimum deviation measurement on a prism made from any infrared optical material. The method is sometimes called the *Litrow mount deviation angle method* because the prism is placed against a rotating mirror. The procedure followed for each

λ (μm)	Ge Low	Ge High	Si Low	Si High	TI 1173 Low	TI 1173 High	GaAs Low	GaAs High
3	330	455		190	58	98	161	206
5	312	428	162	184	57	92	159	216
8	307	426		186	55	87	157	208
10	304	427		187	56	91	148	203
12	302	424		177	56	93	153	210

All values in $10^{-6}/°C$; low = 25 to −197°C, high = 25 up to 100 to 150°C

TABLE 4.3 Thermal Change in Refractive Index for Ge, Si, TI 1173, and GaAs

wavelength point is demonstrated in Fig. 4.9. In this method, the monochromatic light is refracted at the surface so that it strikes the mirror at a perpendicular angle and retraces itself, forming an inverted slit image with intensity varying with the measured angle of rotation across another slit. An advantage of the method is that the light is refracted twice, in and out, while passing through the prism. The prism angle is not as large as with other methods. The method was used at TI to study the change in index with temperature for a number of materials.[4] A conclusion reached in the study was that it could be possible to select a chalcogenide glass composition that would have a zero thermal change in index in the wavelength range of interest. The instrument was also used to measure TI 20 (Amtir 1) and IRTRAN IV (ZnSe).

Right after AMI started into operation, an infrared refractometer similar to the one used at TI was built. A photograph of the instrument

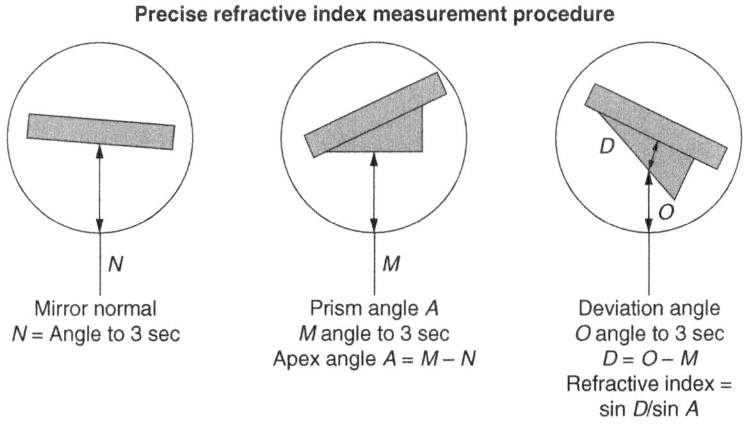

Precise refractive index measurement procedure

Mirror normal
N = Angle to 3 sec

Prism angle A
M angle to 3 sec
Apex angle $A = M − N$

Deviation angle
O angle to 3 sec
$D = O − M$
Refractive index = $\sin D / \sin A$

FIGURE 4.9 The Litrow mirror procedure for measuring refractive index.

is shown in Fig. 4.10. It was used to verify and monitor the refractive index of Amtir 1. Later, when the Army supported the effort for AMI to become a second source of TI 1173 (Amtir 3), the instrument was used to verify that the AMI glass met the refractive index standards. The instrument was also equipped to measure $\Delta N/\Delta T$. As before, a hollow copper chamber was mounted with a flat vertical aluminum mirror attached. But this time, the chamber temperature was controlled by the flow from a Neslab temperature bath. The $\Delta N/\Delta T$ measurements were carried out beginning as low as possible without frost or fogging and up as high as 60 to 70°C.

FIGURE **4.10** Photograph of the AMI refractometer combined with the Perkin Elmer 13 spectrophotometer.

Measurements were made at AMI and published[5,6] in 1990 to 1991 concerning the possible variation in refractive index during production for infrared materials available in the industry. Three prisms were made from different plates of Amtir 1 and measured. Three prisms of zinc selenide from different sources were made and measured. Six prisms were made from three sources of poly- and single-crystal germanium. The accuracy of the AMI method is demonstrated by comparing the results for the zinc selenide prism obtained from Raytheon to the previous results for the Raytheon zinc selenide published by Marilyn Dodge of NIST[7] (then the National Bureau of Standards). Table 4.4 presents the results. Keep in mind that such measurements carried out at NBS took months, time that could not be spared at AMI. The AMI measurements for a glass prism covering 1 to 12 μm are carried out in 2 to 4 weeks. Again, the accuracy of the method is demonstrated in Table 4.5. AMI results for arsenic trisulfide glass agreed with the 1958 published NBS results by the average ±0.0002 measured from 3 to 11 μm. AMI believes the limit of the Litrow mount minimum deviation method when used on infrared materials is accurate and is reproducible to about ±0.0002.

One of the problems with the infrared (IR) refractometer at TI and AMI was using the Bridgeport Rotary Table for measurements. The unit was heavy and difficult to rotate. All measurements were taken in the same direction of rotation. Measurements were slow and tedious. It was decided at AMI to build a computer-controlled unit that used a stepping

Wavelength (μm)	Average	NBS (M. Dodge)	Difference
3	2.4371	2.4376	−0.0005
4	2.4336	2.4332	+0.0004
5	2.4294	2.4295	−0.0001
6	2.4259	2.4258	+0.0001
7	2.4218	2.4218	0
8	2.4173	2.4173	0
9	2.4123	2.4122	+0.0001
10	2.4067	2.4064	+0.0003
11	2.4003	2.4001	+0.0003
12	2.3930	2.3928	+0.0002
Variation	± 0.0007		±0.0002

TABLE 4.4 Comparison of AMI and NBS Refractive Index Results for Raytheon Zinc Selenide: Zinc Selenide IR Refractive Index Measurements, Minimum Deviation, Average Results for Four Measurements, Corrected to 20°C, in Air Sample—Raytheon

	Refractive Index Values for As$_2$S$_3$ Glass		
Wavelength (μm)	AMI Results Two Samples 1991*	Malitson, Rodney & King, 1958†	Δ
3	2.4152	2.4155	0.0003
4	2.4116	2.4112	–0.0004
5	2.4074	2.4073	–0.0001
6	2.4034	2.4032	–0.0002
7	2.3989	2.3988	–0.0001
8	2.3937	2.3937	0
9	2.3883	2.3881	–0.0002
10	2.3810	2.3814	–0.0004
11	2.3736	2.3738	0.0002

*Batch-to-batch variation ±0.0015.
†Values interpolated to the wavelengths of AMI.

TABLE 4.5 Comparison of AMI and NBS Refractive Index Results for As$_2$S$_3$ Glass

motor to replace the indexing table. The diagram of the current AMI IR refractometer is shown in Fig. 4.11. A computer-controlled stepping motor mounted in a vertical position supports the base on which the mirror is mounted. The motor requires 655,360 steps per rotation which means one step is 2 arc seconds. The monochromatic light source is the same prism monochromator originally part of the Perkin Elmer 13. It is equipped with a fused quartz prism which works well over the wavelength range of 1 to 2.5 μm and a sodium chloride prism that covers the range of 3 to 14 μm. The wavelength of the monochromatic light is changed by accurately rotating (drum turns) the mirror behind the quartz or sodium chloride prism. The monochromator is wavelength-calibrated in drum turns versus wavelength for each prism from spectral absorption bands. Polystyrene is used for 3 to 14 μm while didymium glass and a mercury arc are used for 1 to 2.5 μm. The light source for the NIR is a high-intensity light. The MWIR and LWIR use a globar source. The globar light is chopped at the source before entering the slit of the monochromator. A signal from the chopper is furnished to the phase-sensitive amplifier. The detector is a cooled HgCdTe detector for the NaCl prism and a Dexter thermopile detector for the NIR.

Experience has shown that no fabricated prism is perfect. AMI results have shown that making four measurements of each prism averages out any errors. Each prism is measured facing the light right and left, up or down. The mirror is adjusted perpendicular to the horizontal plane of

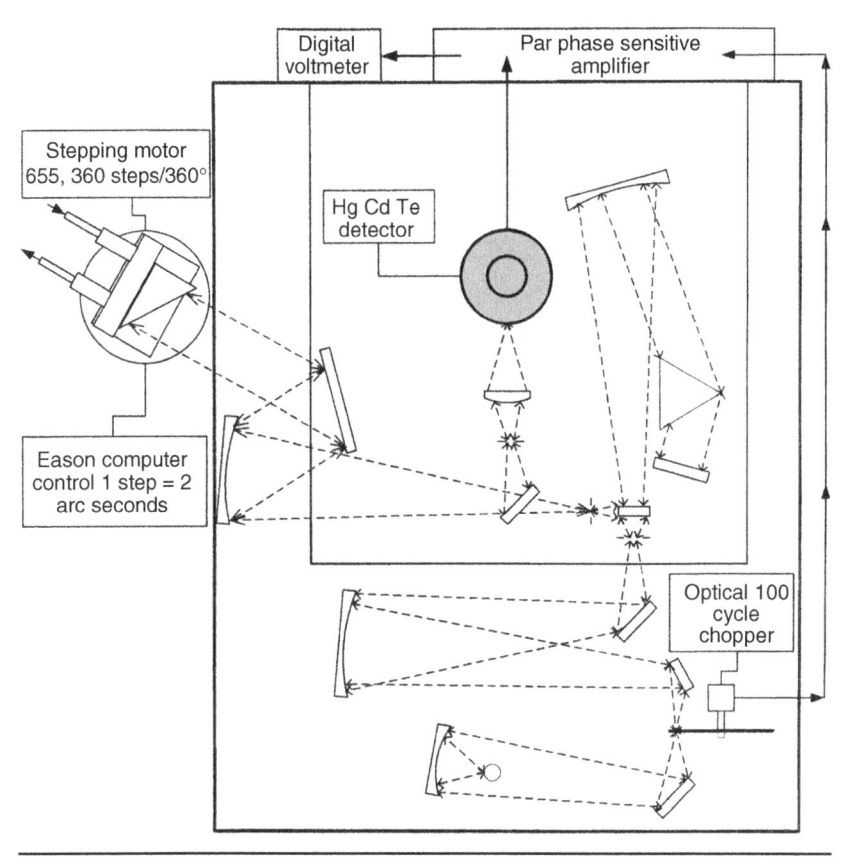

FIGURE 4.11 AMI computer-controlled infrared refractometer.

the instrument. A HeNe laser beam is reflected from the mirror to a wall 25 ft away. A spot is marked. Alignment of the prism in each of the four orientations is accomplished by reflecting the HeNe beam from the prism face and adjusting the prism to the wall spot 25 ft away.

The mirror is rotated clockwise until a switch detects a small magnet attached to its side. The computer records this step as zero. For any angle designated, the computer calculates the number of steps required and sends them to the stepping motor. A single angle input is made to the computer based on the expected angle for maximum intensity. As the mirror is rotated, a slit image of refracted monochromatic light will sweep the external slit, and the detected signal will vary with the angle. From that single initial point, the angle is increased by a specific designated increment, 0.02 degree for instance, and stops at each time so an intensity reading from the millivolt meter can be recorded for each angle. The process is repeated so 10 data points in a row are recorded. The process is adjusted and

106 Chapter Four

repeated until the approximate maximum intensity angle is located. The computer-controlled process locates the approximate angle producing the greatest signal. A scan is then made in exact angle increments three stops on each side of the peak intensity with one near the peak. A drawing of the scan process is shown in Fig. 4.12, marking

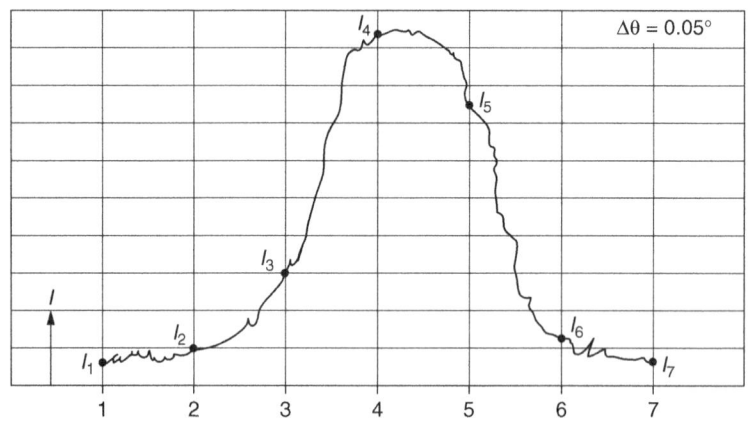

Solving for angle at maximum intensity using 7 sets of data.

Intensity – 3	5	Angle – 3	0.33 angle – 3sq			0.1089
Intensity – 2	32	Angle – 2	0.38 angle – 2sq			0.1444
Intensity – 1	109	Angle – 1	0.43 angle – 1sq			0.1849
Intensity 0	123	Angle 0	0.48 angle 0 sq			0.2304
Intensity + 1	107	Angle + 1	0.53 angle + 1sq			0.2809
Intensity + 2	38	Angle + 2	0.58 angle + 2sq			0.3364
Intensity + 3	7	Angle + 3	0.63 angle + 3sq			0.3969
Sum 1	146	= Bx	1.14	Ax		0.4382
Sum 2	152	= Bx	1.74	Ax		1.0142
3xlo-sum 1	369	= Bx	1.44	Ax		0.6912
3xlo-sum 2	223	= Bx	0.3	Ax		0.253
Add	217	= Bx	–0.3	Ax		–0.323
	440	= Bx	0	Ax		–0.07
	A equals	–6265.714286	2 A =	–12,571.4		
210-sum 2	217	= Bx	–0.3			2030.286
		0.03B =	–1813.29			
		B =	6044.286			
	Angle = B/2A		0.480795	91		91.4808

FIGURE 4.12 A drawing representing the collection of seven sets of intensity-angle data points using equal angle increments.

the intensity at each angle point. The seven sets of data points are input to the computer program to calculate the deviation angle. The increments are equal. The seven sets of data are expressed as seven quadratic equations with intensity I and angle as the variables:

$$I = A\theta^2 + B\theta + C$$

A derivative of intensity with respect to angle is taken and equated to zero for maximum intensity:

$$\frac{dI}{d\theta} = 0 = 2A\theta + B \quad \text{or} \quad \theta = -\frac{B}{2A}$$

With all angle differences the same, the seven equations are added and subtracted to solve for the angle of maximum intensity. The solution is a fraction of 1°. The deviation angle as shown in the diagram is composed of the initial angle, in this case 91, added to the fraction to become 91.4808. The value is then recorded in the proper place in the computer program, so the refractive index is calculated and recorded for that specific wavelength point. An example of how the method was applied to measure the refractive index for a new AMI glass, Amtir 4, is shown in Table 4.6. The orientation is designated LD for left and down. Notice that NIR and numbers 3 to 12 are together. Such data are combined with the results from the three other orientations (LU, RD, and RU), averaged, and then fit by Bill Thompson[8] to a Sellmeier equation or to a Hertzberger equation.

Material Wavelength	Mirror normal 75.72893 Amtir 4 Ref. Angle	Number P Normal	Prism Normal 65.69428 Amtir 4 #1 A + D Angle	Date Apex Angle Sin A + D	Apex Angle 10.03465 4/10/2006	Orientation Sin A	NIR LD Index
1	92.94464	65.69428	27.25036	10.03465	0.45788	0.174244	2.627811
1.064	92.81109	65.69428	27.11681	10.03465	0.455806	0.174244	2.615911
1.25	92.52313	65.69428	26.82885	10.03465	0.451327	0.174244	2.590205
1.5	92.29906	65.69428	26.60478	10.03465	0.447834	0.174244	2.570157
1.75	92.17473	65.69428	26.48045	10.03465	0.445892	0.174244	2.559016
2	92.09728	65.69428	26.403	10.03465	0.444682	0.174244	2.55207
		65.69428		10.03465	0	0.174244	0
3	91.91464	65.69428	26.22036	10.03465	0.441825	0.174244	2.535671
4	91.86354	65.69428	26.16926	10.03465	0.441024	0.174244	2.531078
5	91.83104	65.69428	26.13676	10.03465	0.440515	0.174244	2.528156
6	91.80147	65.69428	26.10719	10.03465	0.440052	0.174244	2.525496
7	91.77626	65.69428	26.08198	10.03465	0.439657	0.174244	2.523229
8	91.74275	65.69428	26.04847	10.03465	0.439131	0.174244	2.520213
9	91.71588	65.69428	26.0216	10.03465	0.43871	0.174244	2.517795
10	91.67937	65.69428	25.98509	10.03465	0.438137	0.174244	2.514508
11	91.64149	65.69428	25.94721	10.03465	0.437543	0.174244	2.511097
12	91.59983	65.69428	25.90555	10.03465	0.436889	0.174244	2.507344
13							
14							

TABLE 4.6 AMI infrared refractive index results for Amtir 4, wavelengths 1 to 12 µm, prism orientation LD.

4.3 Physical Properties Important for Optical Use

4.3.1 Hardness

Hardness is a property important in fabrication of a glass. To produce good optical surfaces, some degree of hardness is required. It has already been mentioned that some physical properties are interdependent. As shown in Chap. 2, the higher the softening point of a glass, the harder its surface. A photograph of the Leitz Miniload Hardness Tester used by AMI is shown in Fig. 4.13. The instrument has an arm with a Knoop pyramid diamond tip that is slowly pressed into the surface of the specimen under a specified load determined by added weights. The load weights may vary from 15 to 2000 g depending upon the material to be tested. For these soft materials, AMI usually uses the 50-g load. The diamond penetrates the surface, leaving a distinctive long diagonal mark

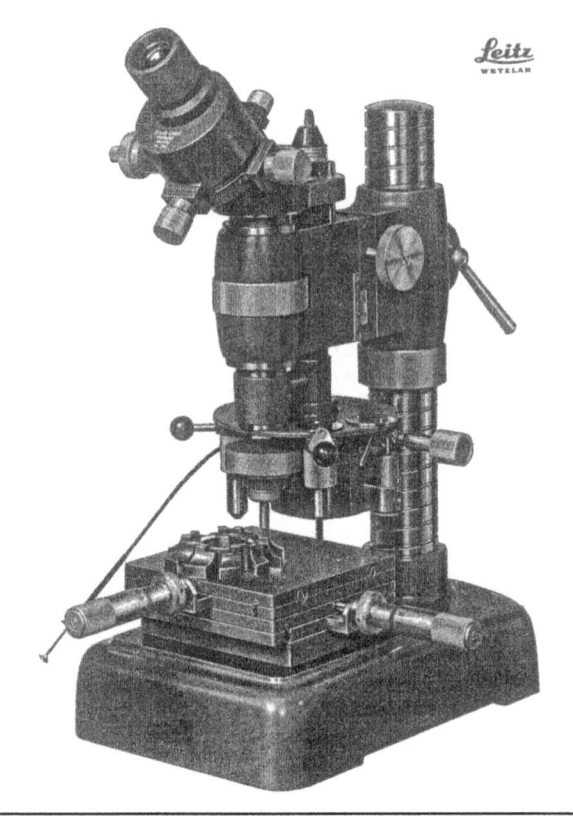

FIGURE 4.13 Photograph of the Leitz Miniload Hardness Tester instrument used by AMI to measure Knoop hardness of chalcogenide glasses.

Characterization of Glass Properties

that is observed, and the millimeter length is measured using the scale in the microscope of the instrument. Several measurements may be made and averaged. The *Knoop hardness* (KH) is calculated from the standard formula.

$$KH = 14.23 \times 10^3 \times \frac{P}{L^2}$$

where P = load in g
L = indention length in mm

4.3.2 Young's Modulus, Shear Modulus, and Poisson's Ratio

These data provide information to mechanical engineers concerned with the physical strength of each component in their systems. AMI uses simple sound velocity measurements to measure Young's Modulus E, shear modulus G, and Poisson's ratio v. Figure 4.14 shows a photograph of the simple device AMI used to measure sound in glass, the Panametrics 25-hp Plus Ultrasonic Thickness Gauge. Also in the photograph, right next to the gauge, is the Thermal Comparator, the instrument used by AMI to measure thermal conductivity of the glasses to be discussed in a later section. The thickness gauge instrument is equipped with a sensing head that uses transducers suited for longitudinal sound wave measurements or a transducer suitable for shear sound wave measurements. The velocity of each wave in the glass is directly measured, and along with the density ρ of the glass, Young's

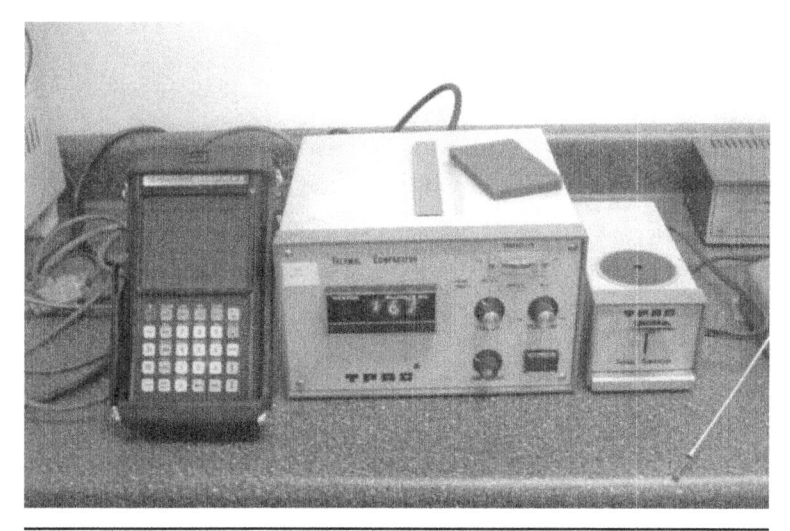

FIGURE 4.14 The Panametric 25-hp Plus Ultrasonic Thickness Gauge and the Thermal Comparator.

modulus E, the shear modulus G, and Poisson's ratio v may be calculated from the following equations:

v_S = sound velocity, shear

v_L = sound velocity, longitudinal

$$v = \text{Poisson's ratio} = \frac{1}{2} - \frac{1}{2}\left(\frac{v_S^2}{v_L^2 - v_S^2}\right)$$

ρ = density

$$G = \rho v_S^2 \qquad E = (1 + v)$$

Perhaps the first application of acoustical methods to chalcogenide glasses was reported from Bell Labs by Krause et al.[9] when they studied sound velocity and acoustical attenuation in TI 20 (Amtir 1) glass. The application of this technique to chalcogenide glasses made at TI and at AMI was the result of efforts by Don Hayes while at TI.[10] Hayes applied this method to characterize some of the TI sulfur and selenium chalcogenide glasses shown in Table 4.7. Later, he published a more complete treatment[11] for Ge-Sb-S glasses. He was the one who selected the equipment used by AMI.

4.3.3 Rupture Modulus

The rupture modulus is an experimentally determined value related to the ability of the glass to resist fracture under the stress of force.

	E (10^6 psi)	G (10^6 psi)	v
Se	1.43	0.545	0.315
$Ge_{17.5}Sb_{7.5}Se_{75}$	2.35	0.92	0.279
$Ge_{21}Sb_9Se_{70}$	2.56	1.00	0.278
$Ge_{28}Sb_{12}Se_{60}$ (1173)	3.16	1.26	0.265
As_2Se_3	2.65	1.03	0.289
#20	3.17	1.26	0.266
#20 (Bell Labs[12])	3.29	1.31	0.261
$Ge_{15}S_{70}As_{15}$	2.01	0.776	0.295
$Ge_{36}S_{60}As_4$	3.05	1.22	0.250
$Ge_{37}S_{60}As_3$	3.37	1.38	0.244
$Ge_{40}S_{50}As_{10}$	4.26	1.70	0.251
$Ge_{35}S_{25}As_{40}$	6.08	2.39	0.271
$Ge_{30}Se_{30}S_{30}As_{10}$	2.72	1.02	0.274
$Ge_{20}Se_{25}Te_{30}As_{25}$	3.00	1.18	0.270

TABLE 4.7 Elastic Moduli of Sulfur- and Selenium-Based Glasses

Characterization of Glass Properties

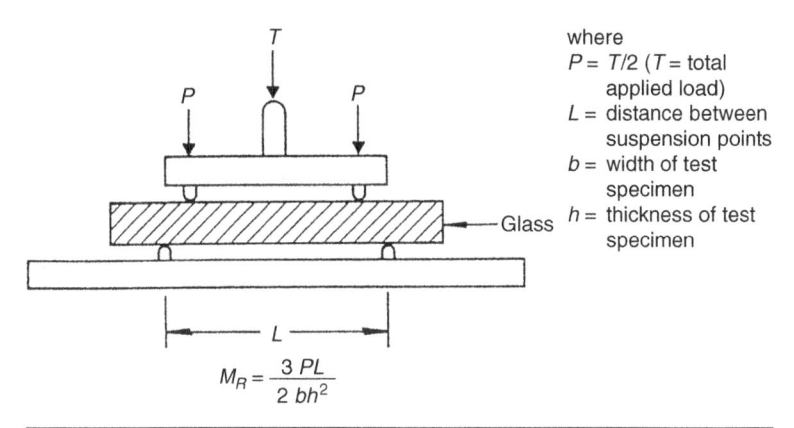

FIGURE 4.15 Diagram depicting the rupture modulus apparatus.

where
P = T/2 (T = total applied load)
L = distance between suspension points
b = width of test specimen
h = thickness of test specimen

$$M_R = \frac{3PL}{2bh^2}$$

The procedure takes time and money in that a lot of good material must be polished first and then destroyed. AMI uses four-point loading on polished bars. The bars must be thick enough to provide good data and long enough to stretch across the apparatus not close to the ends. The bars must be polished carefully with edges beveled and polished. A minimum of 10 bars should be broken. A diagram of the apparatus used is shown in Fig. 4.15. The formula used in the calculation is also given. The preparation for the measurement takes much longer than the actual breaking of the bars. Results obtained for Amtir 4 are given in Table 4.8. Note there are only nine data points. That is

Bar No.	Width (in)	Thickness (in)	Breaking Force (lb)	Rupture Modulus (psi)
1	0.723	0.434	60	2644
2	0.720	0.416	58.5	2817
3	0.725	0.432	57.5	2550
4	0.956	0.414	57.5	2106
5	0.892	0.431	71.5	2589
6	0.894	0.432	56.5	2032
7	0.955	0.433	57.5	1927
8	0.888	0.413	58.5	2317
9	0.890	0.412	56.5	2244
Average = 2358 ± 259				

TABLE 4.8 Amtir 4 Rupture Modulus

so because a bar or two were rejected when examined close to the beginning of the process. All chalcogenide glasses measured at AMI have had a rupture modulus greater than 2000 psi but less than 3000 psi. The highest has been Amtir 1, as you would expect, at about 2700 psi. The value was confirmed twice in a Navy study. A different approach was taken at TI; plates were broken instead of glass bars. A set of metal concentric rings was used in place of bars in determining the rupture modulus. Also, the smaller-diameter upper ring was connected using a swivel joint to ensure force was applied evenly to the surface of the plate. Figure 4.16 shows photographs of a two 5.5-in-diameter plates of TI 1173 after being broken. One was 0.5 in thick and the other 0.25 in. The fracture patterns are clear in the photographs.

4.3.4 Thermal Conductivity

The absolute measurement of thermal conductivity for a solid is no easy task. Such a statement applies strongly for materials with low values. All glasses are considered poor thermal conductors because they are disordered solids. Chalcogenide glasses are even worse than oxide glasses. For many years, AMI has used the Thermal Comparator instrument developed by the Thermophysical Properties Research Center at Purdue University. A photograph of the instrument is seen in Fig. 4.14. The controls and microvoltmeter are the main unit while the thermal probe unit is on the right side. Also shown in the photograph is the acoustical thickness gauge unit to the left. The thermal probe has a thermal reservoir heated to 10 to 30°C above room temperature. The voltmeter is zeroed. A sample is placed over the hole in the sensing unit, and the lever is pushed down, which mechanically brings the probe in contact with the bottom of the sample and the Constantan tube leading from the heat reservoir. Heat begins to transfer from the reservoir to the sample. Chromel thermocouples on top of

Picture taken immediately after fracture, 5.5 in diameter, 0.5 in thick

With top plate removed, 5.5 in diameter, 0.25 in thick

FIGURE 4.16 Rupture modulus measurement of TI 1173 glass plates.

Characterization of Glass Properties 113

the heat reservoir and ones in the tip of the sensor tube begin to develop a voltage difference. When the voltage becomes steady, the reading is taken as a data point. The instrument comes equipped with six samples of known thermal conductivity values certified by NBS. The readings are taken using the standards and plotted on linear 4 log paper as a calibration curve shown in Fig. 4.17. The unknown sample is measured in the same way, and its place on the curve establishes its thermal conductivity value.

4.3.5 Electrical Resistance

Chalcogenide glasses are covalently bonded and show none of the ionic conductance that you may find in oxide glasses. Those based on sulfur are generally insulators, some transmitting visible light which

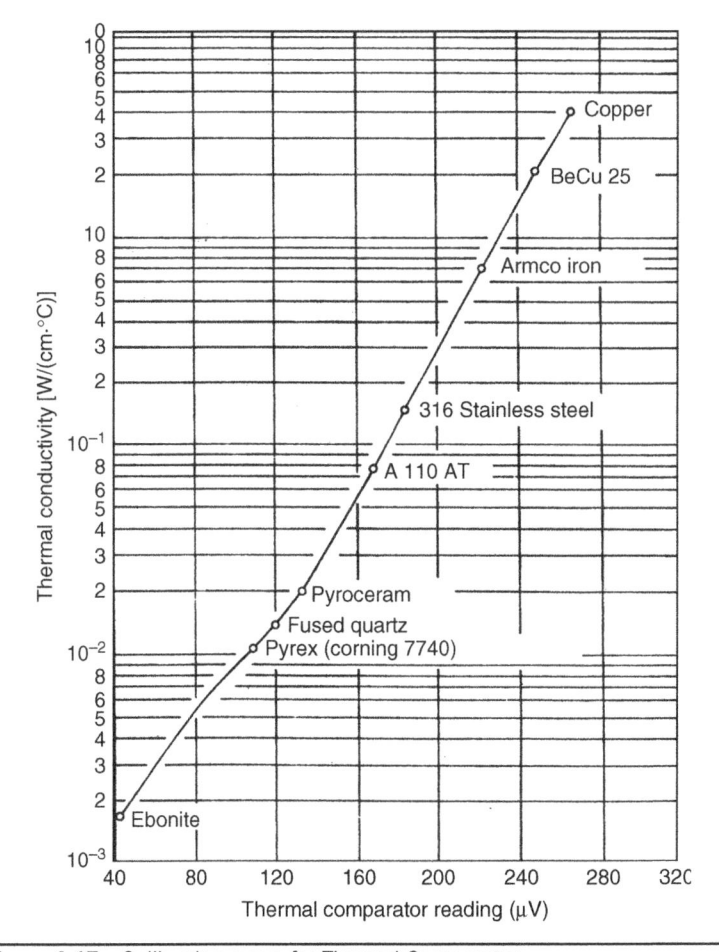

FIGURE 4.17 Calibration curve for Thermal Comparator.

demonstrates an electronic bandgap >1 eV. The bandgap indicates an average bond energy for the solid. Most all selenium-based glasses are semiconductors with bandgaps of 1 eV or less. Just like crystalline semiconductors, they are electronic conductors with electrons and holes but with extremely low carrier mobility. Their resistance is classified as semi-insulating. When they are heated, free carrier absorption in the infrared, as in crystalline semiconductors, if it occurs at all, is too weak to be observed. Glasses containing or based on tellurium may show much higher conductance than selenium-based glasses. Greater metallic character is due to the presence of tellurium. Radiation damages the lattice of crystalline semiconductor materials, knocking atoms out of place and creating defects. For visible light-transmitting solids, color centers may result as impurity metallic elements become activated. For crystalline semiconductors, the conductivity may change due to defects decreasing the carrier mobility. Chalcogenide glasses are already disordered solids. A large amount of radiation would be required to make the solid much more disordered than it is already. Also, most do not transmit visible light, so color centers would not be observed. Carrier mobility is already very low.

4.4 Resistance to Chemical Attack

Glasses generally are less chemically active than crystal materials. Chalcogenide glasses are chemically inert to most common substances with the exception of strong alkaline solutions. Chalcogenide glasses are inert to most common organic liquids such as acetone and alcohol. Exposure for weeks to nonoxidizing acids such as hydrochloric, hydrofluoric, and sulfuric acids at 5N concentration has no effect. Amtir 1 was tested by the Navy in San Diego Bay with flowing seawater with no effect after 3 months. Still, it is always best to test. Glasses based on different elements may show different chemical reactivity. For example, Si-Se glasses are very reactive with water, evolving H_2Se gas, while Ge-Se glasses are stable to water. One test method to follow is to cut and polish a number of small disks, with about 1 in diameter and 0.2 in thick. Weigh each one, identify them, and measure the transmission of each. Expose each one to a chemical to be tested, varying the conditions such as time and temperature. At the conclusion, weigh and run the transmission. Comparison before and after will indicate the resistance to each chemical tested.

4.5 Final Production Procedure

Once the chalcogenide glass composition has been selected, the related physical properties and optical properties are measured accurately, and the decision is made to produce the glass, the process begins to produce the glass in quantity and verified quality. In the AMI process, the reactants are weighed out accurately to 0.1 g using an electronic balance. Sealed in the quartz containers, the finished composition is guaranteed. Each step of the compounding casting process is controlled by a computer program. A unique program in temperature and time is developed and stored for each glass composition produced. Generally, the process, start to finish, takes 48 h.

After the process ends, the quartz container is broken and the plate is removed. Preliminary evaluation consists of an FTIR scan and examination for particles, bubbles, and fractures using an infrared microscope. If the plate is judged good, the anneal cycle is next. The anneal furnace is large with a good circulation of air. The plate is heated up to an anneal point near the Tg and then held at this temperature for several hours. Cooldown is slow, perhaps 1° per hour for the first 20° then faster to room temperature. Anneal cycles vary for each glass, lasting from 3 to 5 days. The thermal history of the glass has an effect on the refractive index in the last three to four decimals. The process should be fixed before the final refractive measurements are made and the numbers distributed. Once established, the anneal process should not be changed.

After anneal, the plate is ground flat and parallel and is polished; transmission is measured and checked against standards and examined again for particles. Results are recorded on the QC sheet for the plate. Each plate has a number assigned in the beginning, and the sheet is kept on file for future reference.

The plate is then examined for the presence of striae using the AMI "striae scope." a diagram of which is shown in Fig. 4.18.

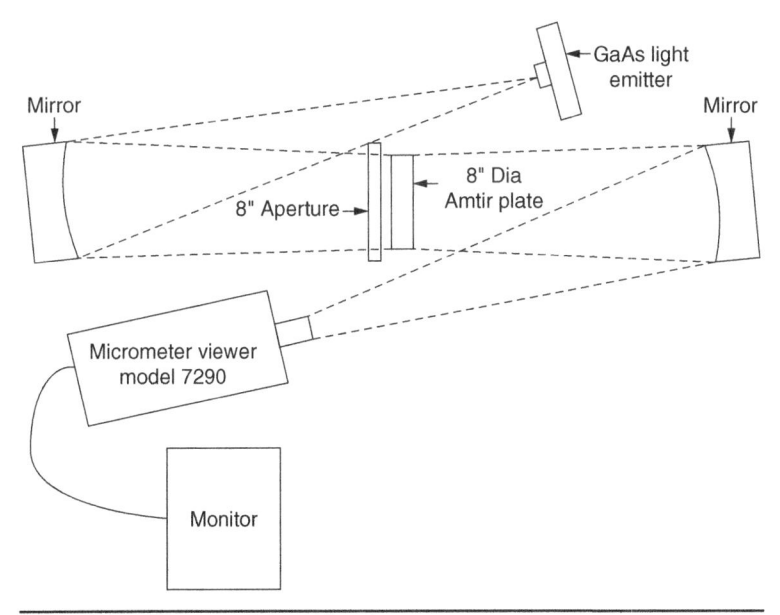

FIGURE 4.18 Diagram of the AMI striae scope.

A 1-µm-wavelength gallium arsenide light emitter is placed off-axis at the focal point of a 10-in telescope mirror. Focal lengths of both mirrors are 48 in. The mirrors are aligned facing each other. A NIR camera is placed off-axis in the opposite direction at the second mirror's focal point. The result is a beam of parallel light passing through an aperture and focused on the camera with the image shown on a TV display. The polished plane parallel plate is placed against the aperture. Any regions in the plate where the refractive index is varying will show up in the image darker than the rest of the plate due to phase cancellation in the parallel wavefront. Also, cracks or large bubbles or particles will be visible. The process is carried out in a darkened room. The operator may mark on the plate with a grease pencil any areas that are not homogeneous and should be avoided. When the plate is sawed or core-drilled for blanks, these marks are used to guide the operator.

In the 1980s, AMI used an image spoiling test as ordered by the Army to verify the optical quality of each lens blank produced. The process was used to measure the modulation transfer function (MTF) in the 8- to 12-µm range of a high-performance FLIR test module and then to remeasure the module with the blank in the optical path. The decrease of the MTF score was used to pass or reject the blank. The process was time-consuming and expensive. Some blanks failed because of the quality of the polish, not striations. After a period of time, AMI was able to demonstrate to the Army that the AMI striae scope method was better than the image spoiling test and the MTF test should be discontinued. The cost of the lens blanks was reduced by 20 percent. To emphasize the usefulness of this technique, two striae scope photographs are shown in Fig. 4.19. The top photograph is taken of an early Amtir 5 plate, number 10. One can clearly see the variation in index in the plate, striae. The second photograph is plate number 19, showing a striae-free plate due to the success of the process adjustments.

Use of the striae scope is somewhat subjective and yields no absolute number useful for comparing the homogeneity of different infrared optical materials. The classical method is to use an interferometer to measure the optical wavefront distortion (OPD) when light is transmitted through a plate of the material. Rosberry[12] reported perhaps the first homogeneity results for infrared optical materials. The materials tested were silicon, zinc sulfide, magnesium fluoride, and calcium fluoride. Although other companies continued using MTF image spoiling tests, AMI decided it would be a good idea to have AMI materials evaluated by an interferometer test. A plate of Amtir 1 was sent to R. M. Ranat of Pilkington. A MTF image spoiling test as well as the interferometer showed that Amtir 1 was about 2 times as homogeneous as single-crystal, annealed germanium. Bill Spurlock of Exotic Materials found a similar value for Amtir 1. Later Spurlock measured Amtir 3 and Sullivan of Exotic Materials measured gallium arsenide for AMI. The results

Characterization of Glass Properties 117

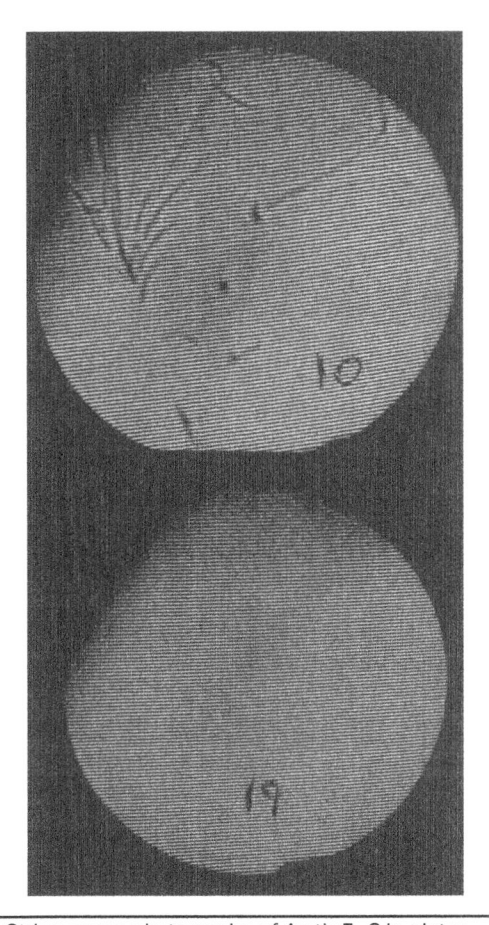

FIGURE 4.19 Striae scope photographs of Amtir 5, 8-in plates.

are presented in Table 4.9 for AMI materials along with other widely used infrared materials for comparison. Also included are approximate values of $\Delta N/\Delta T$ for the materials. The homogeneity numbers are given as $\Delta N/N$ in units of 10^{-6}. Amtir 1 has the lowest value except for ZnSe which was not a three-dimensional test. One explanation for the low value for Amtir 1 is that all three constituent atoms, Ge-As-Se, are on the same row of the periodic table and next to one another. Thus, since the refractive index primarily reflects atomic mass and these three atoms are almost the same, compositional variation is not a factor for this glass. To carry this argument one step further, Amtir 5 is an As-Se glass. Therefore, Amtir 5 should be more homogeneous than Amtir 1. Using a Wyko Interferometer, AMI evaluates the homogeneity of every Amtir 5 plate for use in the Lockheed Martin JSF Fighter. Averaging the results for five plates, the Wyko results showed a value for $\Delta N/N$ for Amtir 5 of 6.1, lower than the 8 value for Amtir 1.

Material	IR Refractive Index	Index Homogeneity $\Delta N/N \times 10^{-6}$	Thermal Change Index $\Delta N/\Delta T \times 10^{-5}/°C$
Germanium (single crystal)	4.0	17*	43
Gallium arsenide (poly)	3.3	20[†]	20
Zinc selenide (poly/cvd)	2.4	3[‡]	6
$Ge_{33}As_{12}Se_{55}$ (Amtir 1)	2.5	8[§]	7
$Ge_2Sb_{12}Se_{60}$ Amtir 3	2.6	19[¶]	10
$As_{40}S_{60}$ (As_2S_3 glass)	2.4	N/A	±0.9

*UK Rarde Results, 1984; Spurlock Exotic Materials, 1990.
[†]Sullivan Exotic Materials, 1991.
[‡]Raytheon number for plane perpendicular to the growth axis. Does not include gradient in growth direction.
[§]Ranat Pilkington, 1991; Spurlock Exotic Materials, 1990.
[¶]Spurlock Exotic Materials, 1991.

TABLE 4.9 Optical Homogeneity of Some Infrared Optical Materials

References

1. Thomas Loretz Computer Engineering Services, private communication.
2. R. J. Patterson, "Research on Infrared Optical Materials," TI Report No. AFAL-TR-66, Texas Instruments, November 1966.
3. J. T. Littleton, *J. Amer. Ceramic Soc.* 10, 259 (1927).
4. A. R. Hilton and C. E. Jones, "The Thermal Change in the Nondispersive Infrared Refractive Index of Optical Materials," *J. Appl. Phys.* 6, 1513 (1967).
5. A. R. Hilton, "Precise Refractive Index Measurements of Infrared Materials," *SPIE* 1307, 516 (1990).
6. A. R. Hilton, "Infrared Refractive Index Measurement Results for Single Crystal and Polycrystal Germanium," *SPIE* 1498, 128 (1991).
7. Albert Feldman, Deane Horwitz, Roy M. Waxler, and Marilyn J. Dodge, NBS Technical Note 993, *Optical Materials Characterization* (1978).
8. Bill Thompson, Optical Services Co., Lucas, Texas. AMI Optical Science Consultant.
9. J. T. Krause, C. R. Kurkjian, D. A. Pinnow, and E. A. Sigety, *Appl. Lett.* 17, 367 (1970).
10. A. R. Hilton, D. J. Hayes, and M. D. Rechtin "Chalcogenide Glasses for High Energy Applications," Contract No. NOOO14-73-0367,DARPA Order No. 2443 (1974).
11. D. J. Hayes et al., *Proceedings of the Ultrasonics Symposium*, 502 (1974).
12. F. W. Rosberry, "The Measurement of Homogeneity of Optical Materials in the Visible and Near Infrared," *Appl. Opt.* 5, 961 (1966).

CHAPTER 5
Conventional Lens Fabrication and Spherical Surfaces

A glass composition has been selected for development. The optical and related physical properties have been determined, and the glass has been selected for fabrication into lenses. A plate of the glass has been compounded, cast, and annealed. The plate has been ground flat and parallel using a Blanchard Mill and polished for final inspection. A serial number is assigned and inscribed on the outside edge of the plate and transferred to the quality control sheet along with all the evaluation data. The plate is turned over to production for lens blank fabrication.

5.1 Lens Blank Preparation

In this particular instance, the plate is a 9-kg plate of Amtir 1, with 8 in diameter and 2 in thickness. The plate will be blocked in an oven on a silicate glass plate using blocking wax. After cooling cylinders of glass, the correct diameter for the lens will be diamond cored drilled out of the plate as shown in Fig. 5.1. The plate is shown along with a glass cylinder. The cylinders have a piece of ceramic epoxied the length of the core to serve as the last material to be sawed through, which prevents chipping of the edge of the lens blank. In Fig. 5.2, we see glass cylinders glued together with epoxy end to end to increase the number of blanks sawed at one time. For greater efficiency, other end-to-end cylinders may be added side to side to increase the number of blanks produced at one time. Figure 5.2 is a photograph of blanks being sawed. Notice the white ceramic glued to the bottom of the cylinder. The controls on the saw are set so that the thickness of each blank is the same. The hole in the ID saw is large, with a 3.5-in diameter which allows use on larger blanks or more ganged together. After the sawing is complete, the sawn parts are placed in a container and soaked in acetone to remove the ceramic and epoxy. Most of the lenses

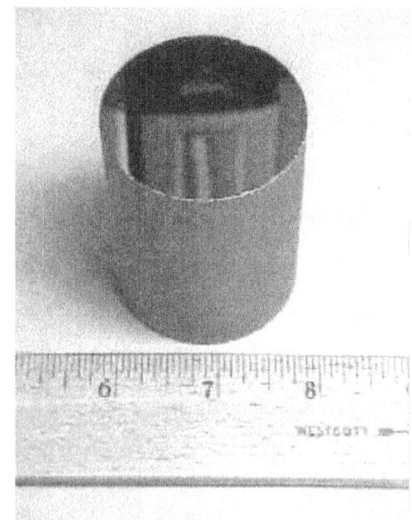

FIGURE 5.1 Core drilling a glass cylinder from a plate.

produced at AMI are small and simple, planoconvex, with 1 in diameter. Production rates as high as 2000 per month have been maintained over long periods. The lenses produced are inexpensive, used mostly for thermal sensing. Total production by AMI since 1984 is well over 250,000. The lenses can be used to form images but are not considered good quality.

FIGURE 5.2 Sawing individual lens blanks with the ID saw from end-to-end glass cores.

5.2 Generation of Spherical Surfaces

Until recent years, almost all lens optical surfaces were spherical. The generation of a spherical surface with the correct radius of curvature was an important step in the production of any lens. Figure 5.3 shows a diagram depicting the process of generating a spherical surface. The blank of glass to be generated is mounted on the lower vertical spindle. The generating tool is mounted on a spindle at an oblique angle to the vertical spindle. Both spindles are spinning. The generating tool is hollow with both its outside and inside edges covered with diamonds. In the diagram, a convex surface is being generated. The inside-diameter edge W is being used. The outside-diameter edge is used for concave surfaces, and W is slightly larger than the inside one. Both of these edge diameters must be greater than one-half the diameter of the part being generated. One may think in terms of the points of edge contact forming a cord on a circle of radius of curvature R of the lens. The angle θ between the axis of rotation for the tool and that of the part being generated can be calculated by

$$\theta = \sin^{-1}(W/2R)(180/\pi)$$

where W is the inside diameter in this case and R the desired convex radius of curvature. In practice, the operators are furnished tables covering the settings for standard tools and desired curvatures by the company that manufactures the generator. This method is practical and has been used for many years.

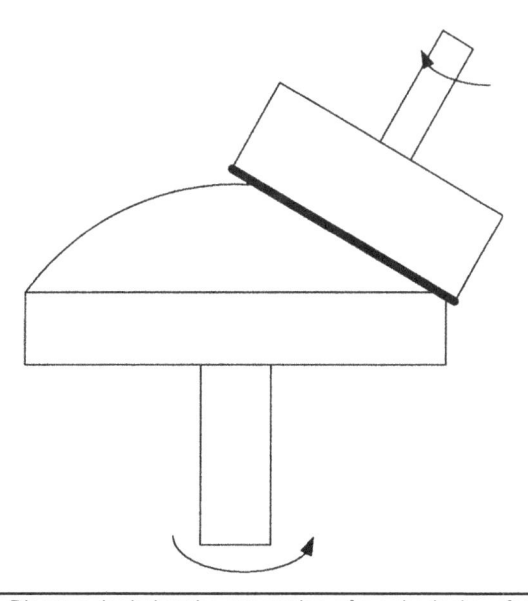

FIGURE 5.3 Diagram depicting the generation of a spherical surface.

5.3 Polishing

AMI has 22 spindles in operation, a small number in comparison to an optics fabrication shop. The spindles are used mostly for polishing. The polishing slurry used is mostly based on aluminum oxide with some silica and zirconium oxide. The slurry is pumped and circulated over the parts during polishing. Single-spindle operation requires a lot of attention by the technician and thus is not very efficient. AMI developed a multiblocking technique to increase production of the small 1-in- diameter planoconvex sensor lenses.

First, one side of each blank is polished and then covered with a protective coating of paint. Next, they are mounted in their respective tool using a soft blocking wax. Figure 5.4 shows two spindles with multiple lens blanks blocked in the recesses on the spherical convex surface of the tool. The smaller tool is for lenses with shorter focal lengths and thus a smaller radius of curvature. Fewer lenses may be made at one time than with the other tool used for longer focal length lenses. Here the radius of curvature is larger, requiring a larger convex sphere. These two tools are placed on the bottom vertical spindles. Not shown is the top concave tools with the exact radius of curvature for their respective lenses. The bottom spindles are rotated and the top spindles moved back and forth while the bottom tools are spinning. Of course a polishing solution is sprayed on all the parts at the same time. After the polishing is finished, the lenses are unblocked in an oven. After cooling, they are placed in an ultrasonic cleaner with a solution that removes all the wax and paint. Each lens is then checked for curvature and center thickness.

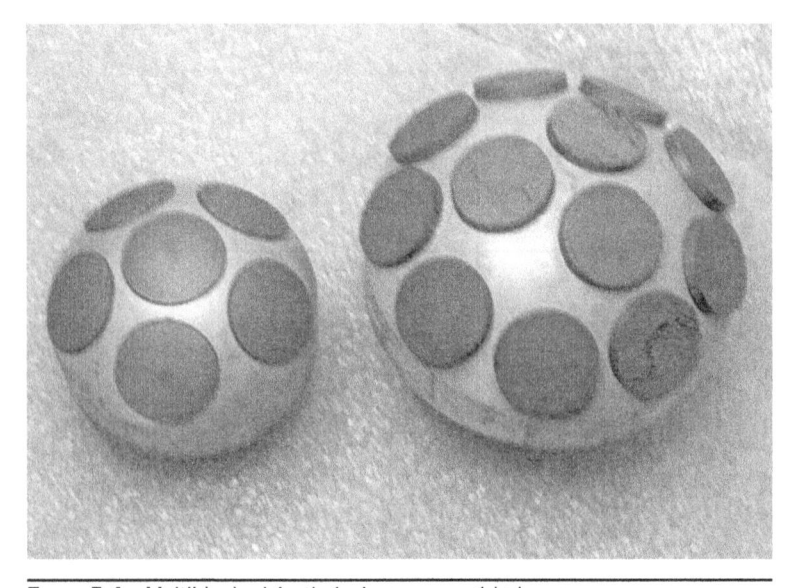

FIGURE 5.4 Multiblocked Amtir 1 planoconvex 1-in lenses.

5.4 Testing

Now that the spherical surface has been generated and polished, how to check it? Again, a simple and practical way is to use a "bell," as illustrated in Fig. 5.5. A linear dial indicator is placed inside a bell that has an opening with a known diameter. The bell with indicator is first placed on a flat to zero the indicator. When it is placed on a part, a sag reading is made. The absolute value may be used to evaluate the radius of curvature, or a reading from a standard may be used for comparison. Concave surfaces may be evaluated in a similar manner. The method is fast and effective for many small lenses.

Many lens fabrication operations use test plates fabricated from optical glass for a specific radius of curvature needed for many parts. The curvature is tested by placing the part in light contact with the test plate under a light source with a specific light emission. Interference fringes results are counted in number and shape to evaluate the finished surface. Others have an instrument capable of measuring the

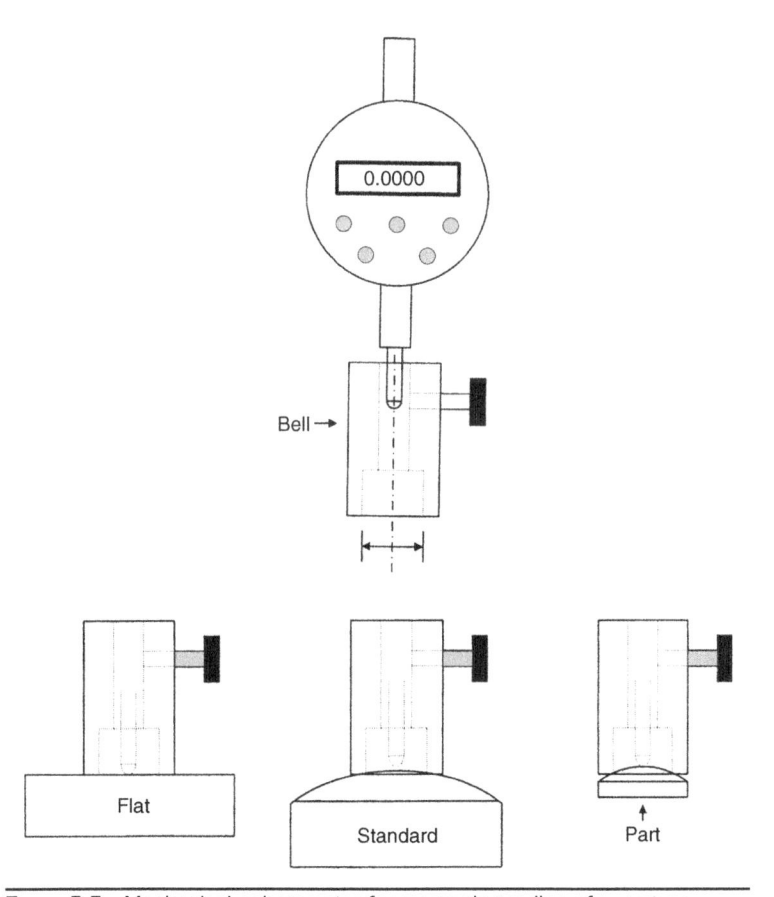

FIGURE 5.5 Mechanical spherometer for measuring radius of curvature.

modulation transfer function (MTF) of the lens using collimated monochromatic light. The focal length of the lens is confirmed, and the quality of imaging in terms of line pairs per millimeter is resolved on and off axis and is calculated and tabulated. In the early years of AMI, the Night Vision Laboratory at Ft. Belvoir (NVL) required AMI to run an image spoiling test on each Amtir 1 blank before shipping. NVL provided the funds for AMI to purchase from Lou Fantozzi of Diversified Optics, a computer-controlled MTF instrument. In the test, the instrument measured the MTF of a high-performance three-element infrared lens designed by Fantozzi and fabricated at Diversified Optics. A carefully polished Amtir 1 blank was then placed in the optical path, and the MTF of the lens was remeasured. The relative MTF score at 10 line pairs per millimeter had to be 94 percent or better for the blank to pass. AMI used the instrument for this test several years until the test was deemed costly and unnecessary because the pass rate was 99 percent. Eventually the instrument was no longer functional and was not replaced by AMI.

In 2000, AMI began molding chalcogenide glass lenses in a joint program with Lockheed Martin in Orlando, Florida. The molds used were of very high quality, and as the work progressed, it became important to have a method to verify the accuracy of the molding process. AMI purchased a ZYGO interferometer instrument, shown in Fig. 5.6 mounted on a stabilized optical table. Not visible in the photograph on the right-hand side of the table is a WYKO interferometer used to evaluate the optical homogeneity of each Amtir 5 plate produced by AMI to be used in the Lockheed Martin JSF Fighter. Figure 5.7 shows a ZYGO evaluation sheet

FIGURE 5.6 Photo of the AMI ZYGO interferometer.

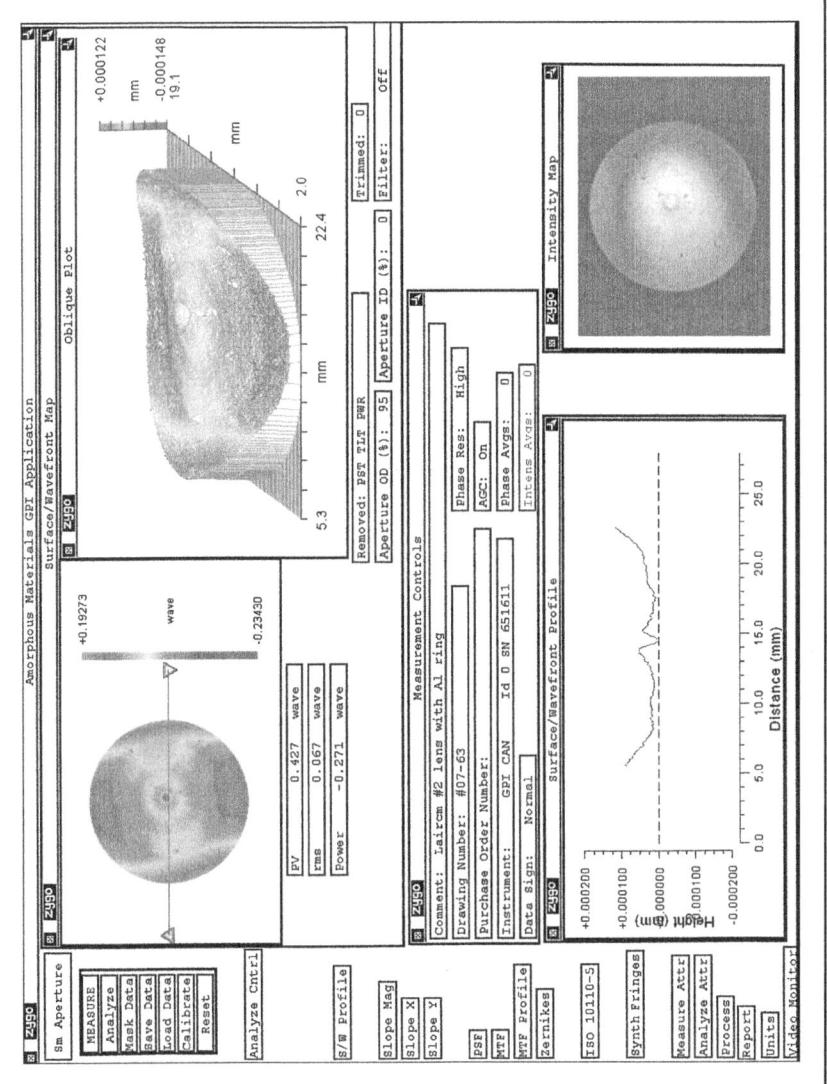

FIGURE 5.7 ZYGO evaluation sheet for a molded Amtir 5 lens.

for a small Amtir 5 molded part. The results on the sheet show a PV of 0.427 wave, an rms of 0.067 wave, and a power of –0.271 wave. The Instrument uses a He-Ne laser emitting at 0.6328 µm. Also not shown in Fig. 5.7 is a mechanical device used to measure the wedge of the molded lens. Each molded lens is evaluated in this manner, and the results are used in improving the molding process. AMI now molds four of its glasses.

5.5 Antireflection Coatings

Previously it was pointed out that glass surfaces are generally more inert than crystalline surfaces. Glasses are more resistant to chemical attack. In applying antireflection coatings, inertness becomes a handicap. Coating designers then have to revert to what they generally call *glue layers*. Each develops her or his own favorite for different materials. The purpose is to apply a very thin layer, too thin to be optically significant, to provide a surface to which the first coating layer will adhere. High melting oxides such as aluminum oxide or magnesium oxide are examples that are favored by some.

AMI for many years was not involved in coatings. The situation changed when in 1997 AMI began a SBIR Phase II program with the Navy to fabricate an infrared imaging bundle from As_2S_3 glass fibers 10 m in length. The program goal was an overall transmission of 50 percent in the 3- to 5-µm band. With an optical path of 10 m to contend with, it seemed obvious that an antireflection coating to reduce Fresnel reflection losses would be required to meet the transmission goals. Ed Carr, a former colleague of the author at TI, was recruited. Carr, although retired, had almost 40 years' experience at TI and other companies in the area designing and producing coatings for infrared optics. He joined the program. From a local company, a rebuilt Temescal unit was purchased with one special modification. Two flanges were welded on top of the chamber. The ends of the flanges were fitted with O-ring seals such that fiber ends could be inserted through the seals to be coated without contaminating the chamber. All coatings designed by Carr used mixed fluorides in place of the radioactive thorium fluoride used by many in the industry.

The results of the three-year effort to produce the 10-m bundle have been reported in the literature.[1] It turns out that the antireflection coatings were not used on the bundle ends because they caused crosstalk problems between the fibers with a loss in contrast of the image. After the SBIR program ended in 1999, efforts turned to providing coatings for all the glasses produced by AMI with special interest in the low-temperature glasses Amtir 4 and 5 developed for use in molding. It became important to be able to coat the lenses after they were molded. New coating designs were developed. A new larger, computer-controlled, more up-to-date coating chamber was purchased. The Cryo pump was backed by a turbo pump so that ion beam assisted deposition could be used to improve coating quality. After deposition, the coatings must be evaluated relative to reflectivity and tested for

Conventional Lens Fabrication and Spherical Surfaces 127

resistance to humidity, adherence, and resistance to abrasion. All AMI films are tested according to MIL-PRF 13830B, Appendix C 3 8.2, the 24-h humidity test, 3841 the severe abrasion test, 3842 the moderate abrasion test, and 385 the adhesion test. The humidity test is carried out in a commercially available environmental chamber. Adhesion is a tape pull test using a designated tape. Both abrasion tests are surface rubs under exact conditions observing damage.

During the coating operation, extra pieces are included to serve as witness samples. Quality tests are performed on these. If the witness samples pass, the coated parts may be approved for delivery. The first step is to verify the reflection from the surface conforms to the design. The Beckman Vis-NIR spectrophotometer has a special reflectance attachment designed for the instrument. Some coatings must be evaluated in the NIR even close to the visible light. Most are confined to the infrared well within the range of the Perkin Elmer FTIR. Figure 5.8 shows the reflection attachment for the FTIR. Notice the sample used is a crystal of high-purity silicon. The reflectivity of silicon is flat at 30 percent throughout the 2- to 14-μm range and is used as a reference. An aluminum, gold, or silver mirror would show much higher reflectivity than silicon, but is not stable with time and offers no sensitivity advantage. When the reference or background scan is silicon, a gain in sensitivity occurs. Since an indicated 100 percent reflectivity is only 30 percent, a horizontal 10 percent line on the scan is in reality only 3 percent reflectivity.

Figure 5.9 shows an actual FTIR scan of a 3- to 5-μm coating on Amtir 4. The 10 percent line represents only 3 percent reflectivity. The point at 5 μm reading 3 percent then is only 1 percent actual reflectivity. Figure 5.10 shows a FTIR reflection scan for an 8- to 12-μm coating

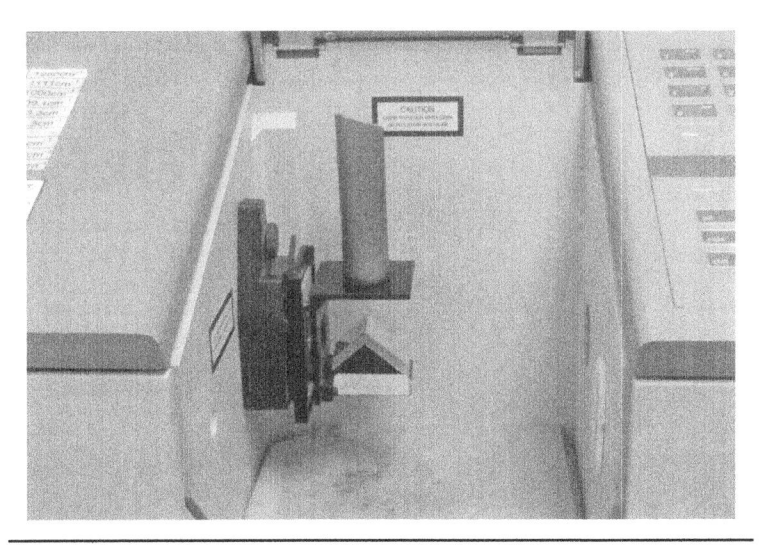

FIGURE 5.8 Reflection attachment for Perkin Elmer FTIR.

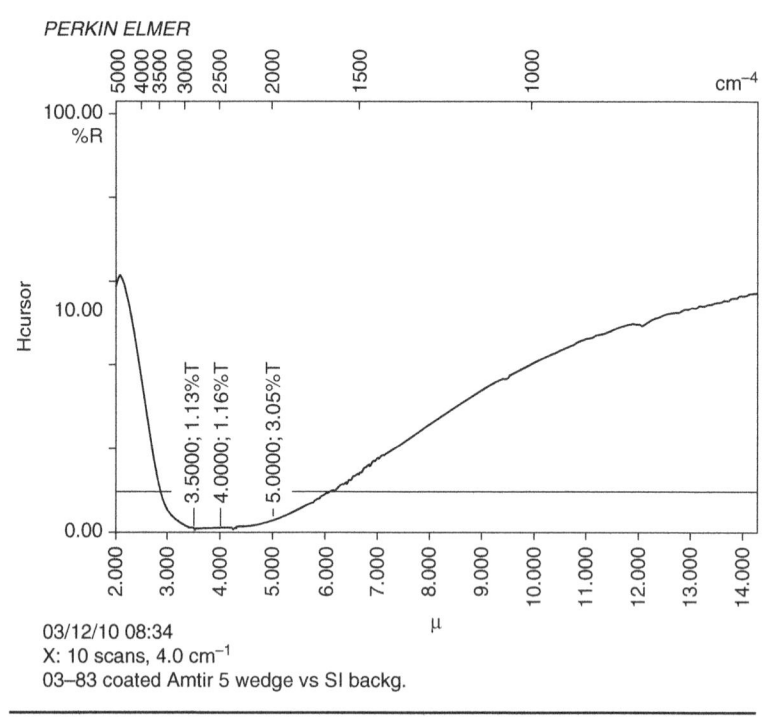

Figure 5.9 FTIR reflection QC scan for 3- to 5-μm coating on Amtir 4.

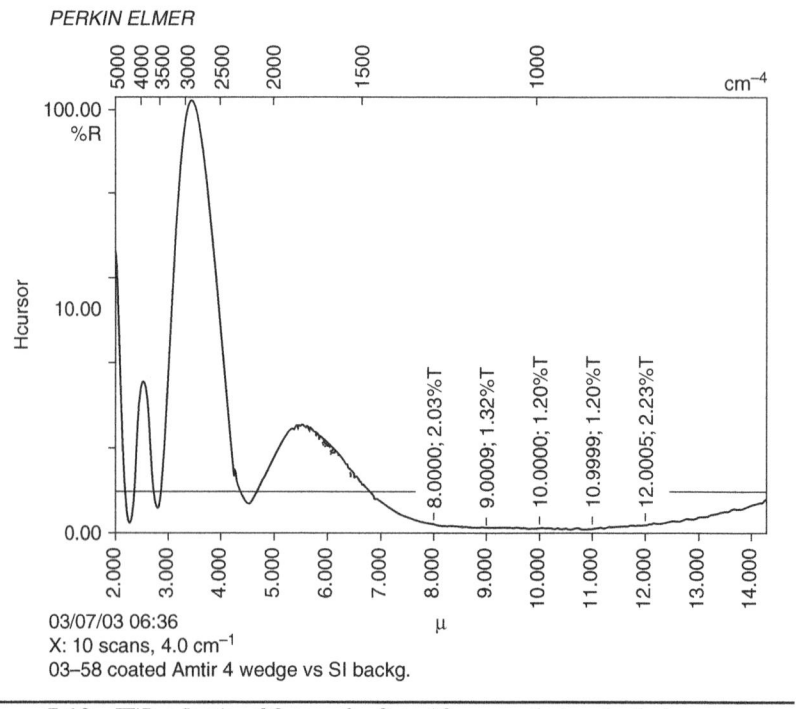

Figure 5.10 FTIR reflection QC scan for 8- to 12-μm coating on Amtir 4.

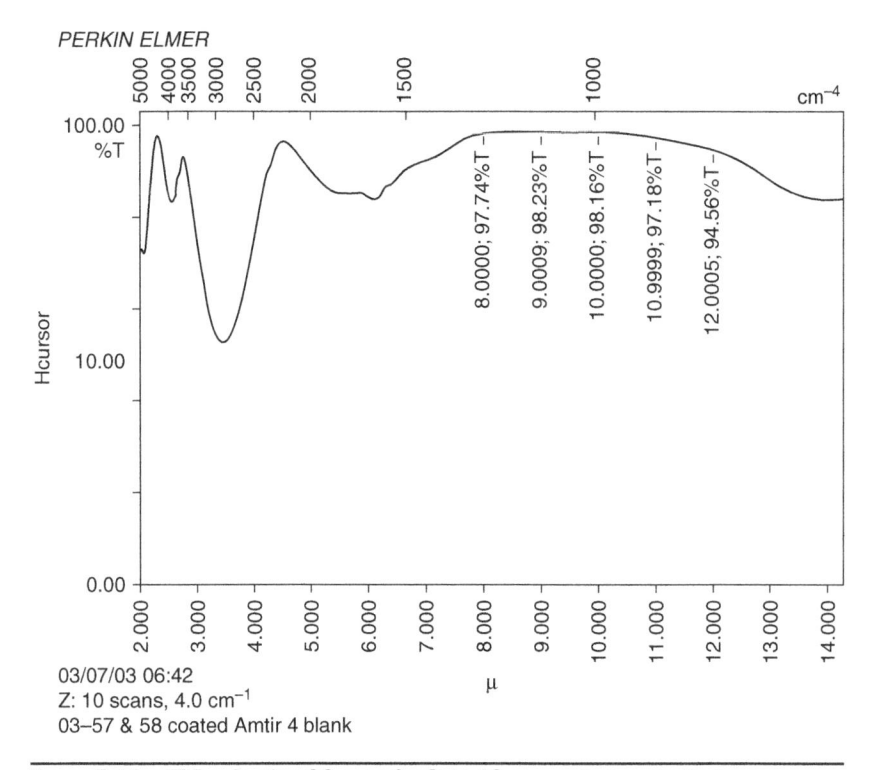

FIGURE 5.11 FTIR reflection QC scan for 8- to 12-μm coating on Amtir 4.

again on Amtir 4. Note all the measured parts were less than 1 percent actual reflectivity. Figure 5.11 shows a FTIR transmission scan again for an Amtir 4 sample. In this case the reference is a standard background. The sample is coated on both sides with the standard coating. The average transmission across the band is 97.18 percent.

One disadvantage for coating the low-softening-point glasses developed for molding is that they must be coated at low temperatures. To produce a good hard coating, one needs heat as the materials are deposited. Only one AMI glass passes the severe abrasion test—Amtir 1. Amtir 1 has the highest softening point and can be coated at a higher temperature than the others. Coatings on other AMI glasses pass the mild abrasion test. One other factor to keep in mind is that chalcogenide glasses have relatively large thermal expansion coefficients. One should keep this fact in mind when selecting the materials to be deposited as layers in the coating design.

Reference

1. Ray Hilton, Sr., Ray Hilton, Jr., James McCord, Glen Whaley, Thomas J. Loretz, and Paul Modlin, "Fabrication of a 10 Meter Length IR Imaging Bundle from Arsenic Trisulfide Glass Fibers," *SPIE* 3596, 64 (1999).

CHAPTER 6

Unconventional Lens Fabrication, Aspheric Surfaces, and Kinos

In Chap. 5 the fabrication methods producing spherical optical surfaces in use for many years were discussed. The methods were applied in the fabrication of visible as well as infrared lenses. Two technical advances brought about a dramatic change. The first was the direct application of computer control of machine tools that led to diamond point turning. The second was the development of inexpensive uncooled infrared detector arrays,[1] the most expensive part of the cost of infrared cameras. Now the most expensive cost of the cameras became the optics. Attention turned to lowering the optics cost.

6.1 Optical Designs

In Chap. 5 an optical element designed by Louis Fantozzi of Diversified Optics was described as a test element in an image spoiling test for Amtir 1 blanks. The lens had three elements, two germanium and one Amtir 1. Each lens had all spherical surfaces. When tested in the wavelength range of operation, 8 to 12 µm, the design produced an MTF score of 70 percent at 10 line pairs per millimeter and 50 percent at 20 line pairs per millimeter. The performance was more than adequate for the then Army common module FLIR. Lens designs for high performance with all spherical surfaces required the use of three or more elements and two or more different optical materials. It is true that aspheric designs could be made to improve performance and produced, but with great difficulty using the old spherical methods. The goal to lower cost became the search for new designs using only two lens elements and still maintaining the required performance.

Better still would be a Petzval design that used only one material in the two-element design. Aspheric surfaces and the addition of diffractive surfaces would improve color correction.

Designers use computer software such as ZEMAX that provides the ability to iterate their designs to produce maximum performance. Positions of each lens may be changed one at a time while the thickness of each lens and the radius of curvature, all may be changed while witnessing the change in performance. Before beginning a lens design for a camera, the designer needs to know the area of the detector array and the number of elements in order to calculate the area of each detector element, the pixel size. The finished lens design must have enough performance to resolve each detector element. The greater the number of detector elements, the greater the optical performance required. Specified field of view and performance off-axis requirements must be met as well.

The final product of a lens design is a *lens drawing* used by the fabricator to produce this specific part. Besides a drawing of the lens element, usually drawn to some scale, the material to be used will be specified along with a callout for the drawing, if available, which specifies the material requirements. Inner and outer diameters may be specified with acceptable tolerances. Spherical radii for both surfaces, if used, will be specified with tolerances. Center thickness is important. Sag for concave and convex surfaces may be given. Alignment of the surfaces with the base in terms of wedge may be specified. Minimum surface figures such as measured with a Zygo interferometer may be listed. Assembly drawings showing all the optical elements in place with spacing between each element specified may be generated by the designer.

6.2 Diamond Turning

The advent of diamond point turning for optics was a major advancement. Optical design is a mathematics-based activity. Computer precise control of a diamond point applied to a circular lens blank accurately located on a round rotating mount produced aspheric or diffractive surfaces (kinoforms) on command. Optical designers use software that allows them to specify optical elements in size and shape, using specific optical materials. The refractive index of each material as a function of wavelength is available to the designer in stored tables. The spacing between elements is a variable as well as the center thickness of each lens. The program can calculate expected performance in terms of MTF with spatial frequency on axis and off axis. The variables may then be adjusted slightly to maximize calculated performance for the final design. If the operator could express the surfaces of each lens he or she required mathematically and load it into the computer controlling the diamond point, it would be generated one surface at a time.

The initial cost of the diamond point machine was great at first. Since the early 1990s the machines have become much more commonplace, resulting in lower initial cost and more general use. Designs previously impossible to consider are routinely produced but are costly. Costs are estimated in terms of labor hours to set up the machine and labor hours for a technician to tend to the machine as the part is generated, machine time. The diamond turning is carried out at high rotation speeds and under the cover of a cooling fluid continuously sprayed on the rotating part. The process is confined in an enclosure in part to protect the operator.

6.3 Slump Molding

Glass has been molded for centuries using simple methods. Generally, the required amount of glass is placed in an open mold and then heated sufficiently for the glass to soften and move under the force of gravity to fill the mold. No pressure is used. Slump molding may be used as a cost-saving measure. Slump molding has been used at TI and AMI to preshape a lens blank to minimize the amount of material removed and wasted during the grinding stage of forming the lens. At AMI, routinely a blank for a deep 7.5-in meniscus Amtir 1 lens has been formed by slumping a flat plate of glass into a concave Pyrex mold. The mold is generated from a Pyrex mirror blank with the concave radius equal to the convex radius of the lens. The plate is placed over the mold and heated in the furnace together until it softens and slumps into the mold under its own weight. After slumping, it is annealed. The thickness of the Amtir 1 plate before slumping is slightly greater than the finished center thickness of the lens. Mostly slump molding is used for shaping glass to dimensions that are not precise.

6.4 Precision Molding

The desire to begin precise molding of lenses was a continuation of the goal to lower the cost of infrared optics. The ability to diamond point turn lenses was a real advancement but too expensive. The expense occurred for each lens ordered. The thinking went that if the expense produced a mold instead of one lens, the mold could be used over and over to make many lenses with only the cost of the infrared glass lens blank. The cost of the diamond turning would thus be averaged over many lenses. In 2000, AMI joined with Lockheed Martin in Orlando (LMCO) in a program to develop the technology required to mold infrared lenses from chalcogenide glasses. At this time, the most experienced and knowledgeable person in the United States regarding molding lenses from glass was Harvey Pollicove. His efforts at Eastman Kodak resulted in the development of a production facility that molded millions of aspheric camera lenses from silicate-based glasses and plastics. He presented a paper at the SPIE OE/LASE Technical Symposium in January 1988 entitled, "Survey of Present Lens Molding Techniques." After Pollicove left Kodak, he joined the American Precision Optics Manufacturers Association (APOMA) at the University of Rochester. Many companies were members along with the U.S. Army. Because of the proprietary nature of the AMI effort, a direct inclusion of Pollicove in the AMI program was not possible.

Shortly after AMI began the molding program, the author met Harvey Pollicove at a NVL (Night Vision Laboratory) meeting held at the Army Picatinny Arsenal. Later, Pollicove visited at AMI and spent some time with the author. He provided a copy of his paper and many glass molding patents. On return from a trip to Japan, he furnished copies of a technical nature describing automatic systems produced by Toshiba. The information he furnished was very helpful to AMI.

The Toshiba units at that time cost $350,000 each, much too expensive for AMI. The units used up to 3500 kg (about 8000 lb) of force, much too severe, we believed, for chalcogenide glasses. The lenses are stamped out. Also, the glass blanks were a sphere of glass heated rapidly in the open. Chalcogenide glasses in contrast to oxide glasses are volatile when heated. Besides being expensive, the method did not seem suited to chalcogenide glasses.

Another approach to be considered was injection molding. Miracles have been produced in the injection molding of plastics. The organic thermal plastics lend themselves quite well to the process. Raw materials are powdered or supplied in small pieces, easily heated to form a flowing liquid. However, they are not volatile and not oxidized when exposed to air. Most contain pigments and do not require optical homogeneity. A few attempts were made at AMI to use the glass extruder unit as a source of glass flowing into a mold without any success. The approach was abandoned.

AMI decided to develop glasses well suited for the molding process. Low-softening-point glasses were chosen. Lower molding temperatures would increase mold lifetime, a cost factor. Selecting glasses not containing germanium would lower the cost of the glass. Molding at low temperature and pressure should minimize the stress of the molding process which in turn would improve lens quality. An arsenic-selenium glass composition was selected, produced, and characterized optically and physically. The glass, designated Amtir 4, was suitable for use in both the 3- to 5-μm band and the 8- to 12-μm band. Because of its very low dispersion, Amtir 4 can be used as a replacement for germanium parts.[1] Amtir 4 has a small negative thermal change in index (-24×10^{-6}/°C) in both bands due to a large thermal expansion coefficient. The index change should be contrasted to the over $+400 \times 10^{-6}$/°C for germanium. As a bonus, it was also found that Amtir 4 could be used to draw small-diameter (50-μm core) flexible, unclad bare fibers that improved the quality of IR imaging

Unconventional Lens Fabrication, Aspheric Surfaces, and Kinos 135

bundles[2] produced at AMI. Later, a second low-softening glass was developed for molding, Amtir 5, similar to Amtir 4 but with a thermal change in refractive index of essentially zero in both the 3- to 5-μm and 8- to 12-μm bands. Also, the glass had a thermal expansion that matches that of aluminum. The glass confirmed an opinion expressed by the author[3,4] many years ago that such a unique composition could be selected. The AMI Amtir 5 glass composition is protected by U.S. Patent 6,984,598.

Early experiments carried out with Lockheed Martin personnel involved slump molding of Amtir 4 glass using Pyrex spherical molds. Results showed that less than one wave rms was easily achieved on one surface. Attention turned to precision pressure molding. The first AMI unit was built, and efforts turned to developing the molding process. A photograph of the first unit is shown in Fig. 6.1. The design is simple and relatively inexpensive, built from parts obtained locally.

FIGURE 6.1 Photograph of the AMI lens molding unit.

The large aluminum chamber is equipped with a door that may be sealed, and the chamber evacuated down to a pressure of 50 μm or less. Inside was placed a standard inexpensive programmable oven equipped with a blower to provide circulating air. The temperatures of the molds are controlled using band heaters. The glass blank is placed in the mold, the door is closed, and the unit is evacuated. The molds and glass blank are heated to the molding temperature, and pressure is applied from the cylinder on top of the unit. A few hundred pounds of force is needed for lenses with 1- to 2-in diameter with less than 1000 lb for lenses with almost 6-in diameter. A linear gauge indicates when the molds are closed, indicating the lens is formed. The molds and lens are cooled slightly, the pressure is released, and the unit is vented. After the door is opened, the oven is turned on and the glass anneal cycle is started.

As results improved, a plan was devised by LMC to prove the quality of molded chalcogenide glass optics was equal to that of those made by diamond point turning (DPT). A 100-mm FL, $F/0.8$, 10° field of view, two-element lens was designed using Amtir 4 for use with a LTC 500 LWIR uncooled bolometer array camera produced by a Lockheed Martin company. The front element had a 4.6-in diameter. The convex surfaces of both lenses were spherical. The concave surfaces were aspheric with kinoforms. The first set of optics was diamond turned and coated at AMI. The second was molded and coated at AMI. Amy Graham[1] of LMCO showed thermal images that demonstrated that optical performance of the camera using the molded lenses was essentially the same as when using the DPT lenses. Actual physical measurements of two sets of lenses supported this statement (Table 6.1).

The only parameter that failed was center thickness CT. The problem was that not enough glass was being removed under current circumstances. The problem was solved by modifying the bottom mold surface so that excess glass was more easily forced out when pressure was applied. Also, the decision was made to use polished blanks that weighed a very small amount greater than the finished lens to minimize glass movement.

Parameter	DPT Part	Molded Part	Delta
R_1	1.369	1.369	0.0000
Sag	0.167	0.1674	0.0004
Center thickness	0.326	0.3568	0.0308
Inner diameter	1.529 ± 0.001	1.5245	0.0045
Outer diameter	1.629 + 0.001	1.6255	0.0035

TABLE 6.1 Molded versus Diamond Point Turned Optical Elements

Unconventional Lens Fabrication, Aspheric Surfaces, and Kinos 137

FIGURE 6.2 Infrared image made using the LTC camera with DPT lenses.

Figure 6.2 shows an infrared image produced by the camera using the DPT lenses. The image was made in the AMI parking lot. Notice the fence wire is clearly resolved and the hot tires on the vehicle are prominent. Presently, AMI has only two infrared cameras for experimental use. The first is an Agema 210, a 3- to 5-μm camera that has proved useful in our infrared imaging bundle work. The second is a Raytheon Palm IR BST uncooled bolometer 8- to 12-μm camera that has been most valuable in evaluating lens performance. Both represent early technology and are not considered high-performance cameras. But they were all AMI could afford for this purpose.

Adapter plates were required for every different lens we wanted to evaluate. AMI wanted to use the 100-mm molded lens with our 8- to 12-μm camera. Unfortunately, the Palm IR has an optical chopper blade in front of the detector array. The design of the 100-mm lens was for the LTC camera; the back focal length of the camera was too long to be in focus, the lens would hit the light chopper. Fortunately, we could use that same mold with Amtir 5 to shorten the back focal length. Our molding skills had improved to the point we did not hesitate to use different glasses with the same mold. So lens number 1 remained molded with Amtir 4 and lens number 2 was molded with Amtir 5. An adapter plate was made and the lens was placed on the Palm IR. A photograph of the camera with lens is shown in Fig. 6.3. One may notice the coated lens does not look bright and shiny. The reason is the fact the Amtir 4 must be coated at a low temperature

138 Chapter Six

Figure 6.3 Photograph of molded 100-mm FL Amtir 4-Amtir 5 lenses adapted to the AMI Raytheon Palm IR camera.

because of its low softening point. Coating temperatures above 100°C yield harder, shiny, stronger coatings. Figure 6.4 shows lens 1 molded with both Amtir 5 and Amtir 6 (As_2S_3) glasses. Note the one on the right is Amtir 5 with an antireflection coating. The one on the left is

Figure 6.4 Photograph of 100-mm FL lens 1 molded with Amtir 5 and 6.

uncoated and shows its natural color. Note the kinos on both are clearly visible.

AMI also worked with Jan Terlow and John Lawson of CBC America to mold four elements used in an IR zoom lens. The lens changes from an $F/0.8$ to $F/1.2$. Three of the lenses are spherical convex with kinoforms on the concave side. The fourth element is double concave with one spherical and one aspheric with a kinoform. The original design used Amtir 4, but as our molding skills improved, we found we could mold the elements with Amtir 5. For the same optical performance, only one element design needed to change going from Amtir 4 to Amtir 5. Again, a mounting plate for the Palm IR camera was constructed so that we could use the zoom with the camera. Figure 6.5 shows the zoom mounted on the Palm IR camera.

Further examples of using the same mold with different glasses are seen in Fig. 6.6. Lens 1 of the zoom is shown molded using Amtir 5, Amtir 6, and C1 glass. The lenses are uncoated. Again, the Amtir 6 lens shows its true color. Kinos are clearly visible on all three.

Two sets of DPT Amtir 4 optics were made during the Lockheed Martin IRAD for a two-element design with a 218-mm focal length. AMI provided one of the lens mounts and antireflection-coated both sets. LMCO kept one set and AMI the other. Lens 1 had a 5.7-in diameter while lens 2 had a 1.8-in diameter. The lenses with mounting are very heavy. An adapter plate was made so the lens could be used on the AMI Palm IR. A support was made for camera and lens. Figure 6.7 shows

FIGURE **6.5** CBC America IR Zoom made with molded Amtir 4 glass lenses mounted on the Raytheon Palm IR camera.

140 Chapter Six

FIGURE 6.6 Photograph of zoom lens 1 molded using Amtir 5, Amtir 6, and the As-Se-Te C1 core glass.

FIGURE 6.7 Photograph of Amtir 4 two-element 218-mm FL lens adapted to the AMI Palm IR camera.

Unconventional Lens Fabrication, Aspheric Surfaces, and Kinos 141

the camera and lens with support mount at the local high school track used by the author each week usually at 5 a.m. The performances of the lenses described with the Palm IR camera are shown by the figures that follow.

Figure 6.8 shows the images produced by the CBC America Zoom IR lens mounted on the Palm IR camera. The top image is in the narrow field mode, and the bottom is the wide field mode image. The object is a person in a darkened warehouse 160 ft from the camera. He is standing between two machines against the back wall. The machinery

FIGURE **6.8** Photographs demonstrating the performance of the CBC America Zoom IR lens mounted on the Palm IR camera.

142 Chapter Six

FIGURE 6.9 Photograph demonstrating the performance of the molded Amtir 4-Amtir 5 100-mm FL IR lens mounted on the Palm IR camera.

is not visible at all in the narrow mode but shows up slightly in the wide field. Figure 6.9 shows a photograph taken of the same scene only using the molded Amtir 4-Amtir 5 100-mm FL IR lens again mounted on the Palm IR camera. Notice the machinery against the wall is now clearly visible. The image of the person is larger and more detailed. Figure 6.10 shows in the bottom photograph a visible image of the high school track. Look closely and you will see a small white image of a person at the far end of the track. The person is about 170 yd from the camera. The top image was taken using the 218-mm IR lens mounted on the Palm IR very early in the morning. The cool background makes the warm person very visible as a white object. The person is the author. In this case, the camera was on a tripod aimed at the distant location. The camera was turned on, and the author walked to the distant point. Being alone made best focus adjustments impossible. Nevertheless, the shape of a person was clearly distinguishable.

The composition of Amtir 5 was precisely selected to provide a near-zero thermal change in refractive index in both bands, 3 to 5 μm and 8 to 12 μm. The unique value of Amtir 5 was the fact that athermal systems can be designed and built without the need for mechanical thermal compensating parts. Such devices were an extra complication and expense to infrared system designers. After the glass was developed, LMCO built a wide field of view airborne system and confirmed by Dan Woody of LMCO a less than 1 percent change in

Unconventional Lens Fabrication, Aspheric Surfaces, and Kinos 143

FIGURE **6.10** Photograph demonstrating the performance of the DPT Amtir 4 218-mm FL IR lens mounted on the Palm IR camera.

performance during temperature excursions. The decision was made to use Amtir 5 in the infrared system for the new Joint Strike Fighter. As part of the IRAD, AMI molded several of the 5.38-in-diameter lenses. The convex and concave surfaces were both spherical. A photograph of one of the molded lenses is shown in Fig. 6.11.

Another question in molding that has not been mentioned is, How do you get the molded part out of the mold without it being broken? It is true the molds are usually coated with a thin film to which the glass part should not stick. However, these coatings develop open spots with use. Also, some areas of the mold are almost impossible to reach while coating. Another factor can be the shape and fragile nature of a lens.

144 Chapter Six

FIGURE **6.11** Photograph of a molded 5.38-in-diameter Amtir 5 JSF lens.

AMI found itself in a seemingly impossible situation when an agreement was made to mold a lens 8 mm in diameter. Early attempts proved fruitless. Another approach was needed. We decided to place a thin metal ring in the mold, the outside diameter the same as the bottom of the mold and the height of the ring the same as the lens edge, so the glass would bond to the ring. The lens could then be removed because the metal ring would not stick to the metal mold. The scheme worked, and the 8-mm lens fell right out. Figure 6.12 shows a fuzzy photo of the 8-mm molded lens encased in the metal ring at the base. Figure 6.13 shows a diagram of the scheme we followed. Another example is shown in Fig. 6.14. In this case, a fragile

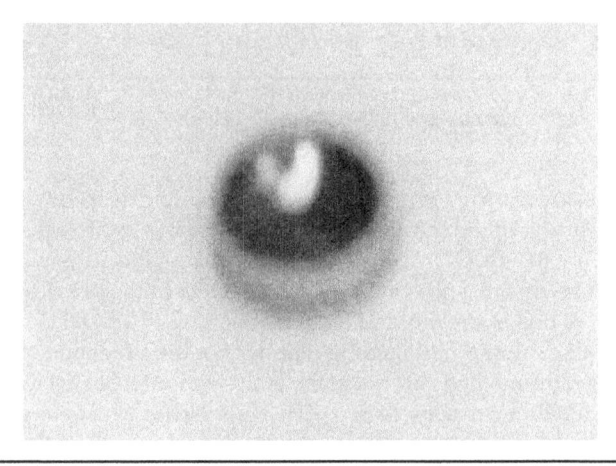

FIGURE **6.12** Photograph of a molded 8-mm-diameter lens in a metal ring.

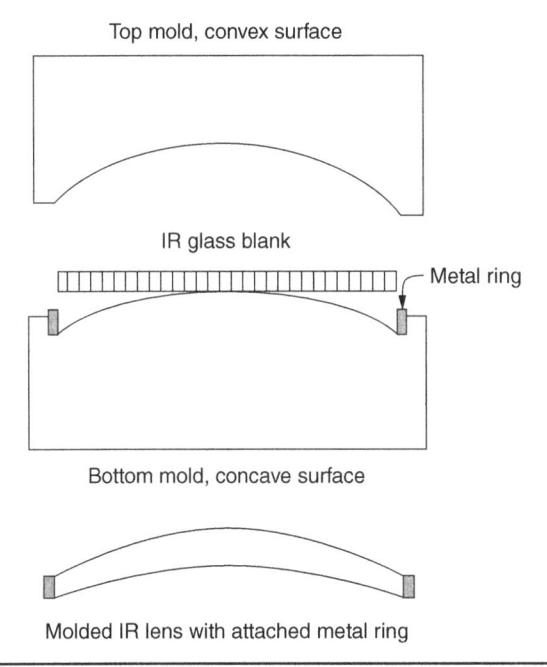

Unconventional Lens Fabrication, Aspheric Surfaces, and Kinos 145

FIGURE 6.13 Diagram of the metal ring procedure used to reduce the difficulty in removing a molded lens from the mold.

FIGURE 6.14 Photograph of a molded 17-mm-diameter lens in metal ring.

tophat-shaped 17-mm-diameter lens was molded in a metal ring. The procedure increased the yield of the molding process substantially. Also, the ring protects the glass from edge fractures.

6.5 Volume Production

From the very start in the molding effort, AMI has been more interested in producing quality lenses based on the glasses we produce. We developed molding units and antireflection coatings compatible with chalcogenide glasses. Rapid stamping out of lenses from chalcogenide glasses was rejected. The AMI molding process uses low pressure and takes time. There are 5 units installed in the production area as shown in Fig. 6.15. All are the same and constructed here with local help and component parts. The goal here was simple inexpensive units. With two operators, each unit may be used twice per day to go through a complete cycle. On that basis, it is possible the 5 units could produce 10 lenses per day, 50 lenses per week, or about 2600 lenses per year. To increase production volume even further, there is enough room in this area to place another 5 or even 10 units in place. One factor not mentioned is molds and their cost. The value of the business would have to support the cost of the new units and the great expense of molds for all the units, sometimes $10,000 per set.

FIGURE 6.15 Photograph of five AMI molding units in the production area.

Unconventional Lens Fabrication, Aspheric Surfaces, and Kinos 147

FIGURE 6.16 Photograph of a mounted mold pair with alignment pins.

Another approach, for smaller lenses, is multimolding. The AMI mounting plates are large and could accommodate several lens molds. Figure 6.16 shows how the large molding plates with alignment pins to ensure top and bottom molds match. Figure 6.17 shows eight molds

FIGURE 6.17 Photograph of mounted multimolds top and bottom.

for the same small lens in place both top and bottom. This approach is much more productive from the standpoint of the number of operators and molding units. Mold cost will still be a major factor. The business volume must be there to support the effort.

Now, AMI is often asked about producing parts in terms of a hundred thousand or more. Our molding approach will lend itself to producing hundreds or a few thousand perhaps, but not a hundred thousand. A small independent company such as Amorphous Materials lacks the resources and the desire for such an undertaking.

6.6 Problem of Refractive Index Change When Pressure Molding

When a glass is pressure-molded, it must be heated above its glass transition temperature. One must remember that below that temperature, glass acts as a solid and will break if pressure is applied to reshape it. As for any liquid, if enough pressure is applied, the liquid is very slightly compressed. Since refractive index is a measure of the concentration of atoms per centimeter cubed, the index increases in a positive sense. In optics, a small change in refractive index is important, The change needs to be measured to see how large a problem exists. Also, a way to avoid the problem may be found.

AMI Infrared Refractometer

Previous discussions concerning the AMI infrared refractometer have pointed out that measurements of a single prism in a single orientation are reproducible to a small number in the fourth decimal place. In each instance of the data presented, the prism has been removed and measured in the same orientation on different days. The data presented in Table 6.2 support the claimed accuracy.

Previous discussions have pointed out that the lack of perfection in the fabricated prism is overcome by measuring the prism in each of

Date	N @ 3 µm	N @ 4 µm	N @ 5 µm
8/4/04	2.7616	2.7565	2.7533
8/6/04	2.7621	2.7569	2.7535
8/9/04	2.7621	2.7565	2.7534
Average	2.7619	2.7566	2.7534
	±0.00016	±0.00016	±0.00007

TABLE 6.2 Amtir 5 Prism 03-1B, Orientation LD

Unconventional Lens Fabrication, Aspheric Surfaces, and Kinos

λ (μm)	Before Molding	After Standard Molding Proc.	Δ	After Modified Molding Proc.	Δ
3	2.7575	2.7622	+0.0047	2.7575	+0.0000
4	2.7530	2.7568	+0.0038	2.7523	−0.0007
5	2.7493	2.7534	+0.0041	2.7488	−0.0005
8	2.7426	2.7466	+0.0040	2.7421	−0.0005
10	2.7379	2.7418	+0.0039	2.7372	−0.0007

TABLE 6.3 Change in Index on Molding Amtir 5

four orientations on the mirror, left or right relative to the beam (L or R) and up or down for the top of the prism (U or D). The four measurements averaged lead to the accurate value, as has been demonstrated a number of times for materials where the resulting values are compared to those published by NBS or NIST.

To demonstrate the change in index on molding Amtir 5, we have taken the before-and-after approach (Table 6.3). A prism was fabricated from annealed glass from plate 04-23 and measured 3 to 10 μm in all four orientations. Next, glass from the plate was used to mold a flat 5-in-diameter plate using our standard temperature and pressure. A prism was fabricated from the plate and measured in an identical manner. Then a second plate was molded at the same temperature and pressure, only care was taken to release the pressure at a point above the T_g in order to relax the compressed liquid.

In conclusion, molding under pressure will increase the refractive index appreciably, 0.0040. However, if the pressure is released before the glass becomes a solid, the change is small, < 0.0010.

In summary,[5] AMI has developed the technology to mold high-quality lenses from glasses produced by AMI. Two low-softening glasses developed were successfully used for molding. Antireflection coatings were developed for these glasses. Lenses made using Amtir 4 and 5 are in several important military systems. AMI has applied the knowledge learned to mold four of the seven glasses produced by AMI shown in Table 6.4. Amtir 1 has not been molded because of its high softening temperature of 405°C. Amtir 2 is a very likely candidate because of its similarity to Amtir 5. Amtir 4, not listed, is not fully developed at this time. Amtir 4 is better than Amtir 1 in transmitting visible light. The softening point of Amtir 4 is 277°C, well over 100° lower than Amtir 1. Amtir 4 is a good candidate for molding. AMI has established production capability with five molding production units.

Property	Amtir 1	Amtir 2	Amtir 3	Amtir 4	Amtir 5	Amtir 6	C1
Composition	Ge-Ae-Se	As-Se	Ge-Sb-Se	As-Se	As-Se	As-S	As-Se-Te
Transmission range (μm)	0.7–12	1.0–14	1.0–12	1.0–12	1.0–12	0.6–8	1.2–14
Ref. index at 10 μm	2.4981	2.7691	2.6027	2.6431	2.7398	2.3807	2.8051
$\Delta N/\Delta T$ (°C × 10^{-6})	72	5	91	−23	< 1	< 1 (5 μm)	31
Knoop hardness	170	110	150	84	87	109	110
Thermal expansion × 10^{-6}/°C	12	22.4	14	27	23.7	21.6	23
Thermal condx [cal/(g · s · °C)10^{-4}]	6	5.3	5.3	5.3	5.7	4	5.2
Specific heat [cal/(g · °C)]	0.072	0.068	0.066	0.086	0.076	0.109	0.062
Density (g/cm^3)	4.4	4.66	4.67	4.49	4.51	3.2	4.69
Rupture modulus (psi)	2700	2500	2500	2358	2400	2400	2500
Young's modulus (× 10^6 psi)	3.2	2.65	3.1	2.2	2.56	2.3	1.8
Shear modulus (× 10^6 psi)	1.3	1.03	1.2	0.85	1.01	0.94	1.03
Poisson's ratio	0.27	0.29	0.26	0.297	0.279	0.24	0.29
Softening point (°C)	405	188	295	131	170	210	154
Glass transition temp. T_g (°C)	368	167	278	103	143	187	133
Upper-use temp. (°C)	300	150	250	90	130	150	120
Dispersion values							
3–5 μm	202	176	159	186	175	155	148
8–12 μm	109	162	110	235	172		196

TABLE 6.4 Comparison of IR Transmitting Glasses Produced by AMI

References

1. Amy Graham, Richard A. LeBlanc, and Ray Hilton, Sr., "Low Cost Infrared Glass for IR Imaging Applications," *SPIE* 5078, 26 (2003).
2. A. R. Hilton, Sr., A. R. Hilton, Jr., J. McCord, W. S. Thompson, and R. A. Leblanc, "Infrared Imaging With Fiber Optic Bundles," *SPIE* 5074, 849 (2005).
3. A. Ray Hilton and Charlie Jones, "The Thermal Change in the Nondispersive Infrared Refractive Index of Optical Materials," *Appl. Opt.* 6, 1518 (1967).
4. A. R. Hilton, Sr., A. R. Hilton, Jr., J. McCord, and T. J. Loretz, "Laser Power Delivery Using Chalcogenide Glass Fibers," *SPIE* 2977, 20 (1977).
5. A. Ray Hilton, Sr., James McCord, Ronald Timm, and R. A. LeBlanc, "Amorphous Materials Molded IR Lens Progress Report," SPIE 6940 69400Q-1 (2008).

CHAPTER 7

Glass Processes for Other Applications

7.1 AMI as Supplier of Chalcogenide Glasses for IR Fibers

The first involvement of AMI in the development of infrared fibers based on chalcogenide glasses was that of a materials supplier. There was a 2-year research agreement made between Texas Instruments and Galileo Electro-Optics in 1982 to jointly develop fibers and fiber devices based on the use of TI 1173 glass. The glass was to be provided by TI and the silicate fiber-based technology of Galileo Electro-Optics applied to the infrared glass. The program was negotiated with the help of AMI colleague Thomas Loretz who at that time was director of materials research and engineering at Galileo Electro-Optics. The program made good progress based on TI 1173 glass and was the first in the industry, 1985, to announce the availability of chalcogenide glass fibers. The TI results from the program were reported[1] in *Optical Engineering* in 1987. The final optical-mechanical properties from Galileo Electro-Optics were also reported[2] in 1987. However, after the 2-year program ended, TI would no longer supply the TI 1173 glass. AMI began to supply glasses in 1985 and 1986 to Galileo Electro-Optics and others in the form of billets and rods of core glass. Results relayed to AMI were not good relative to attenuation at the wavelength of 10.6 µm, critical for use with the CO_2 laser. From the AMI standpoint, our evaluation results for the glasses supplied indicated that much better performance should have been experienced. The reaction at AMI was that the silicate fiber-based techniques they used were not suited for the chalcogenide glasses. As an example, we later learned that silicate fibers were often pulled at speeds up to 70 mph. In the marketplace, another factor was the brittleness of TI 1173 and its tendency to crystallize. Glass compositions other than TI 1173 were selected and turned into fibers at Galileo, but the greatest difficulty in using the fibers was breakage.

In 1984 AMI was approached by Codman Fiber Optics, a company owned by Johnson & Johnson, to join in a research effort to develop a chalcogenide glass to serve as a source of fibers capable of transmitting substantial energy from a CO_2 laser emitting at 10.6 µm. Codman is the surgical division of Johnson & Johnson, and its desire was to be the first with a laser scalpel. Also involved in the project was Kathy Laakman of Laakman Electro-Optics (later known as Synrad) which was to supply the carbon dioxide lasers required for the scalpel. It turned out that Tom Loretz, now a senior scientist at Codman, originated the program. Our chances for success improved because Loretz was not only well versed in fiber technology but had a background in glass science, a Master's in Glass Science from Alfred University. The Codman program manager was Mr. Igino Lombardo. At this time, the FDA had approved the CO_2 laser emitting at 10.6 µm for use in surgery. The original goal of the program was an absorption level at 10.6 µm for the fiber of 0.5 dB/m which corresponds to a bulk absorption of 0.001 cm^{-1} and at 4 µm a goal of 1 dB/m or 0.002 cm^{-1}. One should realize that such low absorption levels in infrared optics were very unusual. Lenses and windows in infrared systems were built using materials with absorption levels of, at best, 0.01 to 0.03 cm^{-1}. Lenses and plates were only 1 to 2 cm thick. Fibers could be 100 cm or meters in length. Silicate fiber people spoke in terms of decibels per kilometer. For this application, the low absorption was important to avoid the fiber heating up and failing while transmitting the laser energy. The chalcogenide glasses have low thermal conductivity. Absorption of the laser energy raises the temperature along the axis of the fiber. The sign of the change in refractive index for the chalcogenide glasses is positive. The result is the formation of a positive lens or thermal lensing. As a result, the fiber melts either at the ends or in the middle.

During the program it became apparent that an absorption level of 0.001cm^{-1} at 10.6 µm was below the intrinsic level for selenium-based glasses. Codman agreed to change its required level to 0.003 cm^{-1} or to 1.3 dB/m. Still later in 1985 after much effort had been expended to lower the 10.6-µm level, Codman agreed to accept core material with measured values less than or equal to 0.005 cm^{-1} or 2.3 dB/m. At the same time, AMI pointed out to Lombardo that the lower level for the glasses at 9.27 µm, the wavelength of another strong emission line from the carbon dioxide laser, had already been achieved. Switching to the other line should be just as useful in surgery as the 10.6 µm. Neither the doctor nor the patient could tell the difference between the two wavelengths. The suggestion was refused by Lombardo on the grounds that there was no FDA approval for the use of the 9.27-µm emission for surgery.

Both core glass and clad glass were to be developed as proprietary and exclusive to Codman & Shurtleff of Johnson & Johnson. The clad glass was to have an index difference from the core glass of –0.01 to –0.03 to provide the desired acceptance angle or numerical aperture (NA). The NA may be calculated from $(n_{core}^2 - n_{clad}^2)^{1/2} = \sin \alpha/2$ where n is the index and α is the total included acceptance angle. All the optical fiber formulas[3] are derived from Snell's law: $N_1 \sin \theta_1 = N_2 \sin \theta_2$, where N_1 and N_2 are the indices in medium 1 (air in this case) and medium 2 (fiber core); θ_1 is the angle of incidence to the surface of the core and θ_2 is the angle of the refracted ray to the axis of the core. Part of the incident ray is reflected from the surface of the core and lost while part is refracted into the core. When the core ray strikes the interface with the clad glass, the situation changes since the clad glass index is less than that of the core index and the ray angle with the normal to the interface decreases to a critical angle θ_C where the ray is totally reflected internally. The angle may be calculated from[3] $\theta_C = \cos^{-1}(N_2/N_1)$. The refracted ray then travels down the core with little or no reflection loss. However, when a ray enters the core at a greater oblique angle (outside the NA) such that the refracted ray strikes the core-clad interface at an angle larger than θ_C, the ray is poorly reflected, partially absorbed in the cladding, and finally disappears.

For laser light transmission through a fiber, a small acceptance angle is desirable with the beam focused using a long-focal-length lens or mirror to a spot size smaller than the core diameter of the fiber. The rays then enter the fiber at angles close to normal and are refracted still closer to normal in the fiber. Loss of energy is minimized through fewer internal reflections off the walls of the fiber. Expansion of the clad glass was to be 95 percent of the core glass to minimize stress between core and clad. After the goals of the program were realized and production was started, Codman would guarantee AMI substantial sales of glass to Codman for 3 years.

Glass systems as candidates for producing a glass meeting the goals, in order of probability, were (1) As-Se-Te, (2) Ge-As-Se-Te, and (3) Ge-As-Se glasses based on Amtir 1 diluted with tellurium. Amtir 1 glasses low in germanium and As-Se-Te glasses were all evaluated relative to lowest absorption level at 10.6 µm. Results[4] supported the best glass would be found in the As-Se-Te system. Considerable information concerning these glasses was found in the work of Joe Jerger in the Servo report.[5] A number of glass compositions were made and tested. They all seemed to have absorption levels at 10.6 µm related to the amount of selenium in their composition. That fact indicated that either an impurity in most all selenium or an intrinsic selenium-selenium vibration overtone was the cause of the limiting absorption level. AMI purchased high-purity selenium from suppliers in Japan, Canada, Germany, Belgium, France, and KBI and Asarco in the United States. Comparison of furnished analysis showed major impurities of Cu, Fe, Si, As, Te, and S. Virtually all selenium is produced as a by-product of copper production. The selenium is chemically separated and then purified by physical distillation.

156 Chapter Seven

In 1982 prior to this program, AMI suddenly had a serious problem in producing Amtir 1 glass due to excess absorption. We found out our KBI source selenium was really produced by Asarco in Amarillo. A visit there revealed that they had shut down their high-purity unit 2 years earlier. Asarco had selected a certain fraction of selenium in the distillation process they had learned produced excellent results for our application. Asarco continued supplying AMI out of their inventory. When that fraction was exhausted, the material supplied to us contained too much iron which lowered the overall transmission of the glass. We switched to the Belgium source marketed in the United States by Indussa.

To verify that selenium was the problem, Glen Whaley of AMI constructed the fractional distillation column, from high-purity quartz, shown in Fig. 7.1. In the drawing, the column is shown in a vertical position inside a two-zone furnace. To begin, Asarco selenium powder was placed inside through the bottom chamber. The chamber was sealed and evacuated. The unit was then placed inside in the vertical position and heated to melt all the selenium. The two-zone furnace was adjusted to produce a vertical thermal gradient with the maximum temperature at the first chamber, number 1. Each chamber had its own thermocouple for temperature monitoring. The column was left for a long period to allow the vapors and reflux to come to equilibrium.

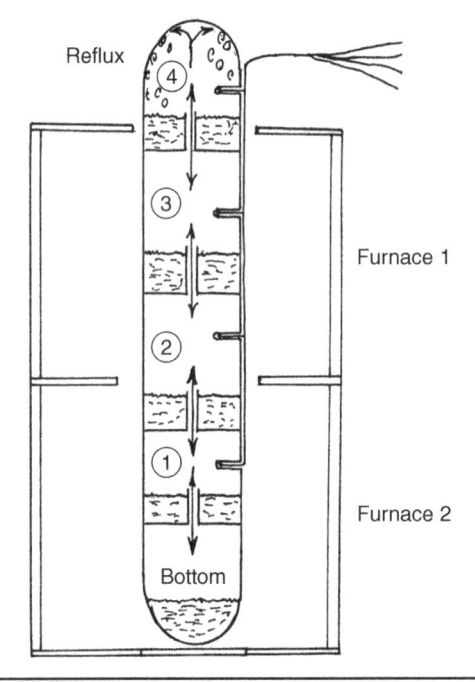

Figure 7.1 Fractional distillation column used on Asarco selenium.

Glass Processes for Other Applications 157

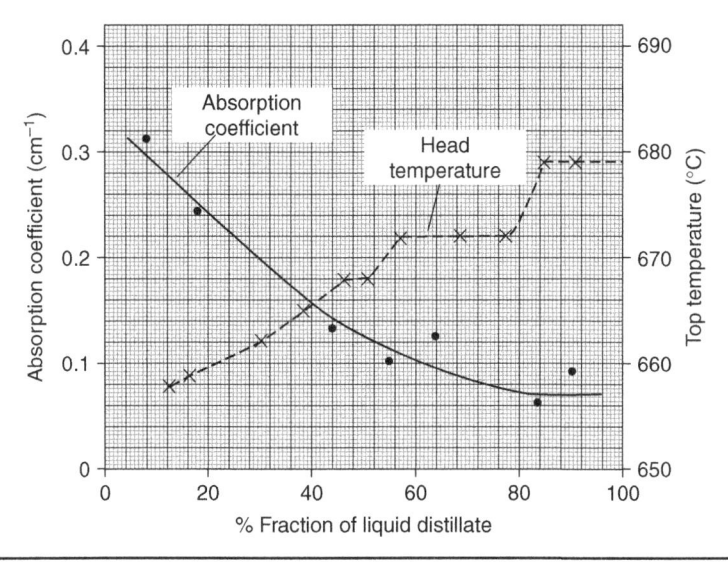

FIGURE 7.2 Absorption at 10.6 μm of selenium rods made from Asarco powder after fractional distillation.

Afterward, the column was rotated to a horizontal position, trapping the liquid selenium in the respective chambers. The heat was removed rapidly by turning off the heaters and opening the furnace to room air. Each fraction of liquid selenium was isolated in its separate chamber. After cooling, the quartz was broken and the amorphous selenium removed from each chamber and weighed to establish the fraction of the distillation in each chamber. Samples were removed from each and evaluated relative to absorption at 10.6 μm. The results are plotted in Fig. 7.2. Samples from all the chambers were sent for analysis. The results are shown in Table 7.1. One-half of the more

Element	Beginning Asarco Se	Still Bottom	First Plate	Second Plate	Third Plate	Fourth Plate
S	217	20	16	28	25	55
Te	27	25	22	23	36	< 5
Si	113	96	33	28	< 5	42
Fe	< 1	< 1	< 1	< 1	< 1	< 1
As	< 1	< 1	< 1	< 1	< 1	< 1
Temperature		573°C	617°C	585°C	538°C	513°C

TABLE 7.1 Analytical Results from Fractional Distillation of Asarco Selenium Powder

volatile selenium was distilled and remained in the lower-temperature chambers 3 and 4. The first half fraction had much higher absorption. The fourth chamber was high in S and Si which may account for the large absorption. The second half was less volatile and at higher temperatures remained in chambers 1 and 2. The second half fraction had absorption at 10.6 µm less than one-third of the more volatile fraction. Chambers 1 and 2 had by far the lowest levels of S and Si. The analytical results for beginning Asarco powder show much higher levels of S and Si. These results indicate these two impurities S and Si are major contributors to the absorption value at 10.6 µm but may be lowered in concentration by distillation of selenium.

AMI supplied glass in billets to Codman so that they could be extruded into rod form. The rods were then drawn into fiber and tested. Also AMI developed methods to cast rods in quartz tubing. There was difficulty in removing the glass from the tubes. We started precoating the tubes with a carbon film before the glass was added. In retrospect, this was not a good idea because hydrogen selenide forms when selenium reacts with carbon. AMI also produced rods by sawing square pieces from flat plates. The square rods were then ground round. Methods were developed to spin-cast thin-walled cylinders in sealed evacuated tubes containing the clad glass. The rods were then placed in the tubes, and the tubes collapsed and sealed to the rod using gentle heat to form a core-clad preform.

Efforts to meet the Codman goals concentrated on improved quartz fabrication techniques, use of high-purity selenium, and good-quality rod preform fabrication. AMI reached the Codman goals and concluded the research agreement in December 1985. Evaluation results of the glass from Codman (Tom Loretz, John Smith, and Gino Lombardo) agreed. Good fiber was produced. Transmission of 15 W of CO_2 laser power through the fiber was reported[6] at SPIE as part of an AMI paper. Glass and preforms continued to be produced and used by Codman. Unfortunately, as time passed, it became apparent that glass produced by AMI and used by Codman later in 1986 was not of constant quality. Purchase of glass by Codman declined and did not reach the amount specified in the agreement. For this and other reasons, the Codman program ended in 1986.

7.2 AMI Fiber Drawing Process

Until late in 1986, AMI was convinced that we had achieved our goals. As we became aware of fiber results at Codman, Galileo Electro-Optics, and Infrared Fiber Systems, we realized our joint efforts had not been very successful. The combination of our methods and theirs was not producing good results. We decided in 1987 to investigate and identify the problems related to our glass producing poor fiber results. The first step was to purchase a carbon dioxide laser. We took samples 5 to 10 cm in length cut from homogeneous glass made using our plate process, and we measured transmission with the infrared spectrophotometer and the laser. The measurements at 10.6 µm agreed with one another. AMI found that glass compounded in a cylindrical tube in a rocking furnace and quenched in billet form was not homogeneous. When slumped and extruded into rods, the rods would not be homogeneous as well. The same problem would exist in fibers drawn from the rod. Striations in the fiber would scatter the laser light, lowering the measured transmission. Glass delivered as specified to Codman, Galileo Electro-Optics, and Infrared Fiber Systems was mostly in billet form. The cast glass rods from AMI were probably not homogeneous as well. The decision was made that fiber had to be drawn from homogeneous glass made in our plate process in order to have a good chance of producing fiber matching the starting glass transmission. In 1987, we notified Rutgers, Galileo Electro-Optics, and Infrared Fiber Systems that we would only deliver glass preforms produced in this manner. We were going to have to learn to draw our own fiber in order to succeed.

One factor holding us back was the cost of a standard silicate fiber drawing tower. Most were 30 ft tall and cost $100,000 to $150,000, much too expensive for a small company with no fiber business dedicated to paying the expense. Presumably, the height was such because the fibers were drawn at high temperatures and at high speeds as already mentioned. The fibers had to be cool before reaching the drum where they were wound. Thinking all this over, we decided in 1988 to build our own. The height was only about 10 ft because the fiber to be drawn was at a relatively low temperature and pulled at a slow rate, a few meters per minute. Extra height was not needed. The tower was constructed from channel iron welded together in our shop. We decided that since the rod drawing process had not produced good results for others, it would not work for us either. We decided to base our process on a cylinder of homogeneous glass from our plate process placed in a metal cylinder with a hole in the bottom, as much as 1 to 5 mm in diameter. The glass would be under inert gas pressure, heated until the glass softened and sealed the bottom hole. At the right glass viscosity, tweezers were used to pull a fiber slowly through the hole out of the bottom. The diameter of the fiber would be much smaller than the hole, so the surface of the fiber would remain pristine, not touched by any surface. The AMI fiber drawing tower is shown in Fig.7.3. The chamber is at the top with the fiber emerging from the hole in the bottom. The fiber is carefully guided through the two open split dies down to the slowly rotating drum and attached to its surface using a small piece of tape. Rotation speed and position on the drum are precisely controlled by the unit made by Sancliff, Inc. As the fiber leaves the chamber, the diameter is continuously monitored by a HeNe "laser mike." The operator adjusts parameters of pressure, glass temperature, and draw speed to produce the core diameter

160 Chapter Seven

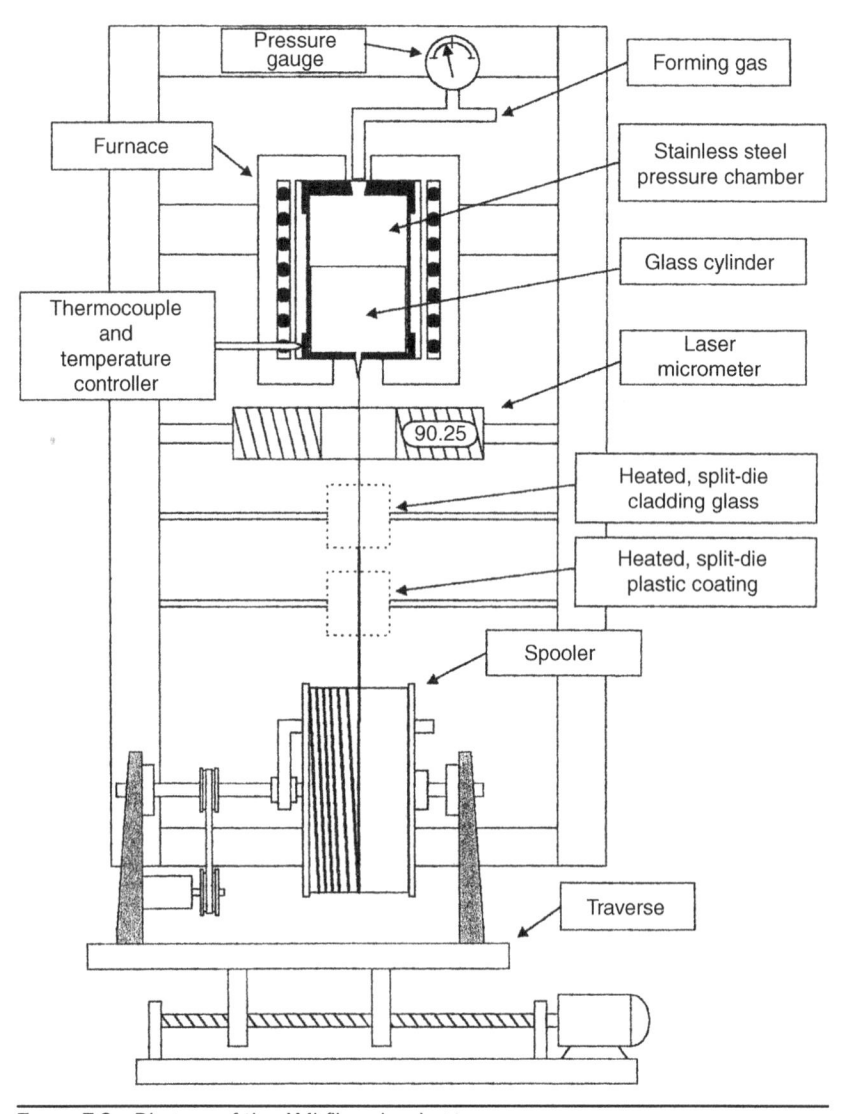

FIGURE 7.3 Diagram of the AMI fiber drawing tower.

desired. The process may continue until many meters later when the glass is used up. Thus the final length depends upon the core diameter. The system is very efficient, leading to inexpensive fiber. Using this system, AMI has drawn bare fibers with core diameters from 50 to 1000 µm. Pull rates are very slow, 0.3 to 2 m/minute depending on core diameter and whether it is being clad or not.

In silicate systems, the cladding is added before drawing the fiber. Coming up with a cladding scheme took some thought, time, and effort.

FIGURE 7.4 A Sancliff split coating die.

An image of a Sancliff split die is shown in Fig. 7.4. The normal use is coating wire with paint. In these units, the fiber passes through a hole, top and bottom, slightly larger than the fiber core. The chamber inside each is filled with a liquid that coats the surface as it passes through. In the first one the liquid is a heated clad glass that coats the fiber surface. In the second the liquid is a heated organic thermal plastic that coats the clad glass surface. The composition of the clad glass was chosen so that it would behave as an inorganic plastic, avoiding strain between core and cladding. The cladding, besides increasing the fiber transmission, protects the fiber, lowering the tendency of the fiber to break during bending. The organic plastic was added as a protective layer for the cladding.

At this time, AMI did not have an FTIR spectrophotometer needed to measure accurately absorption in our fibers. James Harrington, first when he was at Laser Sonics in California and later as a professor at Rutgers University, was kind enough to evaluate our fibers and help us in our progress. His results[4] showed the AMI spectrophotometer measured absorption at 10.6 µm of the homogeneous glass prior to drawing fiber and the FTIR measurement value of the fiber were close in value for fiber from three different glasses. Thus our fiber drawing process was preserving the glass quality. Harrington used his FTIR in the absorption measurement mode. Figure 7.5 shows results obtained for plastic-coated C1 fiber. All the absorption peaks except the one for H_2Se are due to the plastic coating on the fiber, the evanescent wave interaction with the adjacent medium. The dashed lines show the reduced level of absorption that would occur if the

162 Chapter Seven

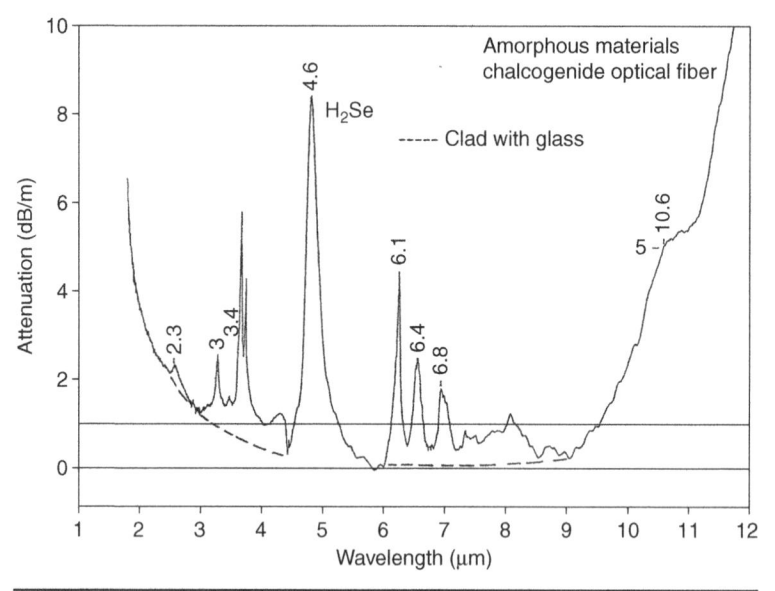

FIGURE 7.5 FTIR measured absorption of AMI C2 fiber.

core were clad with a nonabsorbing glass in place of the plastic. Note the absorption level is below 1 dB/m from 3 to 9.5 μm.

Comparing the bending radius to break a fiber was the next test. The fibers tested were drawn from core glass (Amtir 1 with 2 percent Te), Amtir 2 (As-Se with no Ge), and C2 glass (As-Se-Te with no Ge). Workers at Galileo Electro-Optics reported breaking strength results based on breaking radius, fiber diameter, and Young's modulus. Young's modulus was available from acoustical measurements on Amtir 1 and the value reduced 10 percent for the Te content. Servo results were available for Amtir 2 and C2. The process involved slowly bending the fibers in decreasing radius until they break. Fibers were taped down to the tangent to circles drawn on a table with decreasing radius. The AMI results for the same glass fibers were equal to or higher than the Galileo Electro-Optics results, demonstrating the AMI process produced strong fibers. Figure 7.6 shows a comparison of the bend-to-break radius for the three glasses as a function of fiber diameter. From the figure we find the limiting radius for C2 20-mil core fiber is 0.75 in, 2.8 in for Amtir 2 fiber, and 3.75 in for Te modified Amtir 1 glass fibers. We may conclude that germanium-containing glass fibers are the more brittle than those based on only the As-Se glass. Fibers containing Te along with As-Se in C2 are the most flexible. The addition of Te in the glass composition adds nondirectional metallic bonding to the covalent As-Se glass and at the same time increases longer wave transmission because of increased atomic mass relative to selenium. Addition of cladding to the fibers will improve the strength numbers as well as the

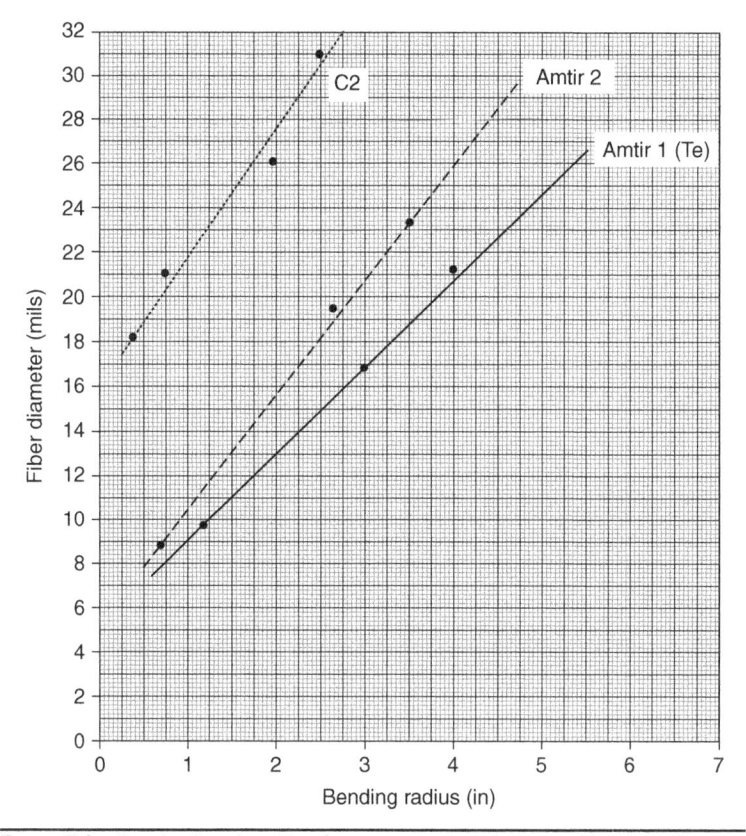

FIGURE 7.6 Bend-to-break radius for some unclad chalcogenide glass fibers as a function of fiber diameter.

transmission. Figure 7.7 presents another FTIR absorption measurement of AMI C2 fiber, again courtesy of Jim Harrington. The fiber was again plastic-coated but measured while straight and then while in a tight coil. The results illustrate how light may be lost through excess reflections from the core surface.

Two As-Se-Te glasses were prepared and evaluated during the Codman program and designated C1 and C2. During the development of the AMI fiber drawing process, six compositions were prepared and fibers produced by each were evaluated. In the end, one composition was identified and designated C1 core glass.[7] The goal from the beginning work with Codman was to produce a fiber that would transmit substantial CW laser energy.

Japanese workers reported[8] transmission of over 100 W of CO laser power through an As_2S_3 glass fiber. AMI turned attention toward preparing high-quality As_2S_3 glass, using our plate process, and drawing fiber as we had with the other glasses.

Earlier, Servo had offered to sell AMI its As_2S_3 glass producing system. But their process was an open process and used technical-grade

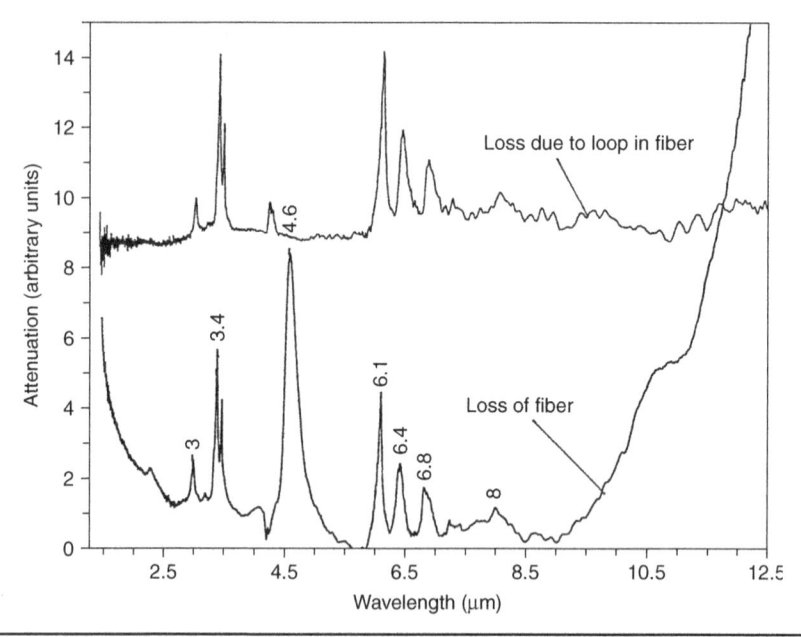

FIGURE 7.7 FTIR measured absorption of AMI C2 fiber straight and tightly coiled.

arsenic and sulfur. We could not use our normal plate process where reactants are placed in the compounding chamber. The heat of reaction between sulfur and arsenic is so great that the heat generated would cause an explosion due to sulfur vapor. So we developed a process[9] in which, under vacuum, sulfur vapor was passed over heated arsenic in a separate chamber to form As-S compounds that were deposited in a third cool chamber as powders. The chamber was sealed off under vacuum, and the powder was melted to form the glass. The glass was removed and then processed in our normal plate process. A cylinder of glass was removed from the plate and fiber was drawn in our usual process. By this time AMI had acquired a Synrad CO laser which emitted at 5.25 μm but only about 5 to 7 W. The energy was easily transmitted by our As_2S_3 fiber. The designation of the As_2S_3 glass fiber was changed to C2 fiber.

The As-Se-Te C1 glass fiber has low attenuation at 5.25 and 9.27 μm. Even with the low absorption, fiber failure occurs at 5 W of laser power. At 10.6 μm, failure occurs at lower power because of greater absorption. Even glass cladding does not help. At that time, the highest CO_2 laser power at 10.6 μm was reported by Nishi et al.[10] as 10 W, slightly better than the AMI experience. But cooling was used. The AMI concluded that absorption was of course important but $\Delta N/\Delta T$, the source of thermal lensing, was probably the most important factor. The author had measured As_2S_3 glass from room temperature down previously and found a very low value, near zero. When it was measured again at AMI from room temperature up, it yielded the same value but with an opposite sign.

Many of the measurements of IR fibers regarding acceptance angle or attenuation when fibers are bent were made using laser light. Laser light is very unusual in that it is extremely well collimated, intense, monochromatic, and to some degree polarized. If the wavelength corresponds to an absorption in the glass or clad glass, the results may be misleading.

The transmission of natural light from a heated source containing all wavelengths can better be used to characterize infrared fiber for normal applications. The experimental arrangement used at AMI is shown in Fig. 7.8. Light is emitted from a heated flat graphite surface. The light is optically chopped before it enters the fiber end. The transmitted light is coupled to a Hg-Cd-Te liquid-nitrogen-cooled detector. The detected signal is fed to a PAR phase-sensitive amplifier, and the detected signal is displayed on an electronic voltmeter. The distance from the end of the fiber to the graphite surface was measured. A variable aperture (iris) was placed at a measured distance X from the end of the fiber. Acceptance angles D were measured by closing the iris and noting the change in detected signal.

Full angle occurred when the signal was first affected. Measurement was made as well at the 80 percent power points. At each point the diameter of the open iris was measured. The angle was calculated from $\tan^{-1} D/2X$.

Conclusions from the measurements were drawn:

1. Clad and unclad fibers had the same total acceptance angles. However, the acceptance angles at the 80 percent points were larger for clad versus unclad, indicating the clad did have some effect. The angles are quite large in comparison to silicate

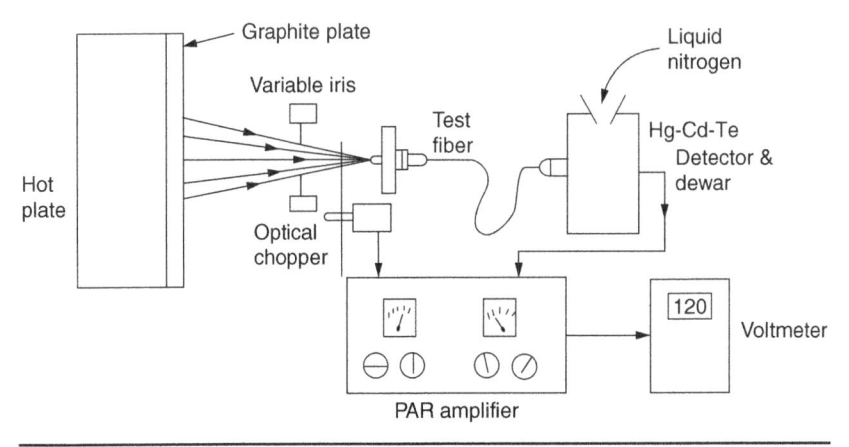

FIGURE 7.8 Experimental arrangement for characterizing infrared glass fibers using natural light.

fibers. Core indices for silicate fibers are < 1.5 while chalcogenide glasses have indices from 2 to 3. Oblique light rays are bent toward the normal when they enter the core surface. The high-index glasses bend the oblique rays much closer to the normal, keeping them in the fiber and away from the core wall.

2. The amount of energy collected was proportional to fiber diameter.
3. Up to a point, the total energy collected increased with distance of the fiber from the heated plate as area within the acceptance angle increased.

In other measurements, the data in Fig. 7.9 show that the C1 As-Se-Te fiber is much better for measuring temperature from 0 to 100°C

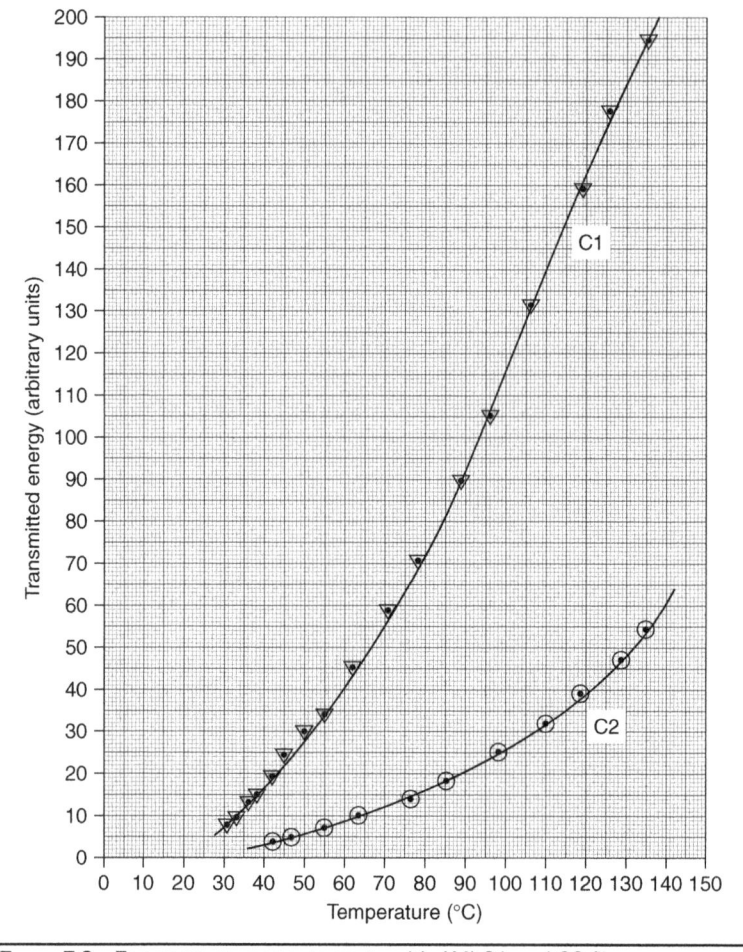

FIGURE 7.9 Temperature measurements with AMI C1 and C2 fiber.

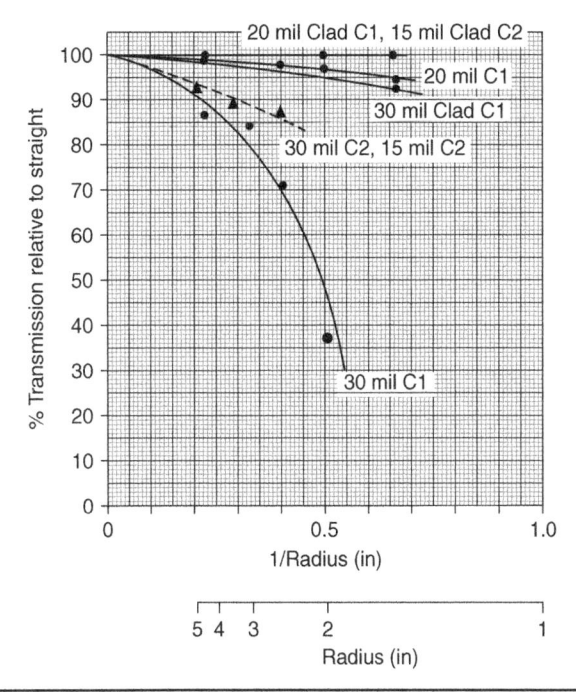

FIGURE 7.10 Transmission of natural light by bent AMI fiber.

than the shorter wavelength transmitting C2 fiber. The signal is in arbitrary units versus degrees Celsius.

Figure 7.10 shows the effect on transmitted energy of bending fiber clad and unclad. The fibers are C1 and C2 fibers with core diameters of 15 mils (375 µm), 20 mils (500 µm), and 30 mils (750 µm). For 30-mil C1 and C2 unclad fibers, the transmitted energy drops off steadily with decreased radius of curvature. Cladding decreases the effect to 10 percent at a radius of 2 in or less. Small-diameter, 10-mil or less, C1 fibers show very little attenuation when bent clad or unclad.

Table 7.2 shows data to compare the properties of core glasses C1 and C2. The thermal change in refractive index for C1 is low but still 3 times the value for C2. The measured absorption for both at laser wavelengths is less than 1 dB/m except at 10.6 µm. The AMI FTIR scans for C1 and C2 fibers are shown in Figs. 7.11 and 7.12. Note the scan for C1 shows absorption less than 1 dB/m from 2 to 9.3 µm. Also notice the absorption as measured with the CO_2 laser at the 9.27-µm wavelength indicates a lower value of 0.4 dB/m. The absorption peak about 4.6 µm is due to H_2Se dissolved in the glass. The other bands originate from the clad glass. The scan for C2 fiber has two absorption bands. The one at 2.9 µm is due to water always present in the sulfur. The second band around 4 µm is due to H_2S, also always

Core Glass	As-Se-Te	As$_2$S$_3$
Glass transition T$_g$ (°C)	136	180
Softening point (°C)	170	208
Thermal expansion × 10^6/°C	23.5	21.4
Refractive index		
∂ 4 µm	2.82	2.41
∂ 10 µm	2.81	2.38
Thermal change in index × 10^5/°C	+3	+0.9
Fiber abs. (laser)		
∂ 5.25 µm	0.7	0.8
∂ 9.27 µm	0.8	—
∂ 10.6 µm	4–5	—
Bending radius		
30-mil diam. (in)	1.5/0.4 Clad	1.0
20-mil diam. (in)	0.3	1.0
10-mil diam. (in)	0.2	1.0
Tensile strength (psi)	15,000 ± 20%	7000 ± 20%
Numerical aperture		
Unclad	0.5–0.6	0.45–0.55
Clad	0.5–0.6	0.45–0.55

TABLE 7.2 A Comparison of the Properties of AMI Fiber Core Glasses C1 and C2

present in the sulfur. The C2 fiber is below 1 dB, except for the two impurities, from the visible to 6 µm. The CO laser value placed on the scan is 0.2 dB/m at 5.25 µm. These two fibers have been supplied by AMI since the early 1990s.

7.3 Chemical Applications of AMI IR Fiber

The original intention of AMI was to develop IR fibers for use with the CO$_2$ laser. For all the reasons already described, that idea turned out to be impractical. The development of inexpensive Amtir 1 sensor lenses for noncontact temperature measurements has also been described. AMI was producing thousands of these lenses for Raytek, the leader in the field at the time. As our business with this company grew, we allowed Raytek to purchase a 10 percent share of AMI stock. The relationship between the two companies became quite close. So

Glass Processes for Other Applications 169

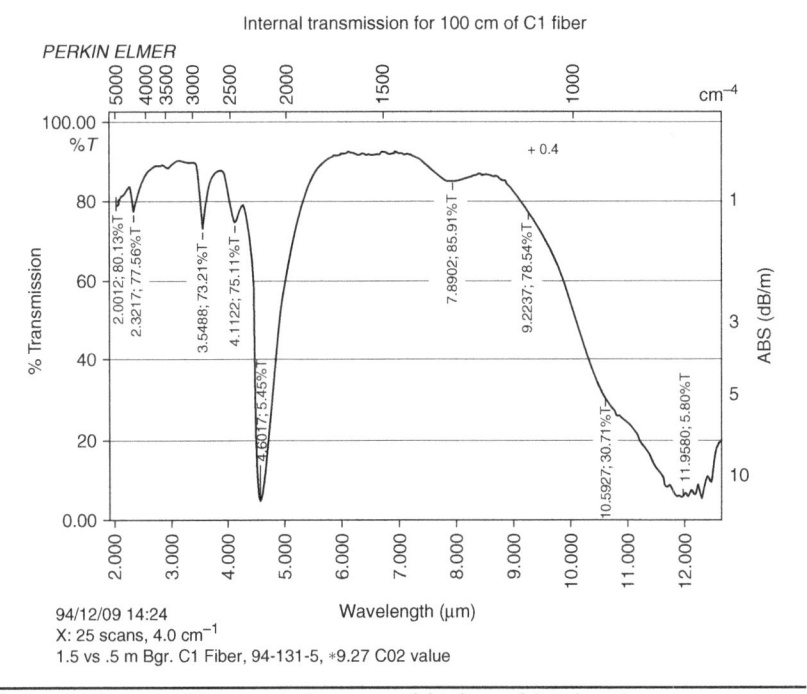

FIGURE 7.11 AMI FTIR transmission scan of C1 fiber of 1 m length.

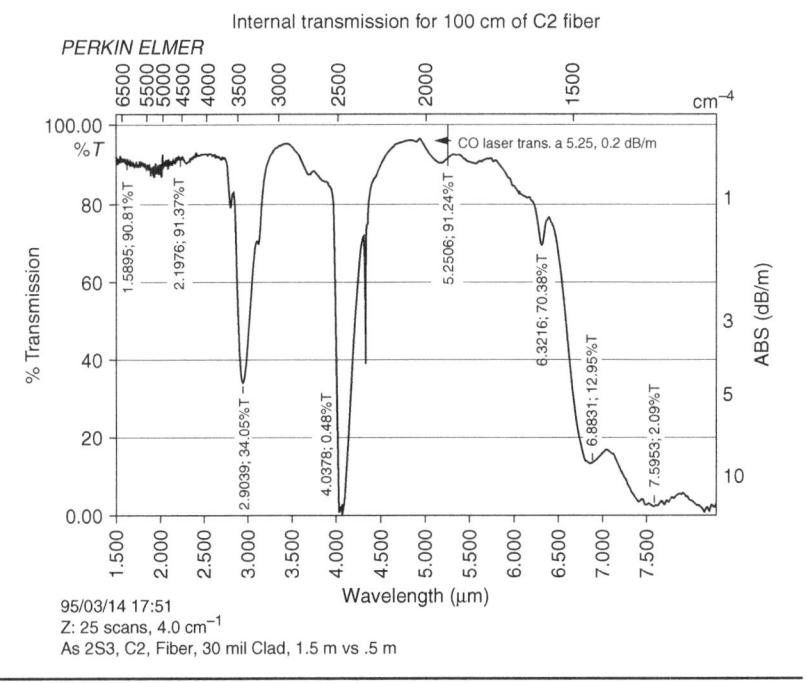

FIGURE 7.12 AMI FTIR transmission scan of C2 fiber of 1 m length.

based on the experimental evidence demonstrating the potential of using fiber to measure temperature, AMI suggested to Raytek that its devices could also be made with an option to use a fiber rather than a lens for direct instant temperature measurement of particular spots. AMI modified several Raytek devices to demonstrate the potential appeal to the customer. In the final analysis, Raytek rejected the idea primarily from the standpoint of the durability of a fiber probe in the field.

One application for which the fiber was ideally suited was chemical analysis. The development and use of Fourier transform infrared spectrophotometers have revolutionized the application of infrared analytical techniques to chemical analysis. The fact that all generated spectra are digital and can be directly compared electronically to a digital reference scan has led to analytical techniques never dreamed of before. The added ability to scan many times the sample and reference before the comparison is made leads to extreme sensitivity. The instrument can be used to measure transmission (absorption) directly through a sample or with attachments, reflection from the surface of a solid sample. All measurements are carried out after a prerecorded reference scan.

An early, very useful FTIR technique involved the use of a flat plate of an infrared material transparent in the wavelength range of interest. Analysis by this method is usually referred to as using an ATR (attenuated total reflection) plate. The plate is polished and most likely will have edges on both ends polished with a 45° bevel relative to the flat surface. The purpose of the bevel is to ensure the light traveling down the plate bounces off the inner plate surface many times before reaching the other end. On each bounce, the evanescent light wave couples with the substance covering the plate surface. An optical arrangement will be provided for the light entering the sample region to strike the plate end and travel through the plate to the other end. On leaving the plate, optical arrangement will be provided to insert the light back into the optical path of the instrument to be converted to an energy versus wavelength digital signal.

First, the plate is scanned a number of times with no liquid in contact or with a liquid containing no solute or sample. Next, the plate is scanned the same number of times in contact with the liquid-sample solution. The two scans are then compared and the results displayed. The absorption due to only the liquid used will be eliminated in the comparison with only the absorption due to the sample remaining. ATR plates are supplied by instrument companies, precisely fabricated and expensive. Several materials are used such as germanium, silicon, zinc selenide, and sometimes Amtir 1.

Figure 7.13 shows a diagram of methods used at AMI with our Perkin Elmer FTIR to perform the same kind of analysis based on AMI fiber and including extruded chalcogenide glass rods. Extrusion will be discussed in a later section. In the diagram, AMI Amtir 1

Glass Processes for Other Applications 171

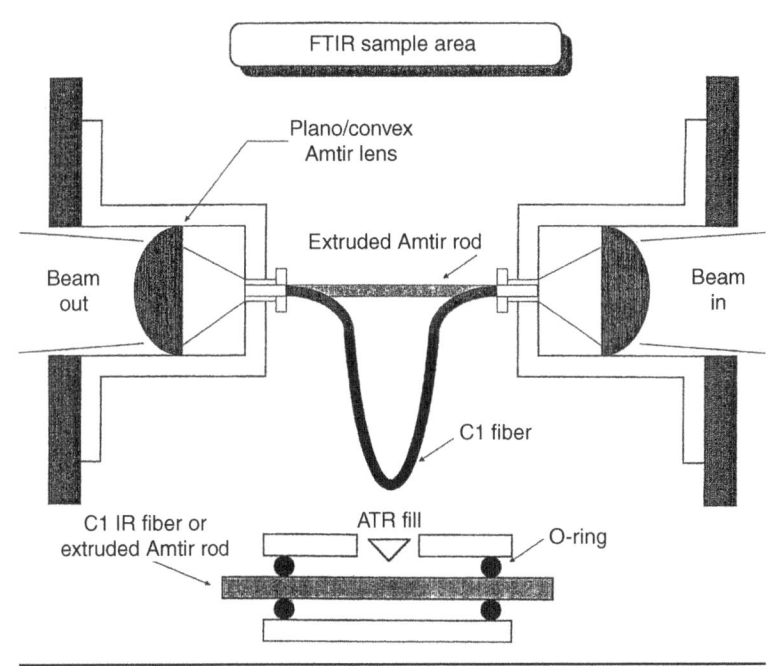

FIGURE 7.13 AMI fiber or extruded rod used with FTIR for ATR analysis.

sensor lenses are facing one another in the sampling area. They are focused on the ends of a C1 fiber or the ends of an extruded C1 rod. In use, the rods have 45° bevels on each end. The fiber is unclad, or cladding is removed in the lower loop area and hangs down into a small chamber containing first the fluid, followed by the fluid with sample. The arrangement is ideal for this operation. The fiber is very inexpensive such that it may be used only once and then discarded.

The extruded rod would be a little more expensive. The rod may be used straight as shown in the cell in the bottom illustration. Another way would be with mild heat to bend the rod in the same configuration as the fiber. In that case, the rod would probably be used more than once. It would cost a few dollars, perhaps $10, while the fiber may cost $1 to $2. The method produced good results. However, the companies that would supply to the users were all instrument companies. Their ATR plates could cost $1000 or more. There was no advantage to the supplier to furnish such an inexpensive replacement. We were unable to develop a market for our units.

First use of chalcogenide glass fibers made possible what is referred to as *remote spectroscopy*. That is, the fiber allows the removal of reactants away from the FTIR, out of the small sampling areas provided in each instrument. Further, complicated optics are not required to move the energy from the instrument to the reaction under study. The instrument energy is easily focused into the fiber to the reactor, passed through the reactor, and returned through fiber to the instrument optical path. Spectra are easily recorded in real time following the course of a reactor. Besides the convenience, the reaction moved away protects the instrument and the operator from hazards. Early examples are described in the literature.[11, 12]

In other applications, a single unclad fiber from the FTIR is placed in a mold away from the instrument and returned to the instrument optical path. The mold is filled with a liquid plastic prior to the liquid undergoing a polymerization chemical reaction. Providing the reactants each have observable absorption bands, the reaction may be followed in real time as the polymerization progresses. An example would be the polymerization of the plastic after a catalyst has been added. The recorded spectra could be used to verify when the reaction is complete so the plastic object can be removed from its mold at the appropriate time.

The largest consumers of AMI glass fibers for chemical applications are Peter Melling and Mary Thomson of Remspec Corporation. Their most unique development has been the multifiber probe.[13] The probe is designed to be used with most FTIR spectrometers; it utilizes 7 fibers in the bundle to input the FTIR energy and 12 fibers to return the signal. The probe has three separate heads: an ATR crystal head, a reflection head, and a liquid transmission. A diagram of the ATR head is shown in Fig. 7.14. The fibers are very flexible, allowing remote collection of spectra from 2 to almost 12 µm. The probe is 7 in long and encased in a ¼-in stainless-steel tube, allowing easy access to a flask or reaction vessel in the laboratory. At the end of the probe is an ATR crystal tip. Again, real-time observation of chemical reactions taking place may lead to better understanding of the reaction and even chemical kinetics data. For an example, the polymerization of isobutylene was studied over 80 min using the probe, and Fig. 7.15 shows the recorded data.

AMI worked very closely with personnel from Foster Miller in Boston in their program to utilize the AMI IR fibers to solve chemical analytical problems. The program included Mark Druy, Roy Bolduc, Paul Glatkowski, Suneet Chada, William Kyle, and Chuck Stevenson.

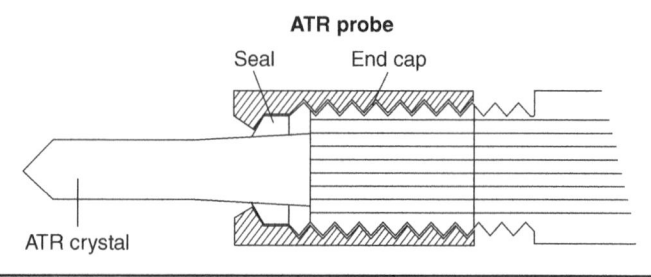

FIGURE 7.14 Remspec multifiber FTIR probe.

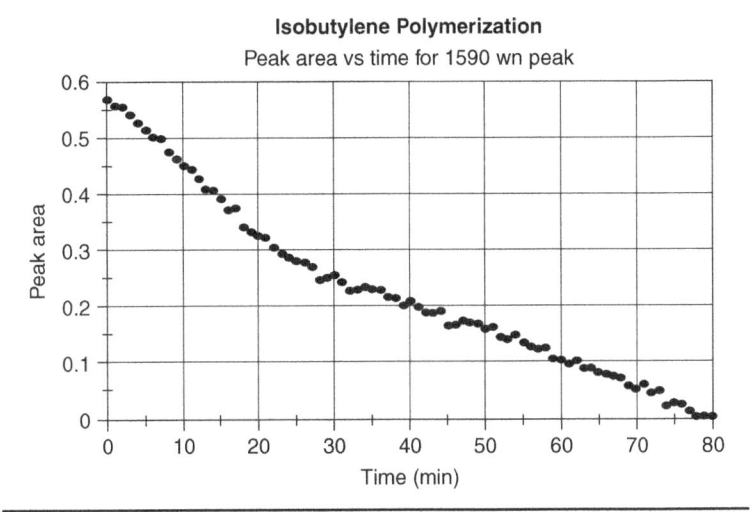

FIGURE 7.15 Remspec multi-FTIR probe results of the study of the polymerization of isobutylene.

Their efforts were concentrated mostly on chemical analysis of hazardous waste in containers. As their work progressed, the group was taken over by the MIDAC Corporation and renamed Sensiv. They developed a barrel "dip probe" for direct sampling of 55-gal drums. Also a portable fiber-optic automated measurement system for field use was developed. Measurements identified benzene, methylene chloride, methyl ethyl ketone, TCE, and toluene in field tests. A portable surface reflectance probe was developed for field use.

Infrared energy from a glow bar is collected with optics and directed to a surface. The reflected light is collected and directed into an IR fiber connected to a FTIR for analysis. Figure 7.16 shows the curing of a urethane paint measured by the Sensiv reflectance probe.

7.3.1 Fiber Summary

The sealed chamber extrusion-pull method developed at AMI for producing optical fibers from chalcogenide glasses has proved to be very reproducible. Continuous draws of 100 m or more are easily achieved depending on the fiber core diameter. The split-die methods developed for glass cladding and plastic coating during the draw process have been very successful. C1 fiber will transmit small amounts (< 5 W) of CO and CO_2 CW laser light.

The broadband As-Se-Te glass C1 fibers are flexible and have low attenuation of 2 to 11 µm. The C2 fibers based on As_2S_3 glass cover the range from the visible to 8 µm and offer the potential of delivering

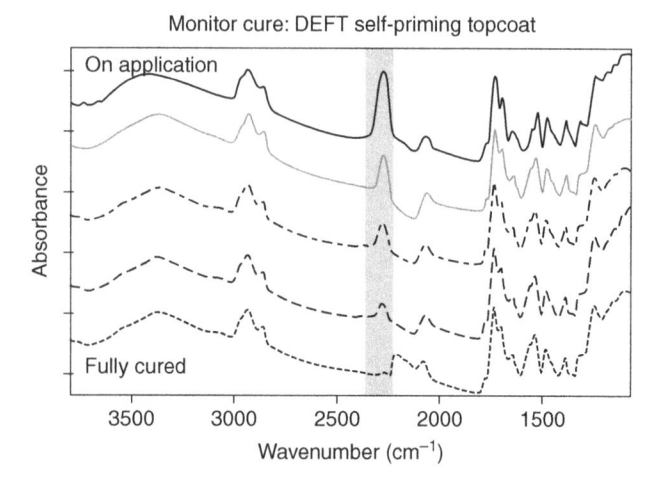

FIGURE 7.16 Cure of a urethane paint as measured by the Semsiv IR reflectance probe. The example shows how the probe can be used to monitor the cure of a coating after its application. This is a urethane paint system, and the curing reaction is followed by the decrease in absorbance of the isocyanate functionality.

substantial energy from the CW CO laser. The main application of AMI IR glass fibers has been chemical analysis and chemical reaction studies. Measurements of this kind were not possible before the development of both the FTIR spectrophotometers and chalcogenide glass fibers covering the wavelength ranges used in analysis. For AMI, IR fibers have been a small but continuous business since the beginning in 1990. Never has this activity been a large portion of total sales, but AMI considers this activity important.

7.4 Extrusion of Chalcogenide Glasses

AMI became involved in the extrusion of chalcogenide glasses in an unusual manner. In late spring of 1993, we had a visit from Jacob Fraden, founder of the company Thermoscan. He had developed an electronic device to measure a person's temperature from the infrared emission of the tympanic membrane of the ear. Figure 7.17 shows the device as advertised by Dillards in the *Dallas Morning News* in 1992. The main market thrust was for use with infants and small children. Sales were already in the thousands. The purpose of Jacob Fraden's visit was to see if we could produce small, inexpensive infrared glass light pipes to replace the light pipe in current use. When a company has a device being sold on the market in quantities of thousands a month, reduction in cost by $1 or $2 of a single element used in its construction leads to a substantial increase in profit. The current light pipe was a short piece of copper tubing that

Glass Processes for Other Applications 175

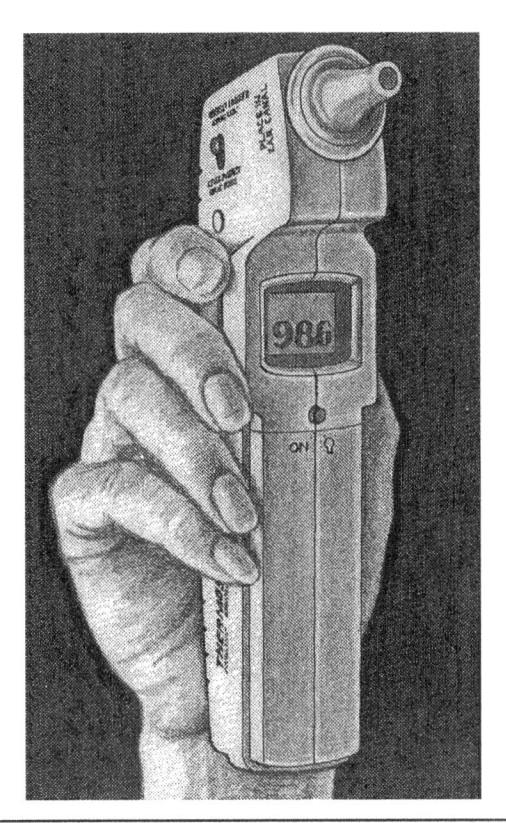

FIGURE 7.17 Thermoscan tympanic membrane thermometer.

was gold-plated. Jacob Fraden is an expert in the design of thermal sensors.[14] His analysis indicated an improvement in performance as well as a cost reduction would occur if AMI could provide a small-diameter (3-mm), short-length (3- to 5-mm) IR glass light pipe. Both ends had to be polished. The problem for AMI, in order to be considered as a replacement supplier, was that we had to demonstrate an ability to produce 50,000 per month. Final production was estimated to reach 1 million per year. The time deadline for us to demonstrate 50,000 was only 3 months away.

Our glass science consultant Tom Loretz of CES was the only person we knew who had any experience extruding glass. We knew he could design and build a first-class unit to produce the rods, or cane, as he called them. But that would take time, and we did not have time. We would have to put together a unit with what we could find locally. Our machinist, Tommy Tyler, produced a stainless-steel cylinder with a piston to match. We purchased a large air cylinder to drive the piston. A heater unit for the cylinder was built. The machinist designed a track and guide for the glass rods as they emerged. A steel framework was welded together and attached to a wall. The units were bolted together and bolted to the framework. The first extrusion took place in July, using scrap Amtir 1 glass. Nitrogen pressure was used to power the piston. The extruded rods were allowed to grow in length almost to the floor before being cut off at the cylinder exit and stacked with the others. The run was ended after the glass was exhausted in the cylinder. Rods were cut into manageable lengths, stacked, and glued together to be gang-cut to required lengths using the diamond ID saw. Individual units had to be polished on both ends. AMI purchased two small planetary polishers for the job. Plastic plates with many holes the diameter of the rods were the polishing gigs. Each small cut piece was placed in a hole on the plate, and then the plate was placed on the rotating polishing pad. One pad was coarse polish, and one was fine polish. The polishing process was very fast. Hundreds of pieces could be produced each day. Tom Loretz was present at the first run. He pointed out the unit was pneumatic, not nearly as good as a unit in which the piston was powered by a jackscrew. Using a pneumatic system would result in poor diameter control. We could see that in volume production with Thermoscan we would need an excellent system. We placed an order with CES to design and build an extrusion unit based on a computer-controlled jackscrew. Delivery would take quite a while.

Meanwhile we continued working with Thermoscan to produce extruded parts from C1 core glass they could evaluate. Using the C1 glass was much easier than using Amtir 1 because of the lower softening point. By late September, AMI could build a case for being able to produce 50,000 parts per month at a price of less than $2 each. Unfortunately, Thermoscan was bought by Gillete. Jacob Fraden lost control of technical decisions. Management used the threat of our glass light pipes to force their present supplier to cut his price considerably. By the end of 1993, it was apparent that there would be no volume rod sales by AMI to Thermoscan.

The system designed by Tom Loretz was delivered in the spring of 1995 and placed in operation. Figure 7.18 shows a diagram of the computer-controlled system. On the left, the system is shown using the tractor attachment that moves small-diameter rods, 1 to 5 mm in diameter, from the die. On the right is the twin rail attachment that both moves and supports the larger-diameter rods, say, 0.5 to 1.0 in. The design and results of applications were described by Loretz et al.[15] at SPIE. AMI continued to develop the production methods, concentrating on C1 glass drawing rods and plates. Figure 7.19 shows a photograph of extruded rods and plate sections.

The rods were made for Peter Melling of Remspec. His intention was to use long rods in place of ATR crystals in order to gain sensitivity because of increased surface contact and many more reflections from the inner surface.

Glass Processes for Other Applications 177

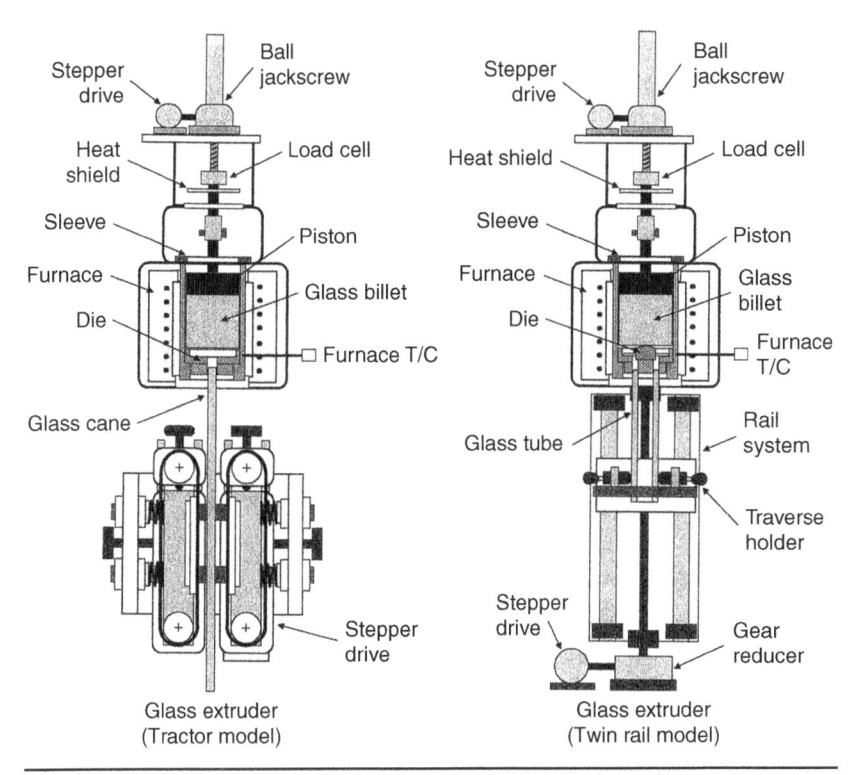

FIGURE 7.18 Diagram of the AMI chalcogenide glass extrusion unit.

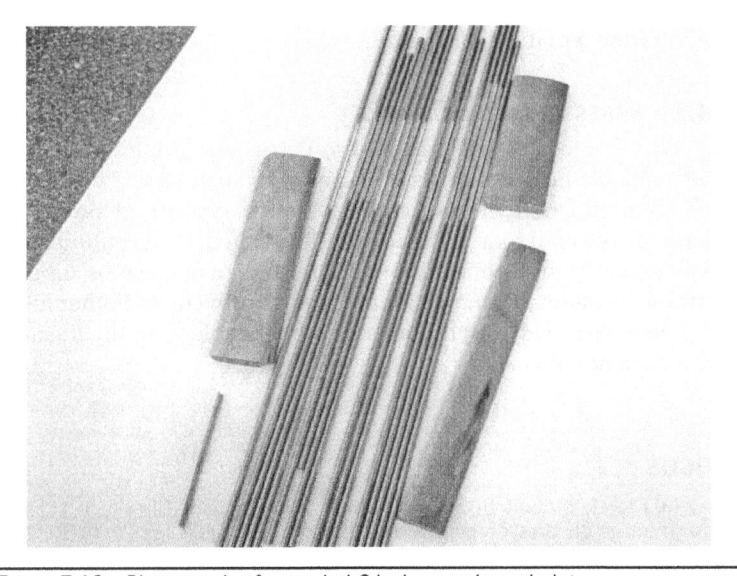

FIGURE 7.19 Photograph of extruded C1 glass rods and plates.

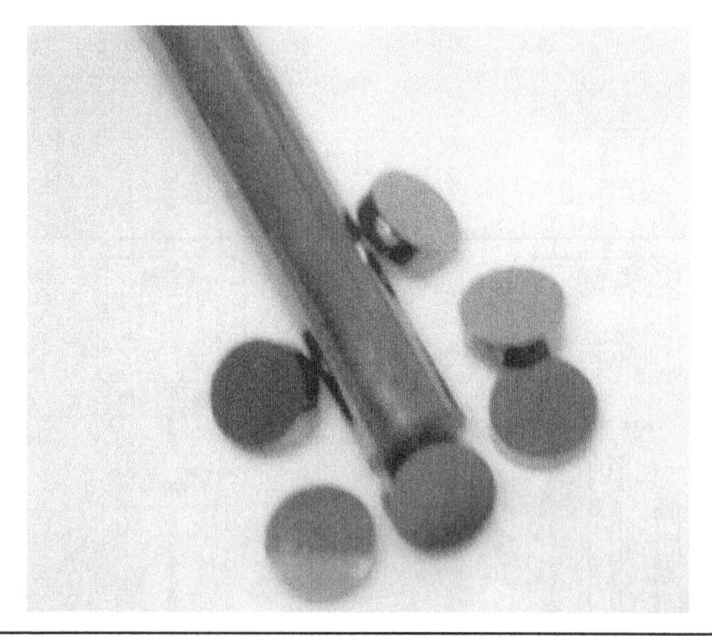

FIGURE 7.20 Disks cut and polished from a 0.75-in-diameter extruded Amtir 1 rod.

Figure 7.20 is a photograph of a 0.75-in-diameter extruded rod of Amtir 1 glass. The illustration shows the potential of extruding large rods and then sawing them into the correct thickness for lens blanks. Material utilization would be greatly increased in comparison to core drilling from a plate.

7.4.1 Glass Extrusion Summary

AMI was able to get started in building extrusion equipment by using local available help and parts in response to an apparent immediate need. A production process was developed capable of producing thousands of small rods. After the need vanished, AMI continued the development of the technology after purchasing a first-class computer-controlled system. Diameter control was improved, and other materials were extruded. Applications of the technology in the business sense have not appeared.

References

1. Paul Klocek, Mark Roth, and R. D. Rock, "Chalcogenide Glass Optical Fibers and Image Bundles: Properties and Applications," *Opt. Engg.* 26, 88 (1987).
2. D. J. McEnroe, M. J. Finney, P. H. Prideaux, and P. C. Schultz, "Optical and Mechanical Properties of Chalcogenide Fibers," 4th International Symposia on Optical Applied Science and Engineering, Netherlands, 1987.

3. T. Katsuyama and H. Matsumura, *Infrared Optical Fibers*, IOP Publishing Ltd., p. 7.
4. A. Ray Hilton, Sr., A. Ray Hilton, Jr., and James McCord, "Production of Far Infrared Glass Fiber," *SPIE* 1048, 85 (1989).
5. J. Jerger, C. J. Billian, and I. J. Melman, Final Technical Report, "Investigate the Properties of Glasses Transmitting in the 8 to 14 Micron Region," Contract No. Nonr 3647(00), 1963.
6. A. R. Hilton and George Cronin, "Production of Infrared Optical Materials at Amorphous Materials," *SPIE* 618 (1986).
7. A. R. Hilton, Sr."As-Se-Te based glass Fibers," *SPIE* 1228 (1990).
8. S. Sato et al., *Appl. Phys. Lett.*, Apr. 14, 1986.
9. A. R. Hilton, Sr., A. R. Hilton, Jr., James McCord, and Glen Whaley, "Production of Arsenic Trisulfide Glass," *SPIE* 3060 (1997).
10. Junji Nishi et al., "Chalcogenide Glass Fibers for Power Delivery of CO_2 Laser," *SPIE* 1228 (1990).
11. William R. Moser, Joseph R. Berard, Peter Melling, and Robert Burger, "A New Spectroscopic Technique for in Situ Chemical Reaction Monitoring Using Mid-Wave Infrared Optical Fibers," *Appl. Spectrosc.* 46 (1992).
12. Jang Heo, Monica Rodrigues, Steven J. Saggese, and George H. Siegal, "Remote Fiber-Optic Chemical Sensing Using Evanescent-Wave Interactions in Chalcogenide Glass Fibers," *Appl. Opt.* 30 (1991).
13. Peter J. Melling and Mary Thomson, "Fiber-Optic Probes for Mid-Infrared Spectrometry," *Handbook of Vibrational Spectroscopy*, Wiley, Chichester, England, 2002.
14. Jacob Fraden, *AIP Handbook of Modern Sensors*, American Institute of Physics, New York, 1993.
15. Thomas J. Loretz, A. R. Hilton, Sr., A. R. Hilton, Jr., and J. McCord, "Fabrication of Chalcogenide Glass Rods and Tubes by Processor Controlled Extrusion Techniques," *SPIE* 2977, 14 (1997).

CHAPTER 8
IR Imaging Bundles Made from Chalcogenide Glass Fibers

8.1 The Stacked Ribbon Method

It is worth repeating that the optical and physical properties of chalcogenide glass fibers are vastly different from those of the silicate-based fibers. Generally, there are two methods for making imaging bundles from glass fibers: (1) the *ribbon stacking method* the original method used to produce bundles, and (2) the method developed for silicate fibers over many years, termed the *leachable bundle method*. In this second method, glass clad rods are clad again with a third glass which is acid-soluble. The rods are then fused in a rigid bundle, heated, and drawn down in size together to a smaller diameter, perhaps one-tenth in diameter, but increased in length. The rods are then cut into convenient lengths and fused again, and the process is repeated until the desired core size is reached. The rods are next cut into imaging bundle lengths. The ends are protected by an acid-resistant coating such as wax, and the bundle is placed in an acid bath to dissolve away the third glass. The fibers are thus separated from each other, providing flexibility for the bundle. Bundles prepared in this manner have a near-perfect fiber arrangement. Image performance is near perfect as well. The bundle characteristics regarding uniformity of core size and clad thickness relative to core are well controlled when the rods are made before the first draw. Unfortunately, chalcogenide glass fibers are not suited for this process due to the relative volatility of the chalcogenide glasses and their chemical inertness to acids. Chalcogenide glasses are only attacked by strong alkali. Also as

182 Chapter Eight

we had already learned, it is very difficult to make chalcogenide glass rods that are homogeneous. The ribbon stacking method is well suited for the chalcogenide glass fibers, particularly considering the AMI method of drawing fibers.

Figure 8.1 illustrates the formation of the ribbons on a 1-m drum as the fiber is drawn from the glass chamber up above. Note that a fiber guide is used above the drum to ensure accurate placement of the fiber. The drum is moved relative to the fiber. As the fiber is wound, the drum may be moved either left or right. All ribbons must be wound in the same direction, right to left or left to right. Otherwise the individual fibers do not match up with one another. For AMI to try making a bundle using this method, it was necessary for Tom Loretz of CES to develop a computer system to control the Sancliff unit driving the drum to provide accurate placement of the fiber on the drum. The pitch between fibers must be controlled, slightly more than the fiber diameter, to form closely wound ribbons. The fiber count for each ribbon is specified and each fiber counted to the final fiber. At the final count, the pitch control moves the drum to set a "gap" to provide space for the next ribbon. The gap move wastes one length of fiber between each ribbon. The drum was previously covered with

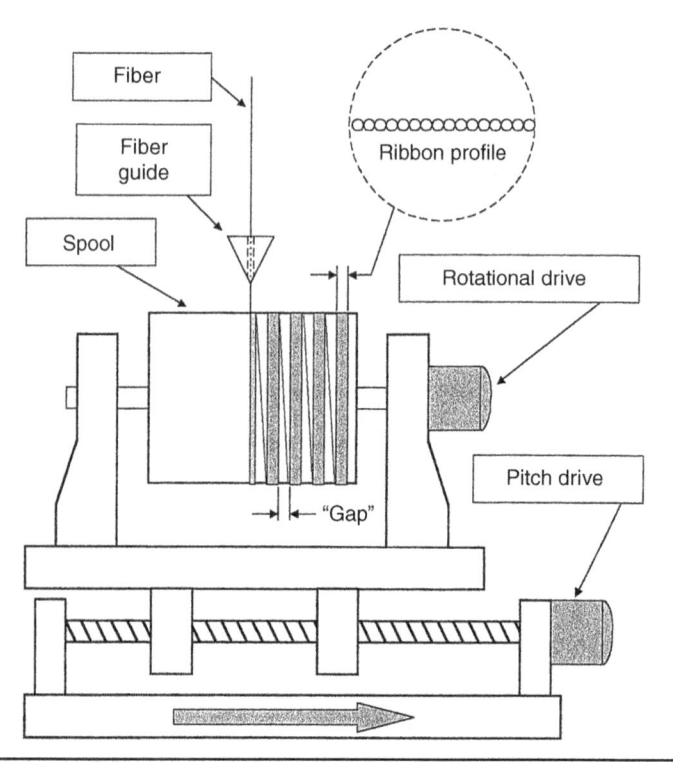

Figure 8.1 Diagram depicting formation of fiber ribbons at AMI.

IR Imaging Bundles Made from Chalcogenide Glass Fibers

a thin sheet of Teflon on which all the ribbons are wound. After the drum has been covered with ribbons, the fibers in each ribbon in a small area are pushed together until touching. A small amount of a thin epoxy is painted on the area to fuse the fibers. After short fused areas on all the ribbons have been made, the Teflon film is removed from the drum without disturbing the placement of the ribbons. Figure 8.2 shows a photograph of fiber ribbons still on a 1-m Teflon-covered drum. The ribbons will be removed one at a time, stacked, and glued on top of one another with the same drum orientation. After drying, the bundle is cut in the center of the fused area. The ends now are optically coherent to one another. For protection, the bundle is usually placed inside a flexible plastic tube. The fused ribbons are carefully aligned during the stacking process and form a rectangular bundle. A bundle made from 100 ribbons of 100 count will have a width of 100 fused fibers, but when stacked, the aligned ribbons will sink down slightly as fibers fit between the fiber of the previous ribbon layer. The bundle will not be square.

Most applications of AMI chalcogenide fibers have involved chemical sensing energy transfer or temperature sensing. In such applications, one is always concerned with obtaining as much energy as possible, always related to core diameter. For this reason, AMI concentrated on fibers with large core diameters in comparison to oxide-based fibers. AMI chose as standard core diameters for fibers, C1 or C2, clad or unclad, 30 mil (750 μm), 20 mil (500 μm), and 10 mil (250 μm). Until the ribbon formation started, AMI had put no effort in trying to draw small-diameter fiber. Besides the fact at that time we had no demand for small fiber, there was a feeling that they would be too weak based on past strength measurements.

FIGURE 8.2 Photograph of fiber ribbons on Teflon-covered drum.

8.2 IR Imaging Bundles of 1-m Length

In the first attempts[1] to make a bundle with C1 glass, the ribbons were only 0.5 m long and made with relatively coarse unclad fiber, 20 mil, 10 mil, and 4 mil. The AMI fiber drawing process was very slow when cladding with glass, so the low-viscosity polyamide plastic coating was used instead. The index of C1 was 2.8 while the index of the polyamide plastic was 1.4 and thus would serve as cladding. At this point the only infrared camera AMI had to use for image evaluation was a pyroelectric vidicon camera sensitive in the 8- to 12-µm band. The camera was imaged through a good-quality FLIR module and was connected to a TV monitor. The input had to be mechanically chopped at 5 to 20 Hz. A diagram of the experimental setup is shown in Fig. 8.3. Sensitivity was not good. For the test, we had a U.S. Air Force resolution target deposited on a 3-in polished piece of As_2S_3 glass. The end of the bundle was moved about the target on the disk imaging resolution patterns with lines of different spacing. A flat, heated hot plate was placed behind the disk at a distance to serve as a light source before any image could be observed due to poor camera sensitivity. The resulting images could not be recorded because of the instability of the image due to the light chopping. Our results based on observed images were as follows:

1. Coarse (20-mil) bundle, limiting resolution was 0.5 lp/mm
2. Medium (10-mil) bundle, limiting resolution was 0.5 to 1 lp/mm
3. Fine (4-mil) bundle, limiting resolution was 1 to 2 lp/mm

In the test without the bundle, only the camera focused on the test disk showed a resolution of 30 lp/mm, demonstrating the camera-FLIR lens combination was not limiting the performance of the bundles.

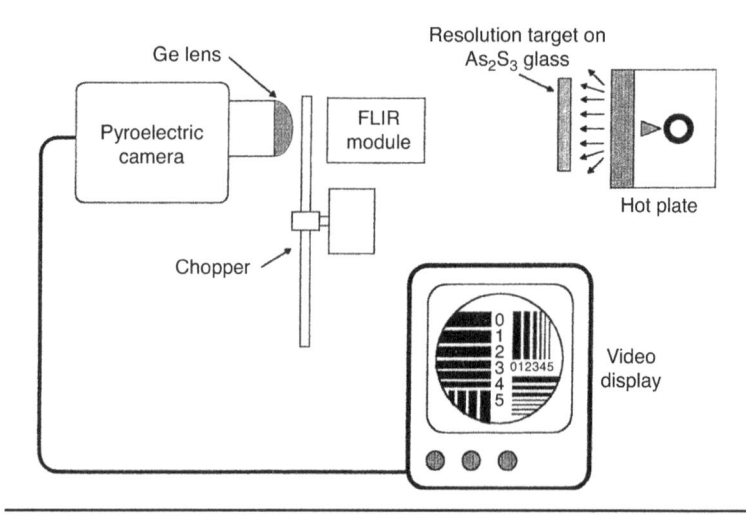

FIGURE 8.3 Experimental setup for viewing image-bundle images.

In considering the use of the polyamide plastic coating as the cladding for the fiber, it was concluded that in the 3- to 12-μm range, the percentage of energy lost in the plastic cladding due to the evanescent wave coupling would be small relative to the total energy in the band. The conclusion was based on measured FTIR transmission of a 500-μm fiber coated with the plastic. However, the FTIR measured transmission of a 1-m-long C1 bundle containing 2500 plastic-coated 100-μm core fibers revealed that almost all the 8- to 12-μm energy had been absorbed by the plastic due to evanescent wave reflection at the core cladding interface. What had not been taken into account was the fact that light traveling down 1 m of a small-diameter fiber reflects off the wall many more times than a fiber with a large diameter. The number of reflections N is proportional to length L divided by the diameter D. It takes twenty-five 100-μm core fibers to provide the same area for the incident light as for a 500-μm core fiber. The total plastic area on the twenty-five 100-μm fibers is 5 times the plastic area on the larger fiber. The polyamide plastic is no substitute for a clad glass when making an imaging bundle with small-diameter fiber. Better the fiber is unclad where the index at the air interface is equal 1. We had also expressed concern[3] that fiber lengths in individual ribbons should be exactly the same, or optical distortion might result. That turned out to be an unfounded fear.

Another misconception that was holding us back was the tensile strength of the fibers. We had previously reported[1] tensile strength of our fibers based on bend-to-break measurements. Gradually we realized that the values were unrealistic and not indicative of true strength. We began breaking fibers in tension and then calculating tensile strength using the fiber cross-sectional area. We developed a value for C1 and C2 fibers calculated from breaking force for all three core sizes (30, 20, and 10 mil) averaged together. The results were

$$C1 = 15,000 \text{ psi} \pm 20\%$$

$$C2 = 7000 \text{ psi} \pm 20\%$$

Our conclusions from these measurements were that we would have a difficult time drawing small-diameter fibers, < 100 μm, based on the small calculated force to break number. However, we tested more C1 glass fiber unclad but plastic-coated and were surprised at the results.[2]

3.5 mil (90 μm)	116,000 psi
10.2 mil (258 μm)	45,000 psi
19.8 mil (502 μm)	36,000 psi

Of course the plastic cladding increased the values. What impressed us was the dramatic increase with decreased core diameter. Our hopes that we could make small-diameter fiber ribbons were increased. We had described C2 fiber as being one-half as strong as C1. New measurements[3] for glass clad C2 fibers, 90-μm core, yielded a value of

122,000 psi, slightly higher than for C1 fiber, although the C1 measurements used plastic-coated fiber. Another surprise was the strength of both C1 and C2 fibers with 90-μm cores was independent of the strain rates used in the measurements. A comparison of 500-μm core C2 fiber results to the previous 500-μm C1 results indicated that C2 fiber was stronger by 70 percent than C1 fiber. We realized we would not have trouble making small-diameter ribbons with both C1 and C2 glass.

The first C2 bundle[3] was made from 31 count ribbons made from glass clad 100-μm core fibers on the 1-m drum. The bundle was about 7 mm² when finished and contained 1054 fibers. The calculated active area was 35 percent. Calculated transmission with Fresnel losses would be only 24 percent. Figure 8.4 shows images of the bundle

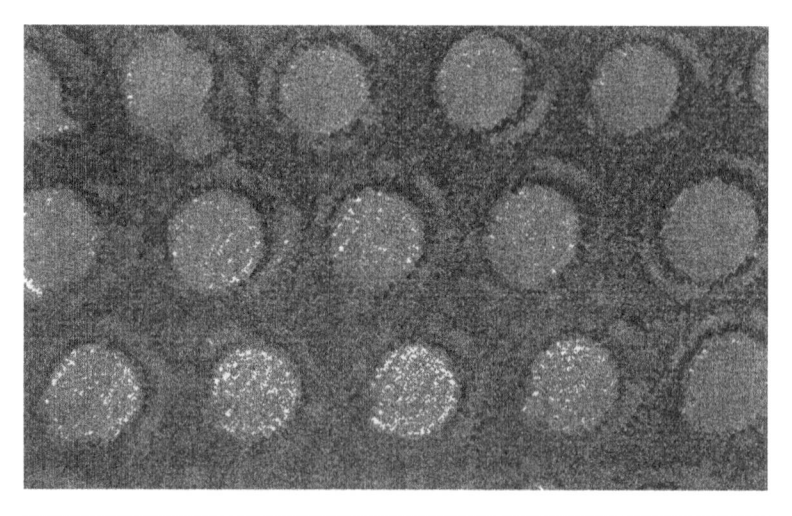

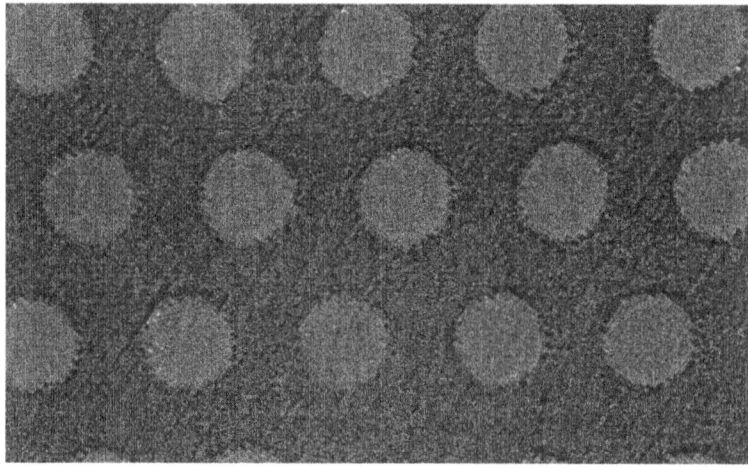

FIGURE 8.4 Microscope photographs of the end of the first C2 bundle.

using a microscope at 5X. The top image shows the cladding rings around each fiber. Note the space between the fibers covered by the cladding. The second image is made when visible light is transmitted through the bundle. The image illustrates how much area is inactive when you use clad fibers, at least when using the AMI method in drawing fibers and making ribbons. The measured spacing between the 100-μm fibers is 160 μm, thus limiting resolution by the bundle. Figure 8.5 shows the transmitted image by the bundle of the "O" pattern as recorded using the 1.5X objective on the microscope. The "O" pattern covers to 1 to 1.8 lp/mm and is clearly resolved. Note the coarseness of the image due to the thick cladding.

It was pointed out earlier that AMI developed an As-Se glass with a low softening point (131°C) and a large thermal expansion coefficient ($\Delta L/L = 27 \times 10^{-6}/°C$) to use for molding chalcogenide glass lenses in a joint program with Lockheed Martin in Orlando. The glass was designated Amtir 4. Fiber drawn from this glass proved our conclusion that chalcogenide glasses with the best physical properties do not make the best fibers. The ir transmission of C4 glass is similar to that of C1 except it drops off a little faster around 12 μm because the composition does not contain tellurium. The refractive index is 2.6 compared to 2.8 for C1 glass. Fibers with a core diameter of 50 μm can be drawn unclad at 10 m/min without breaking. The time required to pull the 4000 to 5000 m of fiber for a bundle is greatly reduced. The finished bundle is flexible enough that it can be placed in the sample area of our FTIR, and the transmission of the bundle can be measured holding each end in place. The results are not quantitative but rather

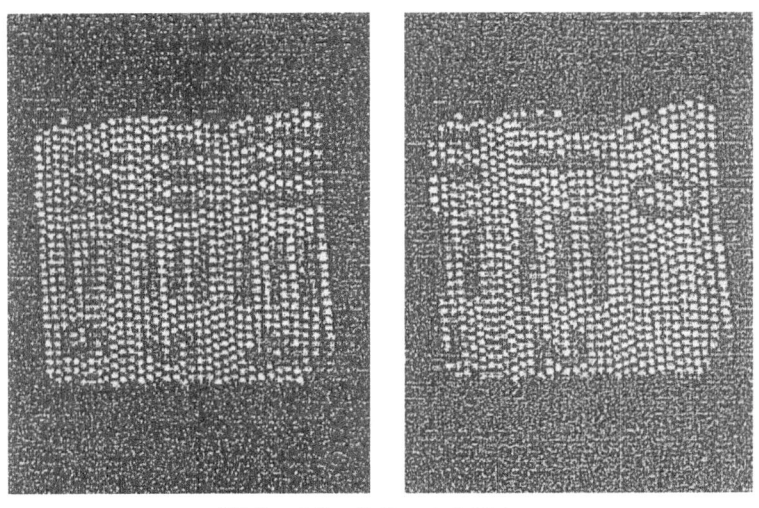

"O" Resolution Pattern 1–1.8 lp/mm

FIGURE 8.5 Transmitted image of the "O" pattern by the C2 bundle.

qualitative, showing the shape of the transmission curve. We found that most all the 8- to 12-μm energy was missing. The thinned epoxy used to cement the ribbons was 1 to 2 in long. The epoxy also served as cladding at each end of the bundle. However, at the end where the light entered the bundle at oblique angles, it underwent many reflections against the epoxy surface that absorbed most all the 8- to 12-μm energy. At this point, AMI had two infrared cameras to use for evaluation of the imaging bundles. The Agema 210 3- to 5-μm camera had a linear PbSe detector array, was thermoelectric cooled, and was made by Magnavox. It was not very expensive or very sensitive. The second camera was Raytheon Palm IR sensitive in the 8- to 12-μm range, which used an uncooled barium-strontium-titanate (BST) detector array. In both cases, the bundles were imaged directly to the camera optics using 2-in-diameter meniscus lenses fabricated from Amtir 1 glass. One was antireflection-coated for 3 to 5 μm for the Agema camera and the other for 8 to 12 μm for the Palm IR camera. A diagram of the arrangement for the Agema camera is shown in Fig. 8.6. The arrangement for the Palm IR was the same. Obviously by now, the Palm IR could only be used when the bundles were made with glass clad fibers.

Bundle no. 26 was 1 m in length made from sixty 60-count ribbons of C4 fibers with core diameters of 63 μm. The bundle was about 4 mm^2 with 3600 fibers and a calculated active area of 65 percent. The first evaluation was carried out using a NIR Electrophysics 7290 tube camera with maximum sensitivity at 1.4 μm. A special relay lens designed by Gary Wiese of Lockheed Martin filled the camera field of view with the bundle image. Intense visible light was passed through the arsenic trisulfide disk which had the Air Force resolution pattern deposited on its surface. The bundle was moved on the disk to different

- **AMI purchased an Agema 210 3- to 5-μm camera**
  Linear PbSe detector array, thermoelectric cooled, made by Magnavox, not a very expensive or highly sensitive camera
- **Camera coupled to 1-m bundle using Amtir 1 f/1 meniscus**
  Antireflection-coated for 3 to 5 μm using AMI facility

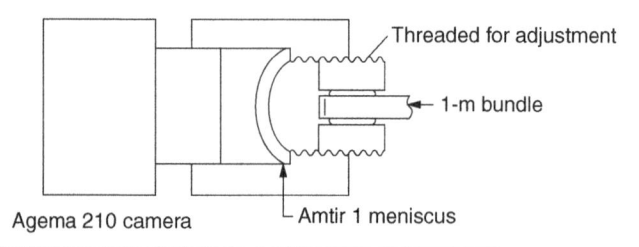

FIGURE 8.6 Attachment for the evaluation of bundles using the MWIR Agema 210 camera.

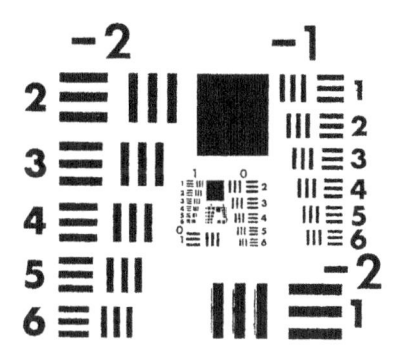

IR Imaging Bundles Made from Chalcogenide Glass Fibers 189

FIGURE 8.7 Air Force resolution target and "O" pattern image transmitted through C4 bundle no. 26.

patterns. Figure 8.7 shows first the Air Force resolution target diagram used in the evaluation and below an image demonstrating the performance of the bundle in the "O" pattern. Notice individual ribbons are clearly resolved. The top image in Fig. 8.8 shows the resulting image when the "1" pattern is transmitted through the bundle. Not shown is that most of the "3" pattern was also resolved, indicating a resolution limit of about 10 lp/mm for the bundle. The lower image in Fig. 8.8 is of a human subject at a distance taken using the Agema 210 camera. A digital video recording was made of individuals in natural light walking away from the bundle up to 50 ft and still discernible. The objective lens on the end of the bundle was a 1-in-diameter

190 Chapter Eight

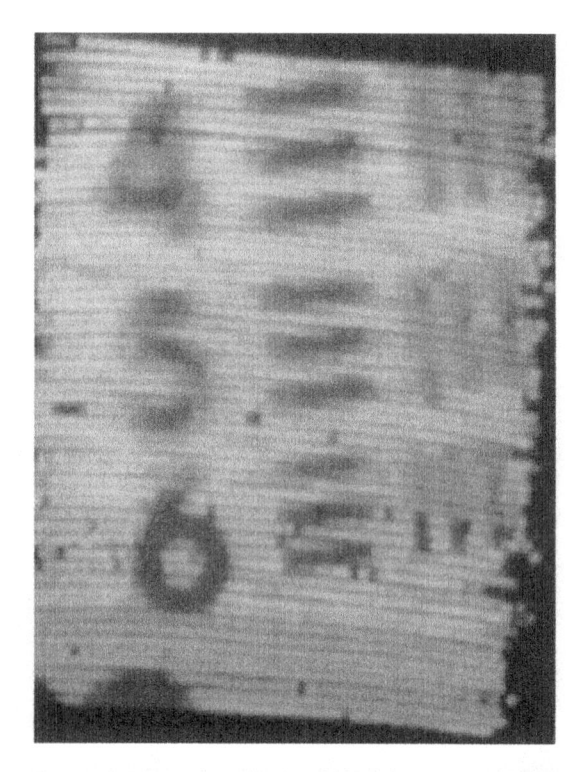

FIGURE **8.8** Agema 210 MWIR images transmitted through C4 bundle no. 26.

Amtir 1 $f/1$ planoconvex antireflection-coated for 3 to 5 µm. The 2-in meniscus Amtir 1 lens serving as the relay lens for the Agema camera did not fill the field of view of the camera with the bundle image. That is why the image has white areas on both sides. The camera is not considered a high-performance camera.

A second evaluation of the bundle[5] was carried out at Lockheed Martin in Orlando by Rich LeBlanc. A high-performance Galileo camera was used. A nearly 1:1 optical relay coupled the bundle to the camera. The acceptance angle (half angle) for the bundle was measured as 30°. Using the good camera and a 1:1 relay, individual fibers were resolved. Some were dark and presumed broken. The limiting resolution was about 10 lp/mm. Resolution target elements were clearly imaged. A clear up-close facial image was made as well as images of vehicles in a parking lot. The effort demonstrated that if the relay lens were chosen to fill the view with the bundle image and the objective lens were of good quality selected for the bundle application, AMI bundles would produce useful infrared images.

Some of this effort to improve bundle image performance was made after the U.S. Navy bundle program began. Improvement of bundle construction techniques would be valuable later on. There was much to be done in the program before any attempt to make a bundle 10 m long could begin. First, we had to wait many months until a drum with a circumference of 10 m could be designed, built, delivered, and put in operation.

In summary, AMI developed methods to produce infrared imaging bundles using the stacked ribbon method. The fibers could be glass clad, but due to the nature of the clad glass, the process was too slow. In addition, the cladding reduced the percentage of active area. Use of C4 glass improved imagery because we could draw flexible small-diameter fibers. However, in fusing the fibers to form the ribbons, a thin epoxy solution was used. As the oblique rays entered the bundle, the 8- to 12-µm energy was lost. AMI bundles were good for use with the NIR (1.5-µm) camera, the MWIR (3- to 5-µm) camera, but not the LWIR (8- to 12-µm) camera.

8.3 Goals of the Navy SBIR 10-m IR Imaging Bundle Program

The goal of this Phase II program[6] was for Amorphous Materials (AMI) to provide for the Navy with an infrared imaging bundle made from arsenic trisulfide glass fibers, 10 m long, to be used to transmit infrared images from optical elements on the surface of an aircraft to a sensitive 3- to 5-µm camera suitably located within. Arsenic trisulfide glass was chosen because it transmits red light, the NIR and the MWIR 3- to 5-µm light used in Navy imaging systems. The method used to form the bundle was to be based on the AMI ribbon stacking method. Modification of the method would be required to accommodate use of a drum with a circumference of 10 m. The drum would have to be designed, constructed, installed, and put in operation before a single 10-m bundle could be made. The fiber-optic imaging bundle was to have a 7-mm × 7-mm cross-sectional format made with fibers with a core diameter of 50 µm. The active fiber area goal was to be 70 percent and with the use of antireflection coatings reach an overall transmission goal of 50 percent in the 3- to 5-µm band. Many problems would have to be solved. Specific objectives are listed here:

1. Reduce the core diameter into the 40- to 50-µm range.
2. Using the ribbon pressing tool, increase the packing density of the fibers to attain the 70 percent goal.
3. Expend considerable effort to develop a reliable method to purify sulfur to be used to produce low-absorption As_2S_3 glass. Our goal is an attenuation of 1 dB/m for water at 2.92 µm and 5 dB/m at 4 µm due to hydrogen sulfide as measured with the FTIR. Attenuation in the fiber will also be measured by laser transmission using the 5.25-µm emission of the CO laser. Another Navy goal is 0.1 dB/m at 2.44 µm.
4. Absorption-free transmission of an arsenic trisulfide fiber is only 69 percent. Fresnel reflection losses must be minimized considering the optical path is 10 m. Establishment of an AMI antireflection coating capability for imaging bundles will be required.
5. The optical performance of the bundles will be evaluated using a resolution target as a critical test. Good-quality relay and objective ir lenses designed to couple the bundles to the ir camera and the image to the bundle will be required to produce useful images. The designs are to be made by Gary Wiese of Panoptics. AMI will fabricate and coat the lenses.

8.4 The Navy Phase II 27-Month Program

AMI developed the technology and built a facility capable of producing for the U.S. Navy ir coherent FO imaging bundles 10 m in length for use with IRFPA (infrared focal plane array) cameras. Results of efforts directed toward the specific goals listed above are discussed below in the order listed.

8.4.1 The 1-m C2 Imaging Bundles

The fabrication of 1-m fiber-optic bundles was used to develop techniques and make decisions suitable to drawing and stacking ribbons on the 10-m drum. The drum was not delivered and put in operation until more than one-half of the 27 months had passed. A total of nine bundles were made during the program, four in Phase I and five in Phase II. The last, bundle 9, was made and evaluated in August 1997, the seventh month of the program. An attempt to reduce core size by changing the exit port size was ineffective. We were able to draw 50-µm fiber, unclad. However, even though the measured tensile strength for C2 is 122,000 psi, when multiplied by the core area in inches, the force to break becomes only a few ounces. This fact is illustrated in Fig. 8.9 which shows force to break as a function of core diameter for C1, C2, and silicate fiber. The same treatment is applied to silica fibers which are about 6 times stronger. Even silica fiber at a core diameter of 100 μm (0.004 in) breaks at less than 10 lb. The calculated force to break for different diameters of C2 fiber is as follows:

$$200\text{-μm core} = 6.1 \text{ lb}$$
$$150\text{-μm core} = 3.4 \text{ lb}$$
$$100\text{-μm core} = 1.5 \text{ lb}$$
$$75\text{-μm core} = 0.86 \text{ lb (28 oz)}$$
$$50\text{-μm core} = 0.38 \text{ lb (12 oz)}$$

Breakage during ribbon drawing would be a serious problem, one we could not tolerate especially while drawing 10-m ribbons. Unclad 50-μm core fiber could be drawn on the 10-m drum but not without risking severe fiber breakage. Each time breakage occurred, the partial ribbon would have to be removed and a new ribbon layer started. An analysis by Gary Wiese of Panoptics showed that MTF performance by the bundle in the infrared was insensitive to fiber

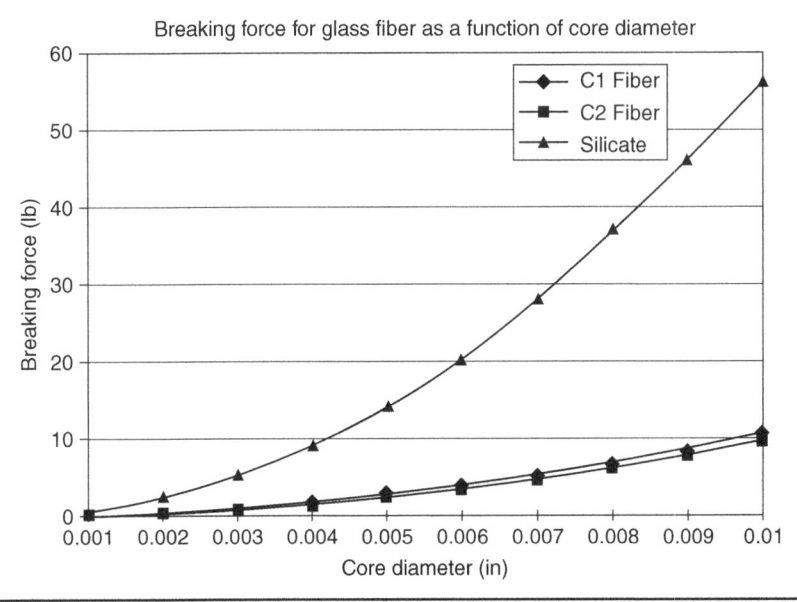

FIGURE 8.9 Breaking force in tension for C1, C2, and silicate fiber as a function of fiber core diameter.

spacing until 150 µm was exceeded. We decided that 100-µm core was our practical limit.

8.4.2 AMI Glass Clad Fibers

In the AMI drawing of glass clad fiber, too much of the space at the bundle ends is occupied by cladding. One approach taken was to have a special tool designed in which the fused bundle section could be placed and heated and pressure could be applied by the tool to force excess clad glass out, to increase the packing density of the fibers. The goal was to reach 70 percent active area. Unfortunately, the fibers moved as well, destroying the spatial arrangement between the fibers. We tried a special ordered split die for applying glass cladding to small-diameter fibers. We ordered a split die with a 125-µm hole to use on the cladding of fiber with a core diameter of 100 µm or less. The glass cladding is a viscous liquid when applied, even when the die hole is much larger than the fiber diameter. When the hole diameters are close, the drag increases. The fiber began to break even at low draw speeds of 0.5 m/min. A draw speed of even 1 m/min would be impossibly slow considering we were to make a 70 × 70 ribbon bundle which would require 49,000 m of fiber. At 1 m/min, it would take thirty-four 24-h days of continuous fiber drawing, if it were possible, with no interruptions. The highest active area achieved with cladding for a 1-m bundle was only 33 percent. The conclusion drawn was that clad fiber was out and unclad fiber was the only possibility of reaching 70 percent and producing a 70 × 70 bundle in a reasonable time frame. A comparison of FTIR results for internal transmission for 10 m of C2 clad compared to 10 m of unclad differs by only a few percent.

8.4.3 AMI Production of High-Purity Arsenic Trisulfide Glass

AMI developed a method to produce high-purity arsenic trisulfide glass. We were able to achieve our goals of attenuation levels for water at 2.9 µm of < 1 dB/m and a level for hydrogen sulfide at 4 µm of < 5 dB/m. Measurement of transmission at 5.25 µm using the CO laser for lengths of 10, 8, 4, and 1.5 m confirmed our level of attenuation was reproducible at 0.3 dB/m when measured for lengths > 4 m. Extrapolation to the 2.44-µm wavelength of the FTIR plot indicated an attenuation level of 0.16 dB/m, close to the program goal of 0.1 dB/m. Figure 8.10 shows an FTIR internal transmission scan for a 100-cm length of 1000-µm core glass clad C2 fiber. AMI developed a process for consistently producing 4- to 6-kg plates of striae-free, low-absorption As_2S_3 glass plates 20 cm in diameter, which was described at the SPIE[7] meeting held at Orlando, Florida, in April 1997.

Figure 8.11 shows a diagram in which, under vacuum, the very high purity sulfur vapor is passed over heated high-purity arsenic, and the resulting As-S molecules are collected on a cool quartz

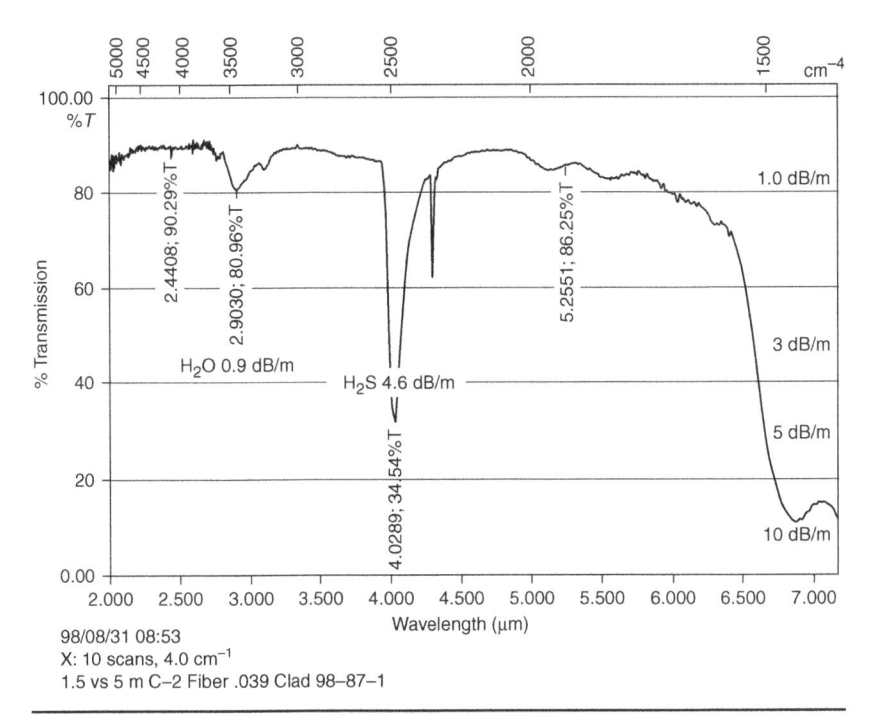

IR Imaging Bundles Made from Chalcogenide Glass Fibers   195

FIGURE 8.10 Internal transmission for 100 cm of C2 fiber.

surface as powders. While still under vacuum, the powders are melted and turned into glass. The glass is then collected and run through the AMI plate process that produces striae-free 8-in-diameter plates. The true test of purity is the measured transmission of fiber made from the glass. Examining Fig. 8.10 again, note the water absorption is less than 1 dB/m while hydrogen sulfide is less than 5 dB/m. The CO laser transmission at 5.25 µm verifies the accuracy of the FTIR

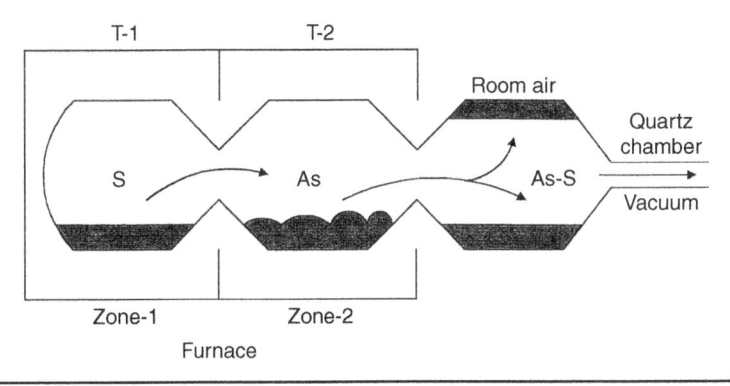

FIGURE 8.11 Arsenic trisulfide glass compounding process.

results. The data presented here represent the present state of the art at AMI and are reproducible using our 4- to 6-kg striae-free plates. Our results have demonstrated that sustained drawing of small-diameter fiber to form 70 count 10-m-long ribbons requires the use of striae-free glass.

8.4.4 The 50 Percent Transmission Goal

To have a chance to meet the 50 percent transmission goal, AMI had to develop an antireflection coating capability. Antireflection coatings were designed by Ed Carr, the AMI coating consultant, and applied by Gail Hanna to our bundles and lenses used in the imaging evaluations. The designs cover the NIR, MWIR 3- to 5-µm and LWIR 8- to 12-µm. In this case, only the MWIR coatings are used. Figure 8.12 shows a diagram of the coating chamber used. The chamber has been modified so only the ends of fibers or fiber bundles are in the chamber to be coated. The rest of the fiber or bundle is wound around a spool

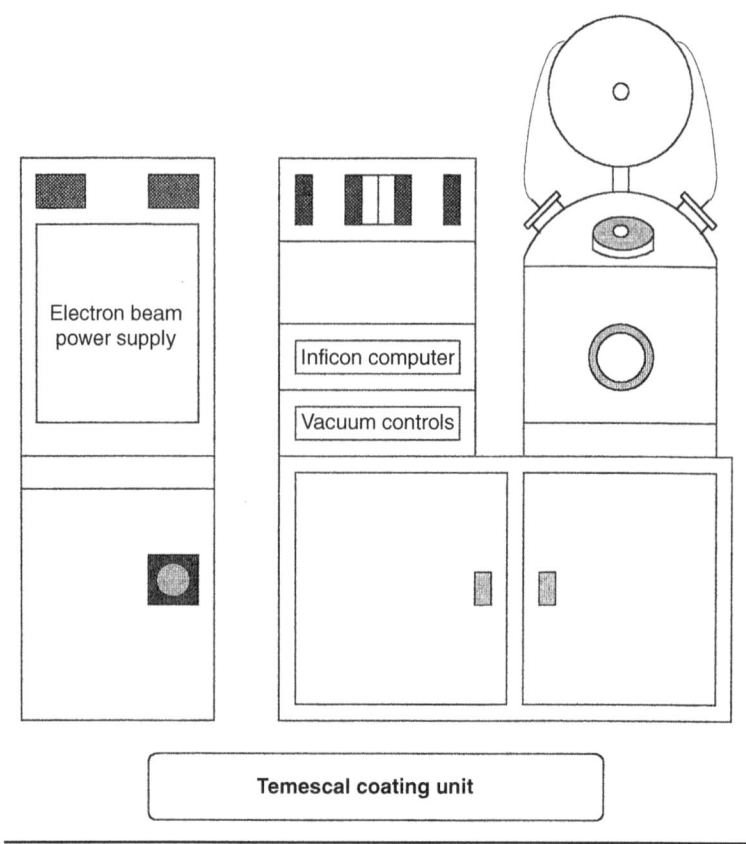

FIGURE 8.12 AMI modified Temescal coating chamber.

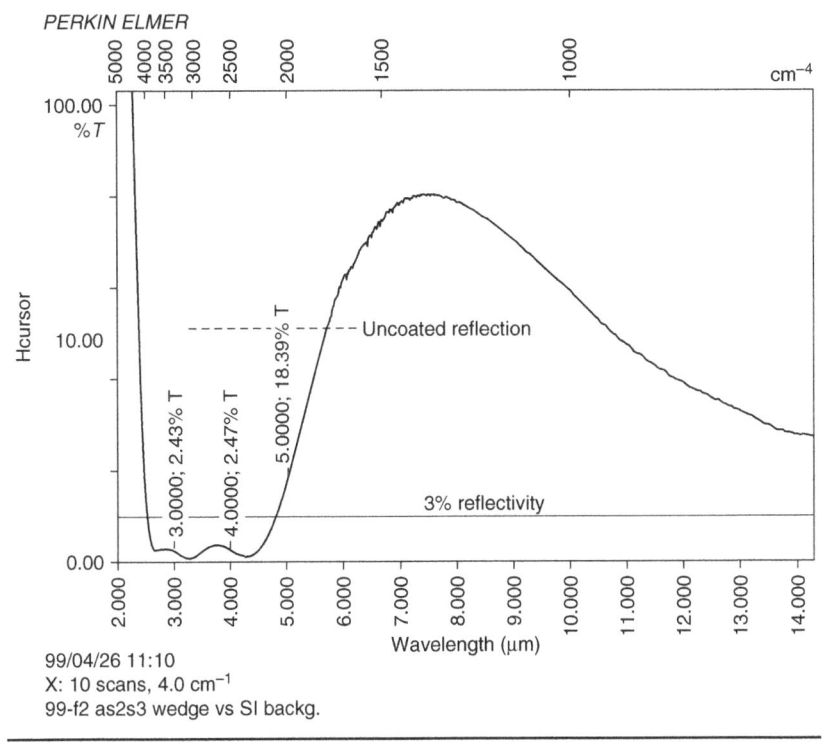

IR Imaging Bundles Made from Chalcogenide Glass Fibers 197

FIGURE 8.13 Measured reflectivity of Navy bundle 10-M-4.

attached above the chamber during coating. Figure 8.13 shows the FTIR measured reflection using a silicon reference background, 100 percent = 30 percent reflectivity, of the coated ends of Navy bundle 10-M-4. Note the 10 percent horizontal reference line marks the 3 percent reflectivity level. The reflection for both ends, 3 to 5 µm, is thus less than 1 percent, increasing the percentage of energy transmitted. The question becomes, How much? And how much will it help for a bundle 10 m long?

Transmission of incident ir energy from the scene through the objective lens to the surface of the bundle through the 10-m (1000-cm) bundle is one of the most critical factors in determining the quality of the image displayed by the infrared camera. The final sum is the product of the percentage of active area in the face of the bundle and the transmission, including Fresnel reflection losses, if any, after antireflection coating, for the 10-m length in the wavelength band of the camera. A discussion of the meaning of the terms appears below. For the exponential bulk absorption coefficient α in terms of decibels per meter, we can say the following:

for 1 dB/m $\alpha = 0.0023$ cm^{-1}
for 0.5 dB/m $\alpha = 0.0011$ cm^{-1}
for 0.3 dB/m $\alpha = 0.0007$ cm^{-1}
for 0.1 dB/m $\alpha = 0.00023$ cm^{-1}

To put these numbers in perspective, the absorption level normally acceptable in infrared optical materials used for lens materials in infrared imaging cameras is 0.01 to 0.02 cm^{-1}. For example, germanium, long the industry standard, is commonly used in the 8- to 12-μm range at 0.03 cm^{-1}. Thus, the absorption level required for this application is 2 orders of magnitude lower than normally required for infrared optical materials, no small order. Fresnel reflection loss is the next factor to consider. With a refractive index value of 2.407, the reflectivity is calculated from

$$R = (n-1)^2/(n+1)^2 = 0.171 \text{ for As}_2\text{S}_3$$

Assuming no internal reflections for fiber, meters in length, the maximum transmission for absorption-free fiber can be calculated as

$$T = (1-R)^2 = 0.69$$

The importance of these facts for a 10-m length of fiber is shown in the following:

Absorption (dB/m)	Uncoated %T	Coated %T	× 70% Active Area
0.1	54.9	79.5	55.7
0.2	43.5	63.1	44.2
0.3	34.6	50.1	35.1
0.4	27.5	39.9	27.9
0.5	21.9	31.7	22.2
1.0	6.9	10.0	7.0

Obviously, the attenuation level must remain between 0.1 and 0.2 dB/m throughout the band for the goal of 50 percent energy to be met. As we shall see, broadening of the impurity bands for long lengths is a serious problem which makes attainment of our goal much more difficult.

Earlier, we indicated FTIR results and laser transmission results agreed if the fiber length was > 4 m. Figure 8.14 shows internal transmission for 10 m of C2 fiber, unclad, which has been marked for absorption levels of 0.1, 0.2, 0.3, 0.4, 0.5, and 1.0 dB/m. Also for reference, the absorption value as determined by CO laser energy transmission for 11 m at 5.25 μm has been added. Note the value of 50 percent, or 0.3 dB/m, falls almost on the FTIR curve, validating the accuracy. Notice the broadening of the water and hydrogen sulfide absorption. In the 3- to 5-μm band, only a small portion meets our criterion of 0.1 to 0.2 dB/m. Also, squares have been marked and

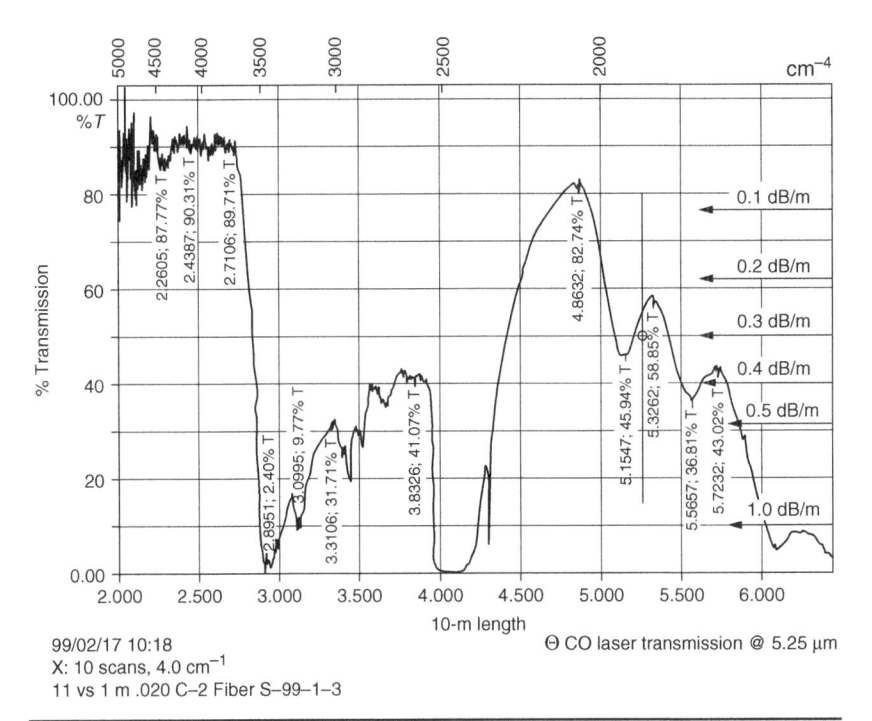

IR Imaging Bundles Made from Chalcogenide Glass Fibers 199

FIGURE 8.14 FTIR measurement of internal transmission for 10 m of 500-μm core unclad C2 fiber.

counted within the 3- to 5-μm band. The calculated internal transmission is 37 percent. Assuming a perfect antireflection coating, when it is multiplied by the 70 percent active area, we arrive at an overall transmission number for a 10-m bundle of 26 percent, one-half of our program goal. Severe broadening of impurity absorption with the length of the fiber reaching 10 m was an unexpected complication.

8.4.5 Formation of Bundles on the 10-m Drum

The 10-m-circumference drum was designed, built, and installed by Tom Loretz of CES and Paul Modlin of Advantek Engineering in January 1999, the 11th month. A diagram of the system is shown in Fig. 8.15. The first thing to notice is the drum is supported by only a single post, not two usually used for a drum of this size. One would be insufficient if the fibers were to be drawn at speeds approaching 70 mi/h. Our intended speeds were only a fraction of that, 5 to 10 m/min. The speed control had to be modified for better control around zero speed. The main reason that the drum was built with only one support column was so that we could remove complete ribbons from the drum for stacking. We were still thinking in our 1-m ribbon stacking mode. At this point we had to face the reality that it

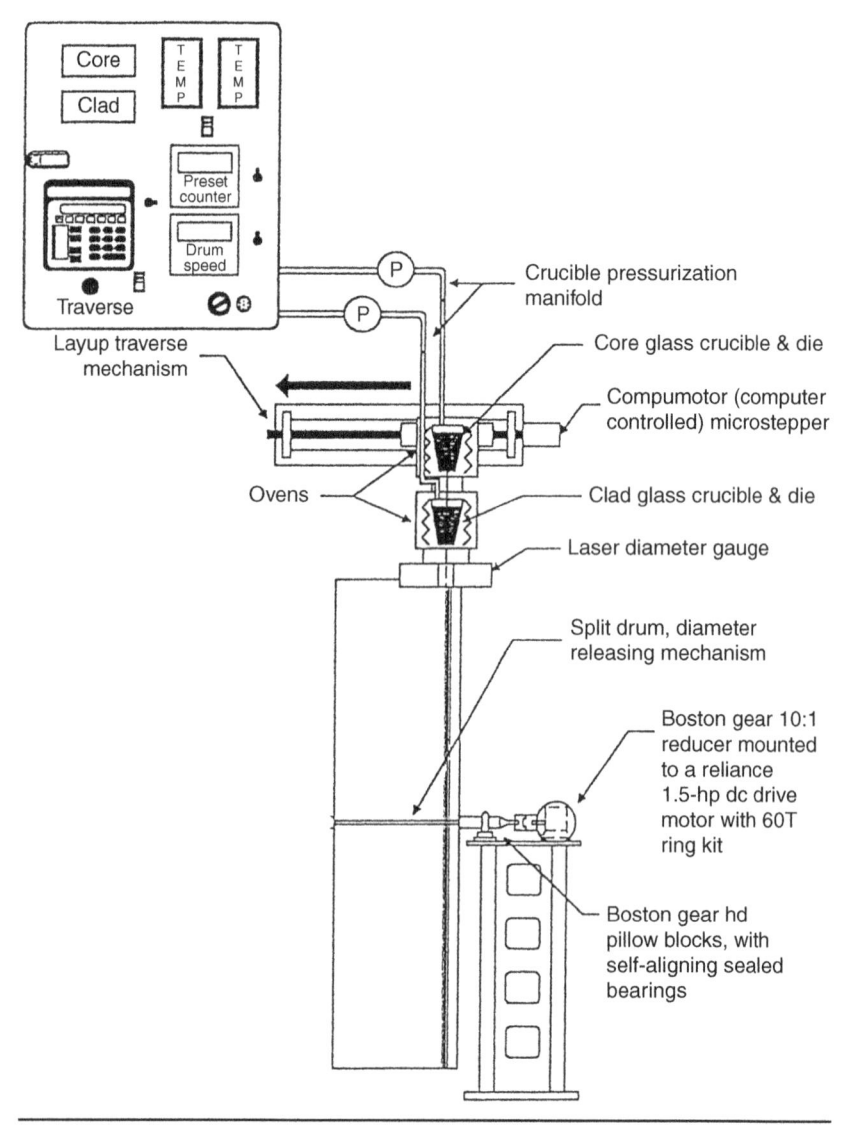

Figure 8.15 Diagram of the AMI 10-m drum assembly.

would be virtually impossible to remove and handle a fragile 10-m ribbon without severe fiber breakage. We decided the ribbons would have to be made on the drum, one stacked and glued to the other without being moved.

To accomplish this, we first obtained some extruded plastic channel and glued it aligned to the surface of the drum one revolution with the open end up. The flat surface inside was 8 mm wide, just right for winding a 7-mm ribbon. Contrary to previous ribbon formation procedures, the furnace, diameter monitor, and split dies are moved relative to the drum. The drum is in a fixed position. A stepping motor is computer-controlled to provide the positional accuracy required to place each fiber precisely in the ribbon being formed. Figure 8.16 shows a diagram of how the unique winding process devised for this program occurs. The computer reference point for the beginning of the ribbon is L_0. The computer reference for the sides of the plastic channel is X_0 and $X_0 + C$, where C is the channel width. The plastic channel glued to the surface of the drum has two notches cut in the sides, ΔN above and $-\Delta N$ below the region in which the layers are to be fused with thinned epoxy. The fiber for each layer (ribbon) is wound in the same direction, with the same pitch and fiber count accurately controlled by the computer. The procedure is started with drum turning; the computer moves the fiber through the notch at $-\Delta N$ to the origin of the first ribbon. When the full count of the ribbon is reached, the computer moves the fiber out of the notch ΔN and then on to point $+\Delta X_0$ where the operator tapes it to the drum surface.

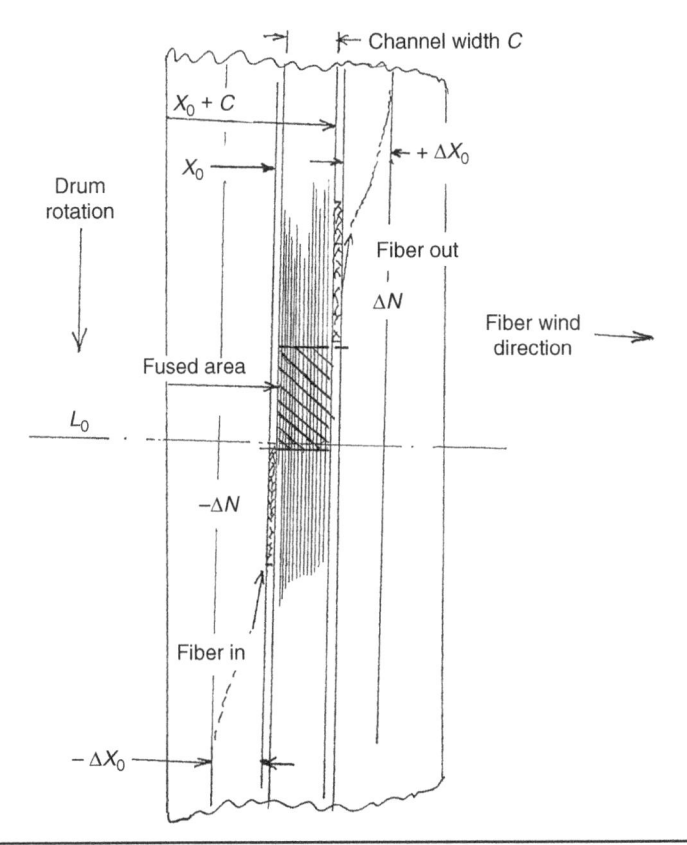

FIGURE 8.16 AMI procedure for construction of an imaging bundle.

202 Chapter Eight

FIGURE 8.17 Photograph of James McCord standing by the big drum.

The operator fuses the fiber to form the first ribbon. After some rotation, the computer moves the fiber across the rotating drum, across the ribbon, and to a position on the opposite side of the channel at $-\Delta X_0$ where the fiber is again taped to the drum. The crossover piece of fiber is cut and removed. The operator then orders the computer to start the next layer. The fiber again enters through notch $-\Delta N$, and the process begins again. Because of the slow pull rate and the high ribbon count, the layers have time to dry before another is ready. Each time the ribbon count is reached, the fiber exits at N on to $+X_0$ and is taped. Meanwhile the ribbon is fused to the preceding ribbon. When enough layers have been made to fill the 7-mm × 7-mm volume, more plastic is added and allowed to dry. After drying, the bundle is cut in the center of the fused area and removed from the drum for polishing the ends. The bundle is placed in a plastic tube for protection. The AMI procedure is unique, resulting in the grant of a U.S. patent in a very short time.[8] Figure 8.17 shows a photograph of the operator, James McCord, standing beside the drum to emphasize the size of the drum. He is holding a 1-m imaging bundle in one hand and one of the early 10-m bundles in the other hand. Notice the bundle in the plastic channel and encased in tubing is flexible in only one direction. This problem was solved on the last bundle produced.

Navy Bundle 10-M-1
During an attempt to try out winding fiber inside plastic channel on the 1-m drum, we realized we could not manually place the fiber

inside the channel with any accuracy. We requested a modification of the software by Tom Loretz and Paul Modlin that was accomplished in February 1999. In our first attempt, we chose to use large-diameter fiber to minimize breakage. Core diameter was 175 µm with center-to-center spacing of 210 µm. The bundle was formed from 24 ribbons, 33 count with a finished size of 7 mm × 4.7 mm which yielded a calculated active area of 57.3 percent. The bundle was coarse, with voids, and had poor resolution, about 2 lp/mm.

Navy Bundle 10-M-2

The second attempt was formed from 42 ribbons, 69 count with 100-µm core diameter. The finished bundle was 7.4 mm × 5.0 mm with a calculated active area of 61.5 percent. The bundle contained more fibers in number, 2898, and more fiber in meters, 28,980, than any bundle we had made at that time. The fused area was large and the fibers were closely packed. The numbers and bar pattern from the −1 group, one step larger than the O group, were transmitted. Limiting resolution was poor, about 1 lp/mm. The bundle showed low contrast, which makes image recording difficult. In the camera image, one can resolve the "O" pattern and the 1 pattern for a resolution limit of about 3.5 lp/mm. We were able to show the low contrast was due to a higher than normal water content in the glass used for the bundle. The fundamental absorption occurs in the glass at 2.92 µm with an overtone at 1.45 µm, close to the wavelength of maximum sensitivity of the NIR camera. The overtone absorption becomes appreciable for a 10-m optical path.

Navy Bundle 10-M-3

The third attempt was designed to increase the active area of the bundle by using a slow-drying epoxy on the layers, which should have allowed the fibers to settle down (we thought) between the fibers of the previous layer. In this way the packing density should increase, which in turn would yield a higher active area. As very often happens, our result was slightly the opposite. We wound 42 layers of 69 count ribbons for a total of 2899 fibers in a bundle 7.7 mm × 5.7 mm with a calculated area of 58 percent, down slightly from 62 percent. The appearance was also no better than 10-M-2 plus there were dark regions in the bundle when viewed with the NIR camera. The dark regions disappeared when viewed with the Agema 210, again supporting the idea that the dark regions were caused by water concentration variation in the fiber.

Navy Bundle 10-M-4

The last 10 m fabricated for the program was by far the best. First, we had recognized that the plastic channel as a complete container for the bundle was not flexible in all directions, leading to crimping and fiber breakage. We restricted our use of the channel to only a small section where the ribbons could be formed and the layers could be fused. At the end, the fused area would be cut in the middle to provide the coherent ends. The bodies of the fibers were wound on a thin sheet of plastic stuck to the drum. When the winding was complete, the edges of the plastic were folded over the loose fibers and sealed to form a thin plastic sack the entire length of the bundle. After the fused area was cut in the middle, the thin sack was placed in a shrinkable plastic tube, heated to size, and sealed on each end to the metal connectors. The bundle was wound from very high purity glass, all from the same plate, homogeneous, striae-free, and low in water absorption (< 1 dB/m) and in hydrogen sulfide absorption (< 5 dB/m). The quality of the glass led to a failure-free winding process with four cores from the plate completing the bundle. The ribbon count was 66 unclad fibers, core diameter 108 µm, 47 layers for a total of 31,020 m. The calculated active area was 71.2 percent, our record (absolute maximum is 78.5 percent). The extra bundle was fabricated to employ all we had learned to prepare a final improved version.

8.4.6 Optical Evaluation of 10-m Imaging Bundles

Evaluation in the NIR using the Electrophysics NIR camera with our microscope played a major role in the program in evaluating the 1-m bundles. Figure 8.18 shows the scheme followed early in the program

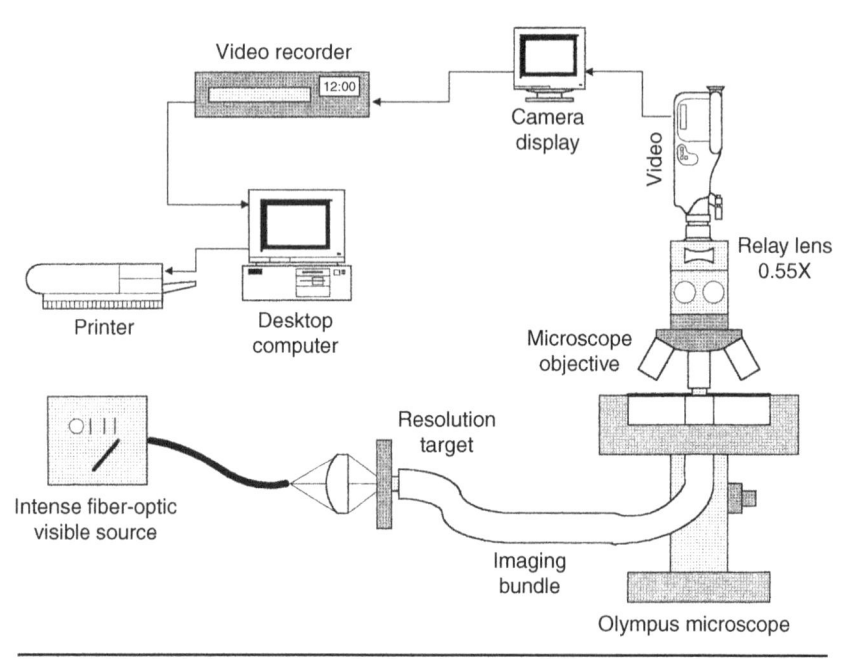

FIGURE 8.18 Optical arrangement for the microscope/NIR camera evaluation of ir bundles.

when the camera, with maximum sensitivity at 1.4 µm, was used with our microscope. The intense light source provided ample light transmitted through C1 fiber bundles for images and red light transmitted through 1-m C2 bundles. However, no red light was visible through a 10-m C2 bundle. Also, the relay lens used on the microscope limited the image to a small area. For the 10-m bundle evaluation, AMI had a special relay lens and objective lens designed by Gary Wiese of Panoptics for use in the evaluation of the 10-m C2 bundles. The NIR camera was removed from the microscope and used directly with the relay lens designed to fill the camera with the image of the 7 mm × 7 mm bundle end. We did not use the Panoptics objective lens. Instead we used the intense visible light source with the AF resolution pattern on the arsenic trisulfide glass disk with the bundle end in direct contact with different patterns. No visible light was transmitted, but enough NIR light passed through to form an image. Video recordings were made for study and record. Most efforts concerned evaluating the resolution of the bundle from the U.S. Air Force 1951 resolution target patterns transmitted by the bundle. For this program, the patterns used have been the −2 (0.25 to 0.45 lp/mm), the −1 (0.5 to 0.9 lp/mm), the 0 (1 to 1.8 lp/mm), the 1 (2 to 3.6 lp/mm), the 2 (4 to 7 lp/mm), and the 3 (8 to 14 lp/mm). The upper image in Fig. 8.19 shows a low contrast number transmitted by 10-M-2 as viewed using the NIR camera. The middle figure shows a number transmitted through 10-M-3 as viewed by the Agema 210 camera. In this case the light source was a globar which results in a small bar image as a background. Notice the high contrast for the number 5. The bottom image shows the uniformity of bundle 10-M-4. Figure 8.20 shows resolution elements transmitted through 10-M-4 as imaged by the NIR camera. A limiting resolution limit of 10 lp/mm was estimated. Figure 8.21 shows the same type of evaluation except using the Agema 210 camera. Also, a flat hot plate was used as a light source. Contrast improved. The rectangular shape of the infrared detector elements used in the Agema 210 becomes obvious in the image. We were fortunate in that we had a visit from Jim Davidson of Thermalscan. Davidson uses a Raytheon Radiance IT camera in his business. He was kind enough to use it in greater evaluation of the bundle 10-M-4. The top image in Fig. 8.22 made by Davidson shows an excellent image of the uniformity of the bundle. We tried without success to form images of human subjects at a distance. Using a 1-m C2 bundle with the sensitive camera, we were able to observe our first human images in natural light. The results led to the work that improved performance of 1-m bundles reported[5] in Sec. 8.2. The second image in Fig. 8.22 is that of a ceramic doll taken a few feet away from bundle 10-M-4 using an AMI Amtir 1 planoconvex lens as the objective. The camera used was the Agema 210 with the hot plate light source.

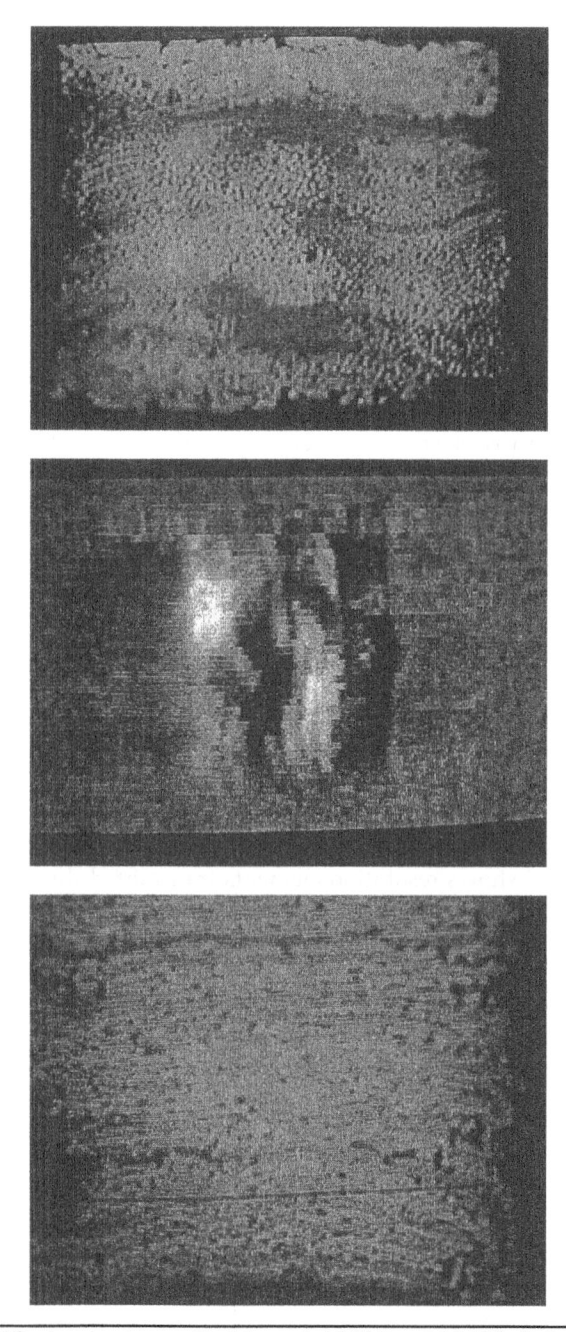

Figure 8.19 Images transmitted through 10-M-2 and 10-M-3. Bottom, NIR camera image of 10-M-4.

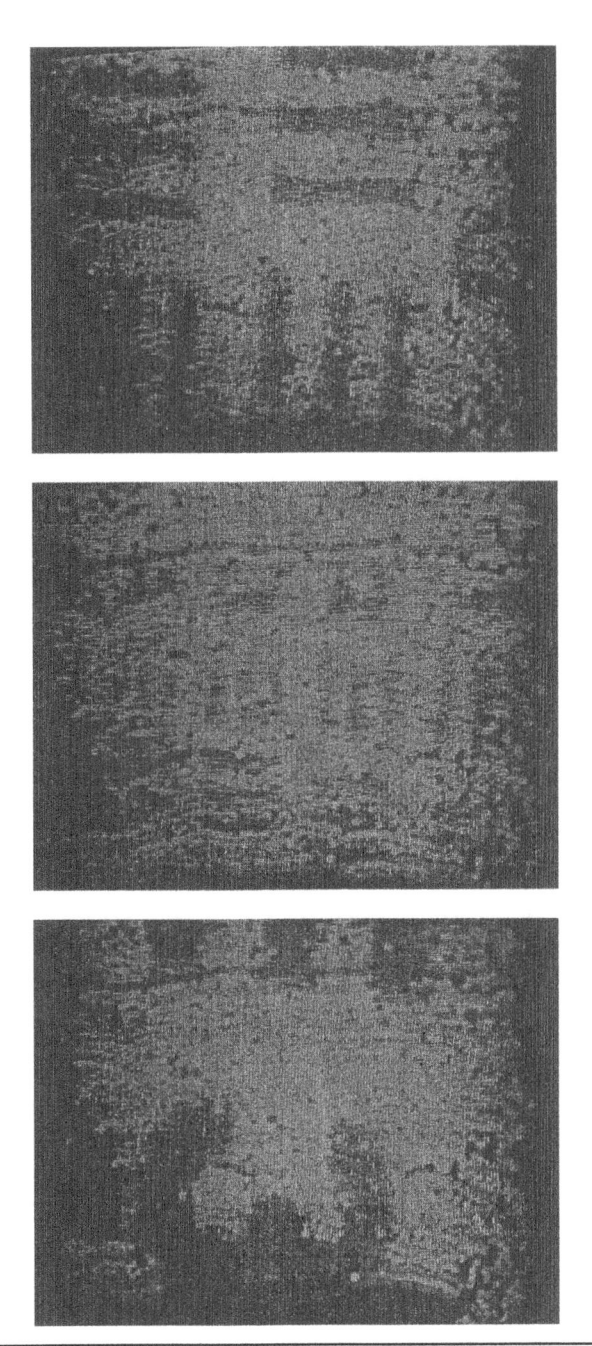

IR Imaging Bundles Made from Chalcogenide Glass Fibers

FIGURE 8.20 Resolution elements transmitted through 10-M-4 as viewed using the NIR camera.

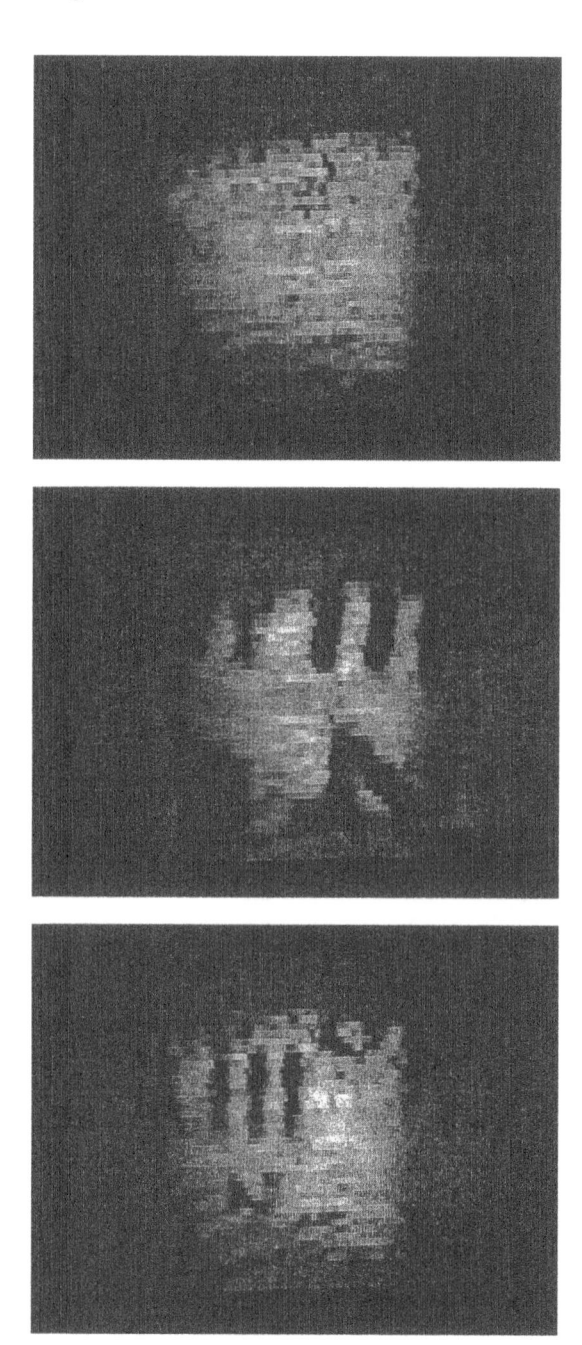

FIGURE 8.21   Bundle 10-M-4 viewed using the Agema 210 camera.

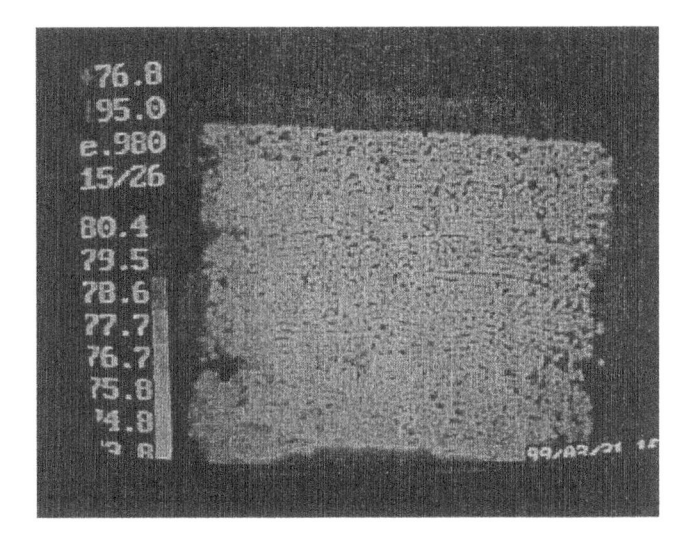

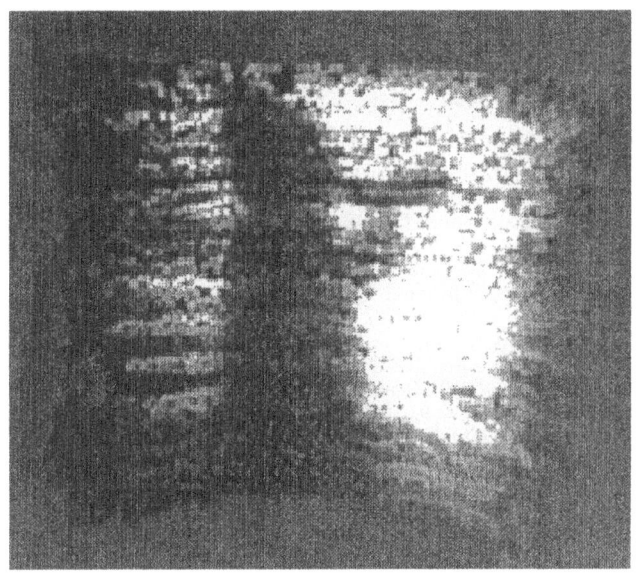

IR Imaging Bundles Made from Chalcogenide Glass Fibers 209

FIGURE 8.22 Images transmitted through 10-M-4 as viewed using a Raytheon Radiance IT camera.

8.5 Summary

AMI has developed in this SBIR Phase II program a method to fabricate an imaging bundle 10 m long from arsenic trisulfide glass fibers, unclad and with core diameters of 100 µm. The bundle is wound inside a plastic channel glued to the surface of a drum which has a circumference of 10 m. A plastic tube is shrunk by heat around the bundle to serve as a container protecting the fragile fibers from breakage. Cladding of the fibers with glass was not used in order to speed up the drawing process, to increase the active area of the bundle to over 70 percent, and to provide flexibility. Crosstalk at the bundle ends did not occur because epoxy used to fuse the ribbons served as cladding. Unclad fibers transmitted almost as well as clad fibers because of the high refractive index (2.4) of As_2S_3 glass. Both ends of the bundle were antireflection—coated to increase transmission in the 3- to 5-μm camera wavelength band. Purification methods used in preparation of the arsenic trisulfide glass have enabled absorption levels at 2.44 μm of 0.1 to 0.2 dB/m to be met, at 2.92 μm (H_2O) below 1 dB/m has been achieved, and < 5 dB/m at 4 μm due to H_2S have been met. However, unexpected broadening of the absorption by the impurity bands for a 10-m optical path led to a decrease in internal transmission in the 3- to 5-μm band to 37 percent. When multiplied by the 70 percent active area, total transmission drops to 26 percent, or one-half of the program goal of 50 percent. Sufficient energy exists, nonetheless, for producing useful images when used with a very good sensitive camera and suitable optics.

References

1. A. R. Hilton, Sr., "As-Se-Te Based Glass Fibers," *SPIE* 1228, 76 (1990).
2. A. R. Hilton, Sr., A. R. Hilton, Jr., J. McCord, and T. J. Loretz, "Preparation of Coherent IR Chalcogenide Glass Fiber Bundles," *SPIE* 2131, 192 (1994).
3. A. R. Hilton, Sr., A. R. Hilton, Jr., and T. J. Loretz, "A Progress Report, Fabrication of IR Coherent IR Glass Fiber Bundles," *SPIE* 2677, 15 (1996).
4. A. Ray Hilton, Sr., "Infrared Imaging Bundle Development at Amorphous Materials," *SPIE* 3849, 61 (1999).
5. A. R. Hilton, Sr., A. R. Hilton, Jr., J. McCord, W. S. Thompson, and R. A. LeBlanc, "Infrared Imaging with Fiber Optic Bundles," *SPIE* 5074, 849 (2003).
6. A. R. Hilton, Sr., "Final Scientific and Technical Report, Fiber Optic Coupled IRFPA," Contract NOO-97-10421-97-C-1046, 1999; and Ray Hilton, Sr., et al., "Fabrication of a 10 Meter Length IR Imaging Bundle From Arsenic Trisulfide Glass Fibers," *SPIE* 3596, 64 (1999).
7. A. R. Hilton, Sr., A. R. Hilton, Jr., James McCord, and Glen Whaley, *SPIE* 3060, 325 (1997).
8. A. R. Hilton, Sr., Method for Constructing a Coherent imaging Bundle. U.S. Patent 5,938,812, August 1999.

CHAPTER 9
AMI Infrared Crystalline Materials

Even though the company was completely dedicated to infrared glasses, at both TI and AMI the author was involved in producing crystalline materials useful in infrared technology. Expertise at AMI was imported when colleagues from TI joined AMI and led efforts to produce crystalline materials needed in government programs.

9.1 Cadmium Telluride

Cadmium telluride first became known as an infrared optical material as one of the hot pressed polycrystalline Kodak Irtran materials. Kodak developed a group of materials, all made using the same process. Fine-grain high-purity powders were placed in a mold, sealed from the atmosphere, heated to a temperature just below the melting point, and molded into a disk using very high pressure. The resulting disk had density very near that of the melt grown material. When polished, they showed high transmission in the infrared and were used to form lenses or serve as windows. They were also strong because of their particulate structure. The designations and materials used to make them were as follows: Irtran 1, MgF_2; Irtran 2, ZnS; Irtran 3, CaF_2; Irtran 4, ZnSe; Irtran 5, MgO; and Irtran 6, CdTe. Optical and physical properties of these materials are described in detail in a Kodak Publication U-72 released in September 1971.

Along about 1970, the U.S. Air Force was funding a great materials effort to develop a window suitable for use with high-energy lasers to be mounted in an aircraft. All infrared window materials were being evaluated for this application. One of the leaders in theoretical studies of infrared optical materials was Marshal Sparks[1] of XONICS. Specific materials were characterized by Stanley Dickinson[2] in an Air Force

Report and in an Airforce Compendium by Charles Sahagian and Carl Pitha.[3] These are only a few of the many such works reported during the 1970s. The understanding of the critical factors related to infrared optical materials use with lasers was advanced markedly.

The TI glass 1173 was under consideration because it could be cast in large window form, something not possible using the Kodak hot pressed ceramic technique. The technology had not developed at this point to produce large plates of melt-formed crystalline materials. Kodak Irtran 6, CdTe, transmitted with no absorption out to 30 µm, farther than most all other materials being considered. As a candidate material early in the Airforce Program, the size of CdTe produced was emphasized. The Kodak limit was a 6-in diameter. The Airforce Materials Laboratory was interested in any scheme to produce large CdTe plates.

9.2 Previous Work at TI

TI at that time was growing single-crystal small-diameter CdTe disks by conventional means to be used as substrates for growth of the new infrared detector material HgCdTe in thin layers by liquid epitaxy. The effort at that time, championed by Dick Reynolds of TI, was to replace mercury-doped germanium as the detector material of choice for infrared systems. The author in 1971 at TI submitted a proposal and won a program funded by the Airforce Materials Laboratory[4] to develop a method to prepare large plates of single-crystal CdTe suitable for application with high-energy CO_2 lasers emitting at 10.6 µm. The method chosen was based on solution epitaxy growth on a mosaic of single-crystal CdTe substrates. An excellent analysis of the theoretical considerations governing solution growth is found in an article by Tiller.[5] In general, a solution of Te saturated with a specific concentration of Cd at a high temperature is cooled and placed in contact with a substrate held at a lower temperature. The layer grows on the substrate but slows as the solution is depleted, depending upon diffusion for continued slow growth. Application to CdTe growth is depicted in the TI process shown in Fig. 9.1. In the diagram, the graphite boat used to hold the CdTe substrate is fitted with a graphite slide that contains a solution chamber. The entire apparatus is contained in a quartz chamber. Because the volatility of cadmium is much greater than that of tellurium, a solution rich in tellurium with cadmium concentration less than 20 percent was chosen. In position 1, the entire apparatus is heated to a liquidus temperature above that for the CdTe solution, above 800°C. Cooling starts and proceeds until the desired solution temperature is reached and the slide is moved to position 2, the substrate is covered by the solution, and the layer begins to grow. Growth is slow and allowed to continue for a period of time. As the solution saturated with Cd next to the substrate becomes depleted, growth depends upon Cd diffusion

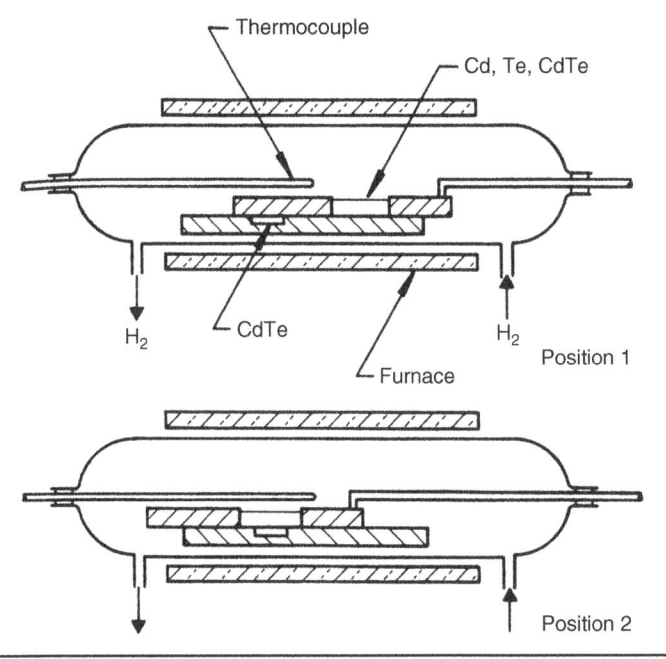

FIGURE 9.1 Conventional solution epitaxial growth process.

through the solution to the substrate. At the end, the slide is moved back to position 1 to prevent the solution from sticking to the epitaxial layer. As described, this is a batch process method, slow, limited by diffusion, and not suited for growth of thick plates. To overcome the disadvantages of the conventional method, the system shown in Fig. 9.2

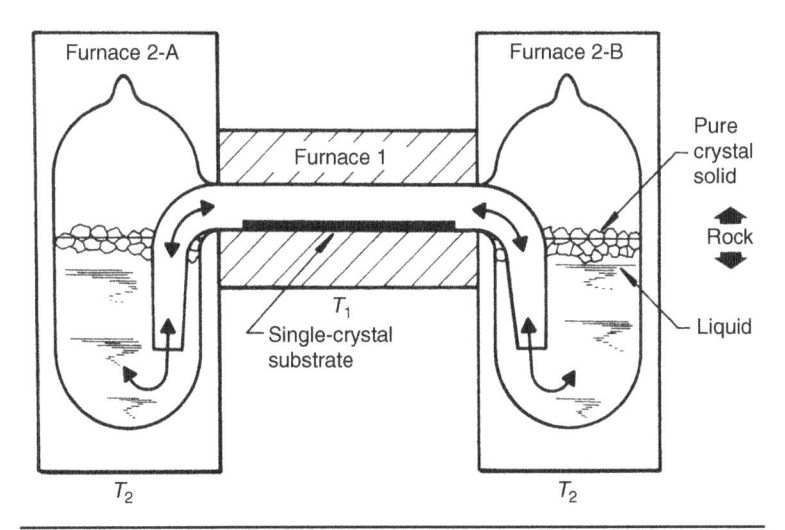

FIGURE 9.2 Continuous solution epitaxial growth design.

was designed and built. The chambers are all made entirely of quartz, assembled in the TI glass shop before placement in the three furnaces. The two source furnaces are controlled at temperatures above that of the furnace around the center reaction zone. The two chambers serving as sources are constructed so the saturated solutions come from the bottom and contain no solid CdTe. The entire apparatus, furnaces and reaction chamber, is placed on a rocking platform and rocked so that solution liquid is moved back and forth across the substrates. As the hot liquid leaves the source chamber, it cools and is depleted of CdTe, adding to the layer on the substrates. At first the rocking rate was 2 times per minute over a period of hours. One reaction run was 18 h with the source temperatures at 900°C and reaction chamber at 700°C. The results were very disappointing. The substrates did not grow into a plate and, in fact, dissolved to a great extent. The conclusion reached was that the temperature differences between substrate and solutions were much too great and the rocking speed was too fast. Small conventional runs were resumed to establish the proper growth conditions and to demonstrate that single-crystal growth across the substrates would occur. Methods used in forming 2 in × 2 in mosaic CdTe substrates are illustrated in Fig. 9.3. Stockbarger formed CdTe crystals squared up by sawing and then cut into squares or long strips. If the beginning crystals are all single, a square substrate is formed in one case or a substrate is formed from long sawed strips. Even if the material is not all single, good results may still be expected if the poly is large-grain. Ideally, if the crystal is completely single, an orientation may be found that that is favorable for CdTe growth. The graphite holder for the substrates shown in Fig. 9.3 has graphite strips at both ends to hold the substrate in place because its density is less than that of the solution and it would float away as the solution moved across. In all the runs, it was found the substrate pieces did not grow together. One possibility to be pursued was to alloy the pieces together before starting. Otherwise, the Te-rich solution flows into the cracks between the pieces, preventing growth.

9.2.1 Conclusions Concerning This Effort

1. High-resistivity layers of CdTe will grow from Te solutions if Cd concentration is 20 percent or greater in the temperature range of 700 to 750°C.

2. The volatility of Te is such that a closed system must be used to avoid Te loss and solution concentration changes.

3. Results indicate it is necessary to alloy mosaic pieces together prior to layer deposition attempts.

4. The temperature difference between solution and substrate must be kept low to maintain equilibrium conditions.

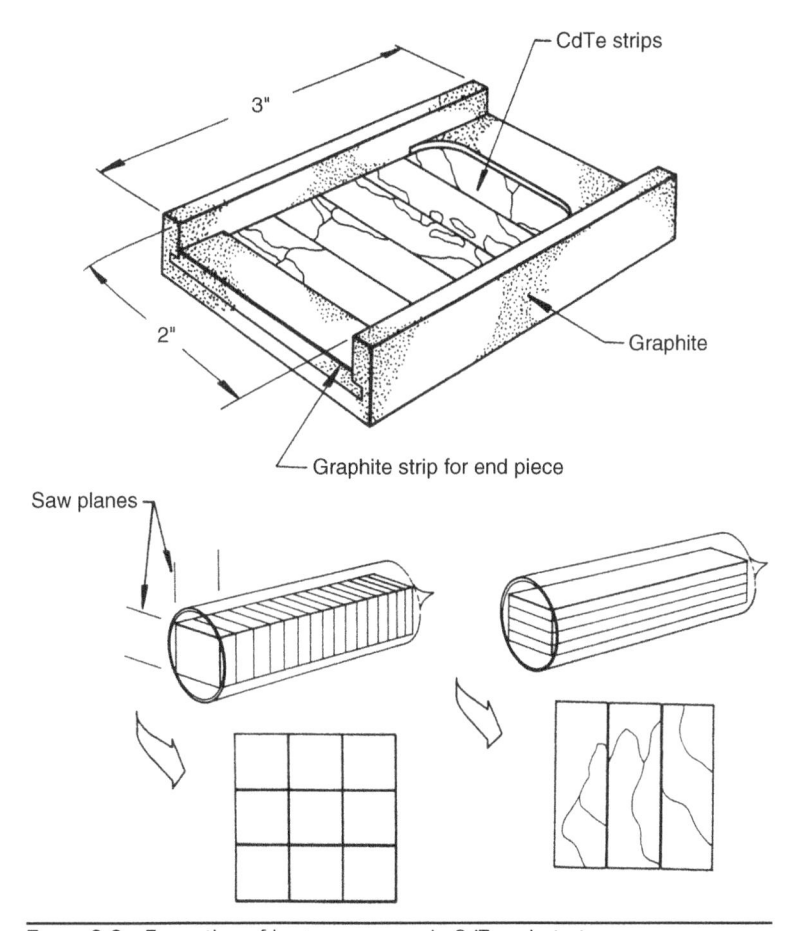

FIGURE 9.3 Formation of large area mosaic CdTe substrates.

9.3 AMI DARPA-Funded Large Plate Process

Some 10 years later, in 1981, the author met with his ex-TI colleague Dick Reynolds for lunch in Washington, D.C. Dick Reynolds had become the deputy director for DARPA and was heavily involved in building support for the production of many thousands of HgCdTe detector arrays that would be needed in the very near future. Conventional methods of producing CdTe for substrates were much too slow. What was needed was a new approach to produce large plates of CdTe. Together they decided that the AMI process for producing large plates of high-purity chalcogenide glasses could be modified to produce CdTe plates with increased size and purity. A plan was made and a proposal written for an effort to produce plates of 6-, 8-, and 10-in diameter. The first contract was DARPA-funded MDA903-82-C-0159 followed by DARPA Contract No. DAAG-29-85-C0004 funded by the U.S. Army covering the period from 1982 to 1985. The author was fortunate in recruiting his good friend and former colleague at TI to join AMI and take over all the crystalline materials work. George Cronin had been a leader in developing methods to produce many crystalline materials including all the III–V materials since joining TI in 1958. Of particular interest to AMI was his work with GaAs and high-purity vacuum float zoned silicon; later we discuss the AMI activity in producing these two materials.

Cadmium telluride is a very difficult material to grow from a melt because both elements are volatile and it has a high melting point where appreciable vapor pressure can exist for both elements even if stoichiometry is near perfect. It has a congruent melting point of 1097°C, which means below that temperature both Cd-rich liquids and Te-rich liquids exist in equilibrium with pure CdTe. Figure 9.4 shows a simple binary, single-compound diagram for Cd and Te.

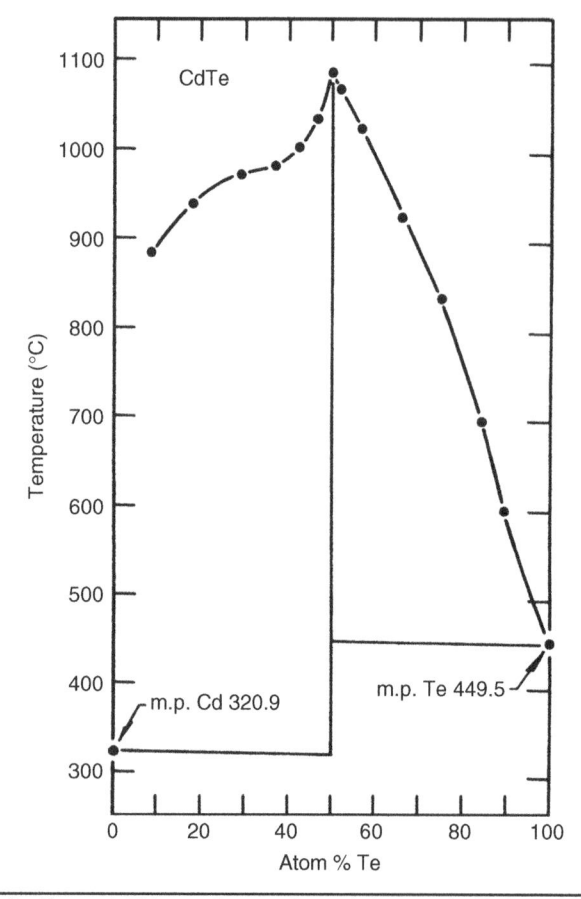

FIGURE 9.4 Elemental binary diagram for cadmium and tellurium.

The solid line represents the liquidus curve for CdTe solutions. Complete discussion of the growth and properties may be found in the extensive work published by D. de Nobel.[6] The vapor growth approach is discussed in detail by Shiozawa and colleagues[7] of Gould Laboratories in Ohio. The AMI approach was confined to modifying our glass production methods. AMI had considerable experience with forming glass plates in round 6-, 8-, and even 10-in quartz chambers. Recalling the AMI glass production process, the elements are placed in the compounding chamber, melted, mixed, and compounded. Then the compounded material is physically distilled through a porous quartz filter to remove particulate matter and is condensed in the round casting chamber. The transfer process also leaves behind most all the impurities found in the beginning reactants. It was natural for us to try first taking precompounded CdTe and transferring it as vapor through a filter to form a plate in the round chamber. However, because of the high melting point of the material, the filters constantly became plugged, thwarting our efforts. We decided it would be better to separate the cadmium and tellurium. We added a third chamber to our standard glass unit, as shown in Fig. 9.5. The reader may recall that the same conclusion was reached in the preparation of arsenic trisulfide glass. The decision regarding CdTe predates the decision on the glass.

Referring to the diagram, the cadmium is slowly transferred to the central chamber over a period of 4 h at a temperature of 750°C. Vapor pressure is about 1 atm. Then tellurium is transferred at 800°C at a pressure of about 150 mm over a period of 8 to 10 h. During the transfer and filtration of the Cd and Te, particulate material, mostly carbon, and other impurities are left behind, improving the final purity of the CdTe plate. After the material is transferred, the temperature of the central chamber is slowly raised above the melting point, 1097°C. The two side chambers are left at 800°C. The center

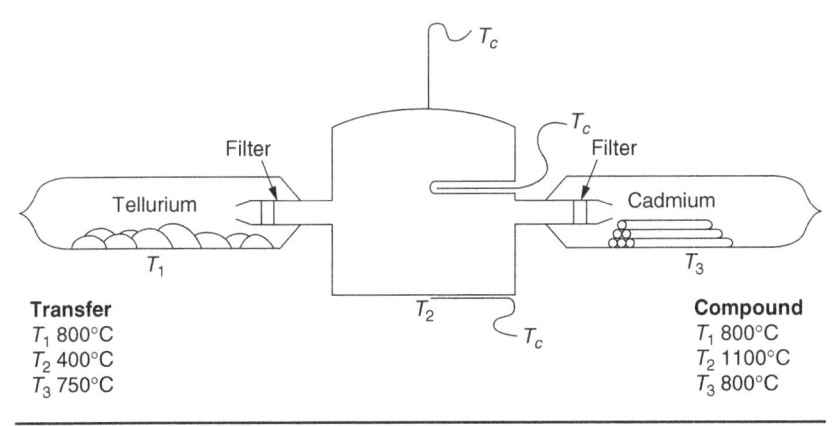

Transfer
T_1 800°C
T_2 400°C
T_3 750°C

Compound
T_1 800°C
T_2 1100°C
T_3 800°C

FIGURE 9.5 AMI three-chamber method for producing plates of CdTe.

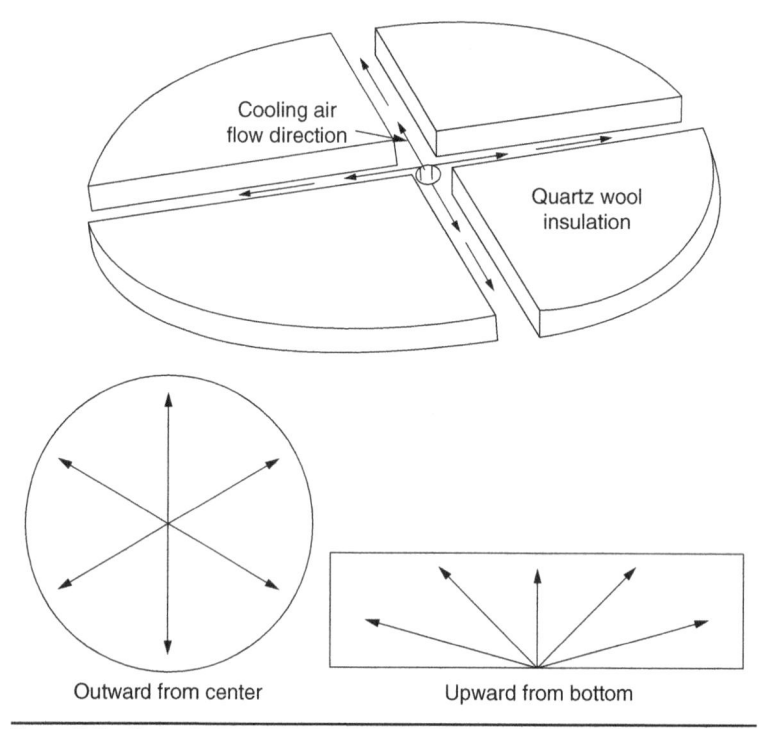

FIGURE 9.6 Flow of air to control nucleation and growth of CdTe plate.

chamber is slowly cooled to promote growth of large-area crystals then quenched, followed by an anneal cycle. During the slow cool cycle, a gentle flow of air is directed at the bottom center surface to ensure that freezing of the liquid is initiated at the center of the plate and crystal growth occurs outward toward the periphery.

Figure 9.6 shows how air is channeled from the center outward through spaces cut in the quartz insulation material on which the center chamber is resting. Airflow continues outward and upward, affecting the crystallization process of the melt in the desired direction. Figure 9.7 in the upper photograph shows plate no. 65 still in the round chamber inside the three-zone furnace while the lower photograph shows plate no. 65 after removal with the top surface lapped. The plate weighed 4 kg. Notice the large areas of the plate that appear to be single-crystal. Plate no. 65 was the best made in this program. Analytical results provided by former colleagues at TI confirmed the impurity level was below any CdTe ever analyzed in their facility. Table 9.1 presents the material evaluation results.

Spontaneous nucleation and growth of a CdTe melt can be made to produce large regions of single crystal in the plate as we have just seen. However, to consistently control the quality of any crystalline material grown from a melt, the interface between solid and liquid

AMI Infrared Crystalline Materials 219

FIGURE 9.7 Plate no. 65 still in the furnace and after being lapped.

Room temperature resistivity:	2.74×10^7 $\Omega \cdot$cm
Liquid nitrogen resistivity:	$> 10^{10}$ $\Omega \cdot$cm
Carrier mobility (+):	51 cm^2/(V·s)
Carrier concentration:	4.5×10^9
Detected impurities:	Cr, Mg, Si, Na, K at < 1 ppm level
Inclusions (Te) concentration:	5.8×10^6/cm^3
Diameter of inclusions:	7 μm
Infrared free carrier absorption not detected:	2–25 μm

TABLE 9.1 Material Evaluation Results

220 Chapter Nine

Cadmium or tellurium

CdTe samples

Evacuated quartz chamber

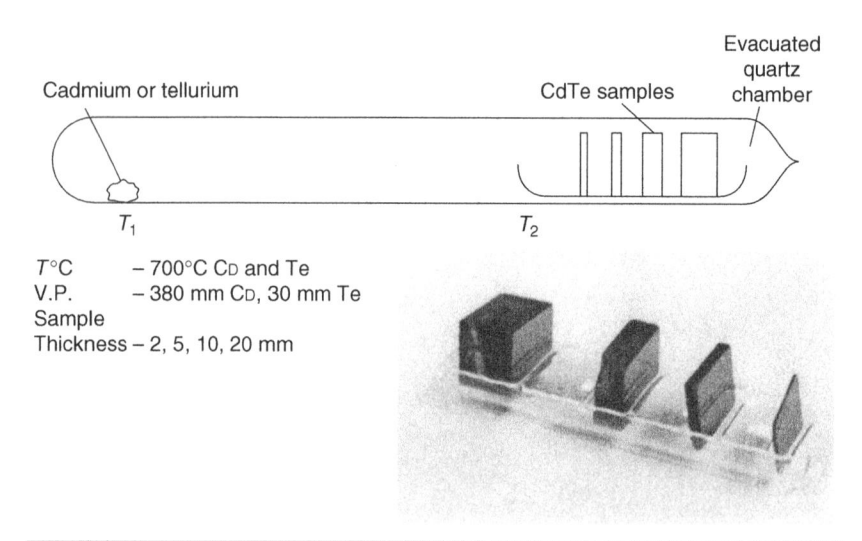

T_1 T_2

T°C — 700°C Cd and Te
V.P. — 380 mm Cd, 30 mm Te
Sample
Thickness — 2, 5, 10, 20 mm

Figure 9.8 Plan to evaluate the effect of Cd and Te vapor on the infrared transmission of CdTe produced using AMI three-chamber method.

must be controlled. Perfect stoichiometry and vapor pressures at the melting point must exist at the melt-solid interface. Growth must be slow to maintain these conditions. This statement will be emphasized again later when the growth of gallium arsenide is discussed.

CdTe grown using the AMI process will have Te precipitants that scatter light, reducing infrared transmission. Measurements made of polished samples of different thickness show an absorption value of 0.3 cm^{-1} across the wavelength band of 5 to 8 µm with no free carrier absorption. The first step in improving the material is to heat-treat with Cd vapor to destroy the Te precipitants. Then follow up with Te vapor to restore stoichiometry. The plan is illustrated in Fig. 9.8. Samples from the plate 2, 5, 10, and 20 mm thick were cut and polished in four sets to be heat-treated for periods of 1, 2, 4, and 8 days at 700°C with Cd vapor at 380-mm vapor pressure and Te vapor at 30-mm vapor pressure. A fifth set of samples was prepared to be a reference with no vapor treatment. The samples were placed in the tubes along with a piece of Cd or Te, sealed, and evacuated. After the tests were complete, the samples were removed, infrared transmission was measured, and results were compared. All the Cd treated samples showed free carrier absorption in the 5- to 8-µm band after only 2 days' treatment, increasing in magnitude with each day of treatment. The untreated did not change, showing no free carrier absorption just low transmission due to Te inclusion scattering. The samples were then reloaded in the chambers only with Te this time, and the procedure was repeated. All samples had a marked increase in transmission with the thickest, 20-mm sample taking nearly 8 days to

Wavelength (μm)	Refractive* Index at 25°C	Absorption Coefficient (cm⁻¹)
3	2.7026	< 0.01
4	2.6971	"
5	2.6922	"
6	2.6886	"
7	2.6865	"
8	2.6846	"
9	2.6825	"
10	2.6797	"
11	2.6766	"
12	2.6749	"

*ν 3–5 μm = 163
†ν 8–12 μm = 173

Thermal Change in Index†
$\Delta N/\Delta T \times 10^{-6}/°C$
+147 @ 1.15 μm
+98 @ 3.39 μm
+98 @ 10.6 μm

*Melt formed crystal.
†R. J. Harris, *Appl. Opt.* 16, 436 (1970).

TABLE 9.2 Optical Properties of Cadmium Telluride

reach maximum transmission with no free carrier absorption. An absorption coefficient of 0.002 cm⁻¹ was measured at 10.6 μm by Laser Power Optics. Table 9.2 lists the data found in the AMI product sheet for melt-formed crystalline CdTe.

9.3.1 Conclusions

The AMI chalcogenide glass compounding process was modified by adding a third chamber so plates of large-grain polycrystalline cadmium telluride could be produced. Plates with as large as 10-in diameter were tried, but failure of the quartz became a problem. Most efforts were concentrated on using our regular 8-in chambers. The slow freezing rate and the air-controlled directional freeze appear to be a reliable method for producing relatively large volumes of randomly oriented single-crystal regions. No single preferred growth orientation was observed. In general, the plates were well compounded, free of voids with very little, if any, dendritic growth.

A successful method for heat-treating the CdTe from the plates with first Cd vapor followed by Te vapor was developed that dramatically improved infrared transmission. A prism was fabricated from a melt-grown plate, and the refractive index measured 3 to 12 µm. The measured values were slightly higher than those measured on Kodak hot pressed Irtran 6. That result would be expected since Irtran 6 only has nearly 100 percent density.

9.4 Vacuum Float Zoned Silicon Detector Material

The normal method for growing crystals of silicon, the Czochralski (Cz) method, in which the crystal is grown from an open melt, suffers from impurities getting into the melt during the growth process. Chief among the impurities is carbon coming from graphite parts of the crystal grower and oxygen from quartz liners containing the silicon melt. A great advance occurred when TI developed and used the extreme high-purity vacuum float zone technique. TI used this method to produce the high-grade silicon required for the "paveway detector" sensitive to the Yag 1.064-µm laser radiation that made possible the first laser-guided bombs used by the military. The supervision of the growth of this material was one of the responsibilities held by George Cronin when he was at TI. George Cronin felt that production of such a specialized silicon was an ideal product for a small niche company such as AMI. In 1983, he searched around and located two intact units in California that had been sold off by TI as surplus equipment. AMI purchased the units and had them moved to the new portion of the AMI building under construction. Figure 9.9 shows a diagram depicting single-crystal growth using vacuum float zoned purified silicon inside the machine under an atmosphere of argon. The diagram shows that a rod of high-purity polycrystalline silicon is passed through a copper coil connected to a high-power RF generator. The energy coupled into the silicon produces a molten zone between two solid pieces of silicon. The molten zone is held in place by surface tension. The silicon crystal is rotated from the pedestal at the bottom. As the crystal passes through the RF coil, the top is melted and the seed is inserted. Growth rate is very slow as indicated. The seed crystal, <111> in this case, may have other orientations. Figure 9.10 shows in the left photograph one of the units with the door open being inspected by Mitchell Jones of AMI. The stainless-steel inside surface is shown along with the pedestal in the bottom. The copper RF coil is visible at the same height as the observation window. The chambers are water-cooled during use to minimize contamination from the inside walls due to the heat. The right photograph shows James McCord, Ray Hilton, Jr., and Mitchell Jones observing a purification run. All three operators were capable of carrying out purification runs. Only James McCord, who was brought to AMI from TI by George Cronin, grew the paveway-grade single

AMI Infrared Crystalline Materials 223

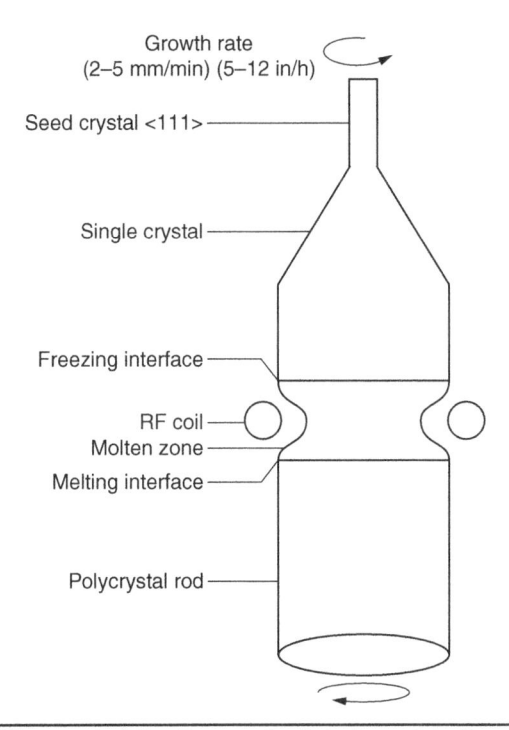

FIGURE 9.9 Diagram depicting the growth of single-crystal silicon in the AMI vacuum float zone machine.

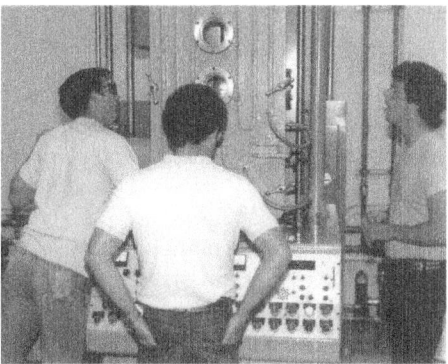

FIGURE 9.10 AMI silicon vacuum float zone chambers.

224 Chapter Nine

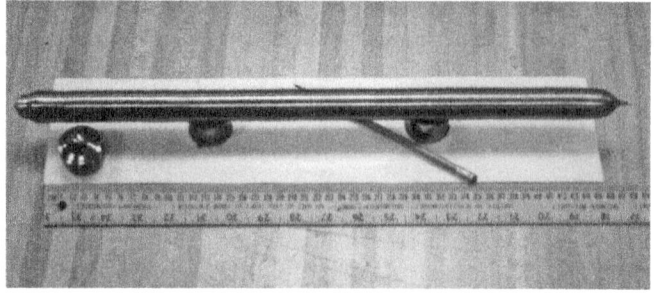

FIGURE **9.11** Photograph of AMI float zone machine and finished "paveway grade" silicon single crystal.

crystals. Figure 9.11 shows a front photograph of the machine including the attached turbomolecular pump used to evacuate the chambers down to 1 to 10 torr during the purification processing. The bottom image in Fig. 9.11 shows a finished float zoned paveway-grade silicon single crystal.

The purification of the silicon uses the same machine setup under vacuum. No seed is involved. A rod of special high-grade, high-purity polycrystalline silicon, such as produced by Hemlock Semiconductor, is the starting point. As the crystal rotates, a narrow zone is slowly swept through the crystal from the bottom to the top, carrying the impurities in the zone to the top of the crystal. Travel time is 3 in/h for the crystal 10 to 14 in long. All purification runs follow the same procedure, always zoning in the same direction relative to the rod. On completion of a run, the crystal is removed and its surface sandblasted. The crystal is evaluated relative to resistivity using a conduction method. A dc value of 0.1 A is run through the crystal. An electric probe with contacts exactly 1 cm apart is applied to the surface to measure and record the voltage drop across each centimeter segment, 1 cm at a time for the full length of the crystal. The numbers are converted to resistivity. After measurement, the crystal surface is acid-etched prior to the next run. The process is repeated 8 to 10 times, depending on the beginning purity of the silicon bar, until the measured resistivity falls within the range of the specification. Then the crystal is ready to be grown in single-crystal form. The top of the crystal is cut off to remove the impurities, and the ends are switched so the growth will be in the same direction relative to the crystal as in the purification process. The crystal is attached to the pedestal for rotation during growth. The crystal growth is controlled by the operator observing the process through the window. The growth rate again is very slow. On completion, the crystal resistivity is measured to ensure all regions meet the resistivity specification, 10,000 to 30,000 Ω·cm. On meeting the standard, the crystal is ready to be sawed into slices that will be polished and sized in thickness and diameter. AMI developed a good paveway silicon business having substantial sales in the period from 1984 to 1992. Eventually, European competition at lower cost created a paveway market in which AMI could no longer compete. Even though government policy usually calls for domestic sources for critical materials, government production programs often ignore the policy and put price first.

9.5 Silicon as an Infrared Optical Material

Semiconductor materials have free carriers of current, both positive (holes) and negative (electrons). Intrinsic levels depend upon temperature and the bandgap of the material. Silicon has a bandgap of only 1.04 eV (electronvolts) which means intrinsic carriers are thermally generated at relatively low temperatures. However, at room temperature, the intrinsic resistivity of silicon is 2×10^5 Ω·cm. Infrared radiation interacts with free carriers and is absorbed. The magnitude of absorption increases with increased wavelength to at least the power of 2. With increase in temperature, mostly the impurity atoms in silicon contribute free carriers, referred to as extrinsic, reducing the resistivity. TI results[8] showed that free carrier absorption at room temperature for N-type silicon, 1 Ω·cm resistivity or less, was significant at 10 µm. Silicon produced in the normal Cz process was never pure enough to have a resistivity over 25 to 50 Ω·cm. When heated, N-type silicon with a resistivity of 50 Ω·cm begins to show free carrier absorption at a 4-μm wavelength around 200°C due to thermally generated intrinsic and extrinsic carriers.

The value of silicon as an infrared optical material was recognized in the 1960s. Processes were developed at TI to spin-cast silicon domes for heat-seeking 3- to 5-μm missiles. A spinning graphite convex-shaped dome mandrel was dipped into a crucible containing molten silicon that stuck to the surface, slowly forming a dome. For this application, the absorption level was not critical because the dome thickness was not great. Silicon was selected because of its outstanding physical properties. Silicon is characterized by a high melting point, high thermal conductivity, and small thermal expansion, indicating resistance to thermal shock. The inherent strength of the material is indicated by its surface hardness and Young's modulus. Table 9.3 compares pertinent physical parameters for commonly used infrared optical materials (3 to 12 μm) including silicon. Examination of the numbers makes one conclude silicon is harder, stronger, and more resistant to thermal shock than any other material listed except sapphire. In fact, silicon, sapphire, and glassy quartz are in a class all by themselves. Considering cost, availability in high-quality form, and physical properties, silicon would have become the universally used infrared optical material (1 to 12 μm) except for two reasons: low use temperature and a lattice absorption band at 9 μm, right in the middle of the passive infrared night vision band at 8 to 12 μm. For that reason, in spite of higher cost and inferior physical properties including lower use temperature, germanium became the FLIR night vision standard optical material. Silicon was favored for 3- to 5-μm applications. During the 1970s efforts to find materials capable of withstanding very high intensity laser radiation, the very low absorption level of silicon from 3 to 5 μm, were reported.[9]

There was great interest in materials that could be used with high-intensity CW gaseous lasers such as the HF emitting at 2.8 μm, the DF emitting at 3.8 μm, and the CO emitting at 5.25 μm. Another laser of interest was the pulsed Er:Yag laser emitting at 2.94 μm. At this point, the selection of the lasers for weapon systems had not been made. Great interest still remained in the CO_2 laser emitting from 9.3 to 10.8 μm. Although at that time AMI did not have one of these lasers, the very low absorption of vacuum float zoned silicon was confirmed in a conventional manner. Taking a P-type single crystal 12 cm in length and 2.5 cm in diameter, the infrared transmission was carefully measured and the absorption coefficient calculated as a function of wavelength as far out as 8 μm. The results are shown in Fig. 9.12.

A value below 0.001 cm^{-1} was found out to be 5.5 μm. An absorption level of 0.001 cm^{-1} corresponds to 0.5 dB/m in fiber language. The low value and resistance to radiation damage was confirmed at Quantronix by Z. Drozowicz and M. Cohen using a pulsed Er:Yag

Material	Trans. Range (μm)	Refractive Index N at 10 μm	Thermal Change in $\Delta N/\Delta t°C \times 10^6$	Upper Use Temp. (°C)	Thermal Expansion Coefficient $\Delta L/L°C \times 10^6$	Surface Hardness Knoop	Young's Modulus $\times 10^{-6}$ psi
Si	1.2–10	3.42	+186	200	4	1150	19
Ge	2–12	4.00	+427	100	6	700	15
GaAs	0.9–12	3.34	+207	500	6	750	12
CdTe	0.9–30	2.67	+100	500	6	45	3
ZnS	0.6–10	2.20	+46	500	8	200	11
ZnSe	0.5–12	2.38	+52	500	9	100	10
Amtir 1 (glass)	0.7–12	2.49	+72	300	12	170	3
Quartz (glass)	0.2–3	1.44	+4	500	0.5	460	11
Sapphire (Al_2O_3)	0.2–4	1.73	10	500	6	1370	50
KRS-5	0.5–40	2.36	234	300	58	40	2

TABLE 9.3 A Comparison of the Physical Properties of IR Optical Materials

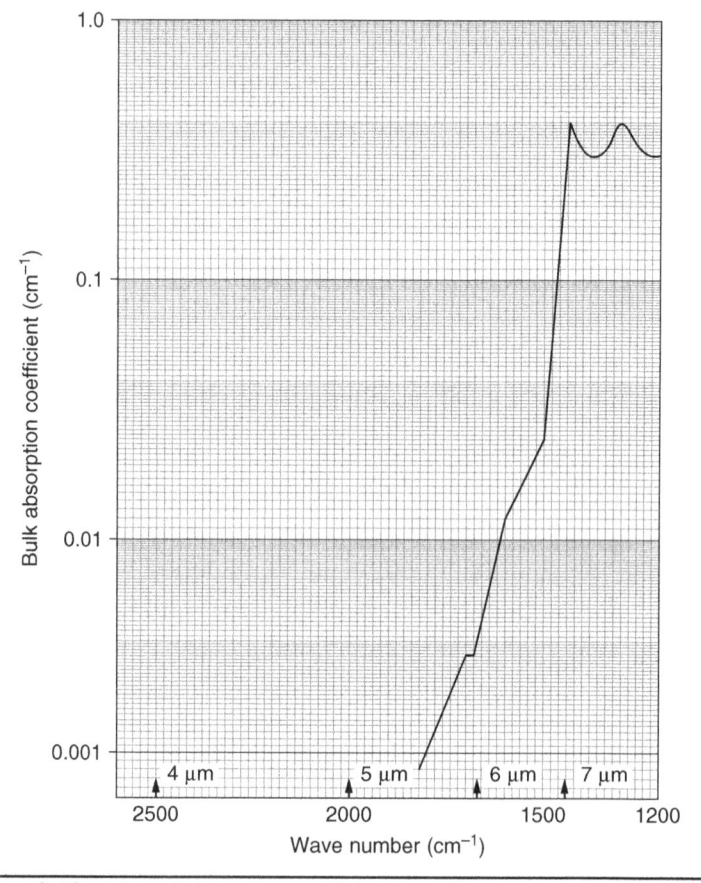

FIGURE 9.12 Infrared absorption coefficient for AMI vacuum float zoned silicon as a function of wavelength.

laser emitting at 2.94 μm. No measurable absorption for the 12-cm crystal was observed. No damage resulted from the intense radiation estimated to be 4 MW/cm^2 after 10 min of irradiation. Results show high-purity single-crystal silicon was suitable for use with high-intensity NIR lasers pulsed or CW.

9.6 Single-Crystal Silicon Fibers

There was great interest in developing flexible infrared fibers to be used with these high-intensity lasers. Silica fibers were extremely low in absorption in the visible region, but because of water absorption cut off before 2 μm. Attention shifted to fluoride glasses with some success. Polycrystalline materials were extruded but suffered from light scatter and breakage at grain boundaries when bent.

Single-crystal fibers had been drawn from sapphire using a carbon dioxide laser as a heat source.[10] The results were encouraging. At about the same time, AMI had been working to produce flexible single-crystal silicon fibers[11] using the float zone technology. The feed crystal was reduced to an 8-mm diameter, the seed diameter was reduced to 2 to 4 mm, and the RF coil reduced to a single turn with a 1.5-cm ID. Linear growth rate was increased and top or bottom spin reduced or eliminated. Single-crystal rods were grown 1 to 2 mm in diameter and 30 to 40 cm long. The sides were irregular due to spin or facet formation. Figure 9.13 shows, in the top photograph, typical small-diameter single-crystal silicon rods. A pencil is included in the photo for comparison. The lower photograph shows a rod being grown from the top of a feed crystal.

Besides being grown with the seed in the <111>, the <100> and <110> orientations were tried in hopes of eliminating the facet effect so the surfaces would be smooth. None seemed better than the other.

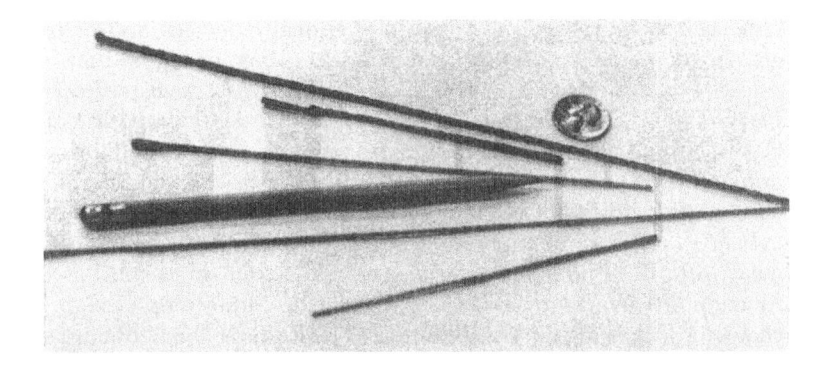

FIGURE 9.13 Small-diameter single-crystal silicon rods and rod growth in process.

When AMI had a rod 1.5 to 2 mm in diameter and 8 cm long tested at Quantronix, 50 percent of the energy was lost by scatter due to the lack of a smooth surface. As evidence of that fact, transmission varied greatly with angle of incidence of the beam relative to the face of the rod. Rods were etched to reduce diameter and smooth their outside surfaces. The quality of the etched surface was affected by seed orientation. Even so, rods in this diameter range had, on average, a bend-to-break radius of about 1.75 in. There was wide variation depending on the seed orientation. The silicon fibers were not flexible. Absorption and resistance to the intense radiation test were passed. Figure 9.13 shows typical results for single-crystal grown silicon rods.

9.7 Gallium Arsenide as an Infrared Optical Material

The effort to develop the compound semiconductor gallium arsenide (GaAs) began in the late 1950s at TI. As mentioned earlier, George Cronin was deeply involved in the effort from the beginning. The idea was to replace silicon for high-frequency devices and enjoy greater temperature stability due to a larger bandgap than that in silicon. Production of GaAs in purity and quality as a device material that could be fabricated went on for years. The technological difficulties associated with crystal, growth, preparation, and device fabrication had been greatly underestimated. The author was only slightly involved as part of his assignment in the analytical chemistry group as head of physical methods applied to semiconductor materials, e.g., measuring the photoconductivity associated with impurities as a function of wavelength in GaAs crystals. The author measured and characterized, using the Perkin Elmer 13 instrument, the light output from the first TI gallium arsenide light emitters. From the measurements, conversion efficiency of current to light was calculated for the first time. Several TI GaAs transistor task force teams were organized and operated over several years, trying to solve all the production problems.

The author first became interested in GaAs as an infrared optical material during the high-energy laser window programs[12,13] of the 1970s when GaAs became considered as an optical material rather than used only for electronic devices. On the periodic chart, Ga is on one side of Ge and As is on the other side. GaAs is said to be *isoelectronic* with Ge. Many of the physical properties of the two materials are almost identical. As one of the candidate window materials, GaAs suffered[14] from not being produced in large enough size, high enough purity, low enough cost and was shown to be subject to high-speed rain erosion damage. Over silicon and germanium it had an upper use temperature of 400°C. At TI in the 1960s, George Cronin and his colleague Bob Haisty[15] developed a method based on chromium doping to insure gallium arsenide would be semi-insulating. Cronin and Haisty were issued a U.S. Patent covering the use of chromium doping. The author and colleague Charlie Jones, developed an absorption method[16] to measure chromium levels in GaAs using the Perkin Elmer 350 instrument. Later a method for measuring the depth of mechanical damage in polished GaAs surfaces was developed[17] using the far infrared Perkin Elmer 301 spectrophotometer. When still at TI, the author and George Cronin started a program to begin casting GaAs plates. This effort led to a TI proposal with the help of Bob Crossland, an optics manager in the TI EO division, submitted to the Air Force Avionics Laboratory at Wright Field. The program was based on a simple casting plate approach. Results for the first phase were reported in 1975.[18] Harold Hafner of TI worked with George Cronin on the program. The author was not involved directly, only in supplying the infrared refractive index data. The program continued into 1977 when it ended after the Airforce Materials Laboratory chose the Raytheon ZnSe/ZnS window as the Airforce Avionics Window. The author took one last trip in 1977 for TI to Wright Field in an unsuccessful attempt to save the TI program. Later, work on casting GaAs windows continued at TI for several years.[19] The TI process involved placing precompounded material in a large round chamber, of up to 30-in diameter, containing a plate mold. The chamber was filled with an inert gas (usually argon) under pressure. The GaAs was heated to the melting point and allowed to form a homogeneous melt. Sometimes the melt was covered with boron oxide to suppress the loss of arsenic. Cooling was slow to encourage large-grain formation.

9.8 Production of GaAs at AMI

While in the process of building up the crystal silicon growth capability, AMI also acquired equipment used for GaAs growth. The expertise of George Cronin was used in getting production started. The first unit was a surplus Czochralski TI silicon puller modified for growing GaAs crystals. The method used was designated LEC (for liquid-encapsulated crystals). The method is by far the most developed and widely used today. A diagram depicting the method is shown in Fig. 9.14. First, GaAs has a melting point of 1238°C which occurs under a pressure of 0.7 atm of As vapor. Arsenic sublimes at 613°C and gallium boils at 2237°C. The chamber is sealed except for an opening where an inert gas is allowed to enter. The gas is either helium or the less expensive argon. The boron nitride (BN) crucible is loaded with the required gallium and/or mixed with GaAs scrap. Pure boron oxide (B_2O_3) is placed in the top of the crucible. The required mass of arsenic is placed in the top of the chamber. The gallium essentially is not volatile. When the chamber is heated, the gallium in the boron nitride (BN) crucible is heated, and the heated arsenic produces vapor that combines with the gallium and continues to be absorbed

232 Chapter Nine

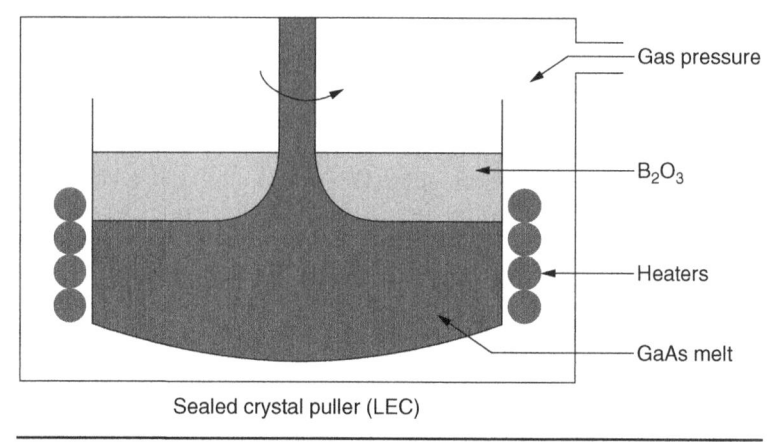

FIGURE 9.14 The Czochralski sealed crystal puller LEC method.

by the melt during the compounding process. When the material in the crucible reaches the melting point of pure GaAs, stoichiometry is complete under an atmosphere of over 500 mm of arsenic pressure. As shown in Fig. 9.14, the melt is covered with a layer of liquid boron oxide the purpose of which is to suppress the vaporization of arsenic from the melt. The compounding process is very slow, taking hours. The next step is to grow the crystal. A rotating seed crystal is inserted through the melted B_2O_3 and makes contact with the GaAs melt, and the growth begins. From start to finish, the process is very difficult and may take more than 12 h. After cooling, the grown crystal is covered with a layer of boron oxide that is easily removed. The AMI operator of this process, Bob Harp, a former colleague of George Cronin, retired from TI, was very experienced. Crystals with 3- to 4-in diameter were grown in this machine, smaller than what was currently produced and used for devices in the industry, but useful for small infrared windows and lenses. The process was expensive and less efficient if not run every day.

Another unit, referred to as a *vertical Bridgeman*, was built by and purchased from a local company. A diagram of the unit is shown in Fig. 9.15. The vertical Bridgman contains a melt in a quartz crucible or one made from BN, in a vertical position. The top chamber holds the required arsenic for compounding and pressure control. While the chamber is spinning, the heater is moved up so that cooling and growth proceed from the bottom up. A seed can be placed in a receptacle in the bottom of the crucible to produce a single crystal of a specified orientation. Instead of moving the heater, a method developed later by Kelly Burke and Russell Kremer, of Crystal Specialties,[20] controlled the energy in each turn of the heater individually. In this way, the thermal gradient can be moved up the crystal without any physical movement disturbing the melt, producing a crystal of

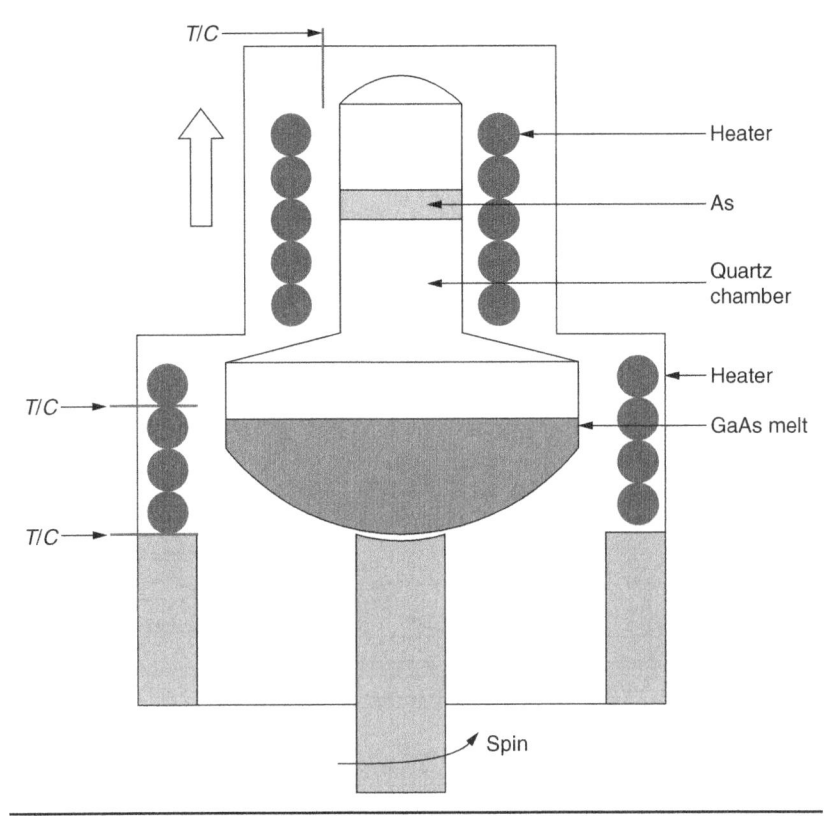

FIGURE 9.15 Vertical Bridgman crystal grower.

improved crystal perfection. Again the crystals produced are not large in diameter. A comparison of the advantages of all the different methods from a device standpoint was also discussed in their paper.[20]

9.9 Horizontal Bridgman Production of GaAs Plates at AMI

AMI purchased a horizontal Bridgman machine from Morgan Semiconductor in Garland after the company was bought out. The most experienced operator from Morgan, Dale Welt, came along with the machine and joined AMI. The machine was very heavy and 12 ft long. The mounting was direct to the concrete floor. Figure 9.16 illustrates the horizontal Bridgman method for growing GaAs crystals. The top diagram is a simplistic illustration of the horizontal Bridgman GaAs crystal growth process. In the connected two evacuated quartz chambers, there is elemental arsenic on one side and elemental gallium in a boat. The remoteness of the arsenic chamber allows accurate control of the arsenic vapor pressure. The boat is usually one-half of a cylinder

234 Chapter Nine

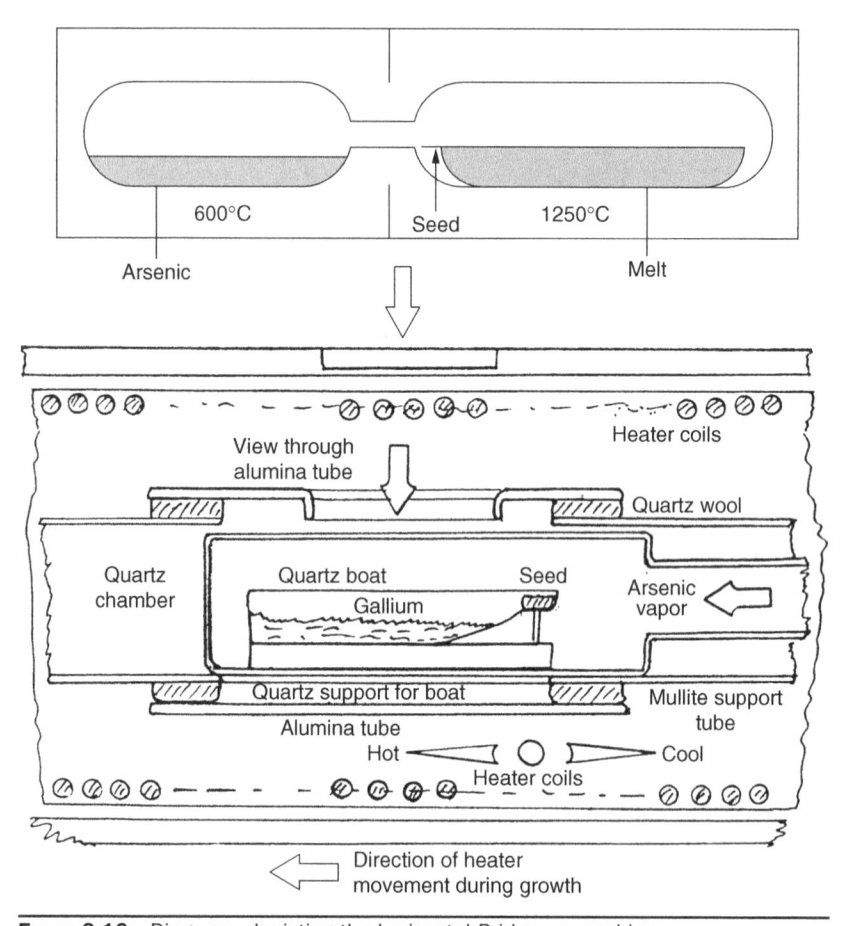

FIGURE 9.16 Diagrams depicting the horizontal Bridgman machine.

of quartz in shape and lined with quartz cloth. The boat may be made from boron nitride (BN) to avoid Si as an impurity in the melt. AMI purchased an expensive BN boat 4 in × 10 in × 1 in for plate production. The process simply described is thus: as arsenic is heated, it sublimes into the chamber containing the gallium and is absorbed, reacting with the gallium. The gallium temperature is slowly raised to the melting point of GaAs, 1238°C, and the arsenic continues to be absorbed until the compounding is complete. A small amount of arsenic in excess maintains the required vapor pressure that also prevents the collapse of the quartz chamber on the melt side. When compounding is complete, the heater begins to move, slowly lowering the melt temperature beginning at the seed end. Thus, the crystal growth is controlled at the melt-solid interface. This fact is very important. In the plate casting process, control at the liquid-solid interface is not possible, so the end plate may have very large grains but they are randomly oriented. We experienced that same problem in producing plates of CdTe. Since gallium is not volatile, only the arsenic pressure must be in equilibrium with the stoichiometric crystal at the melting point. For CdTe, both Cd vapor and Te vapor must be in equilibrium with the stoichiometric crystal at the melting point. Extremely small crystallites of As in GaAs may not appreciably affect ir transmission measured with a spectrophotometer. However, their presence may be verified by measuring transmission with a CO_2 laser. The inclusions heat up locally, causing an immediate drop in infrared transmission. The bottom drawing in Fig. 9.16 illustrates exactly how the real machine looks. The quartz chambers are placed inside an alumina tube for support. The alumina tube rests inside a very long Mullite tube supported at both ends. The furnace is much larger in diameter. Operator viewing windows are cut in both the furnace wall and the alumina tube. The operator watches the compounding process, and when it is complete, he or she raises the crystal end of the Mullite tube so the melt contacts the seed. Growth begins then as the heater is moved slowly, so the growth moves down the boat away from the seed. Crystal length is limited by the total travel for the unit of 18 in.

Material grown in the AMI machine was fabricated into small wafers for use as substrates in the production of second-generation night vision goggles produced by the local company NYTEK in Garland. AMI had a computer-controlled instrument designed and built by Steve Boldish, formerly of NYTEK, to make Hall measurements to verify the resistivity of single-crystal, polished substrate slices. As part of the effort, GaAs slices undoped, lightly doped, doped moderately, and heavily doped were measured to establish resistivity as a function of doping level for AMI GaAs. Some of the more than 50 samples used to form this plot were from outside vendors, which allowed direct comparison of AMI results with those of other suppliers. The agreement was good. The results plotted in Fig. 9.17 for N-type Si doped GaAs cover 12 orders of magnitude in carrier concentration and 7 orders of magnitude for resistivity. Note the plot is almost a straight line until it reaches a carrier concentration of 10^{16} where the electron mobility begins to decrease. This plot was used to guide AMI in producing Si doped GaAs of specific resistivity, which was very important later in trying to produce windows to be used with infrared systems in aircraft while also providing EMI protection.

After the first Desert War, attention at AMI was turned again toward using GaAs as an infrared optical material. First, the ZnS/ZnSe avionics window did not survive well in the desert environment. Second, the infrared system in the U.S. tanks lost sensitivity in the high temperatures of the desert due to thermal-generated carriers in the germanium optics. The systems used in the tanks were complex with many lens elements. The total optical path for germanium was

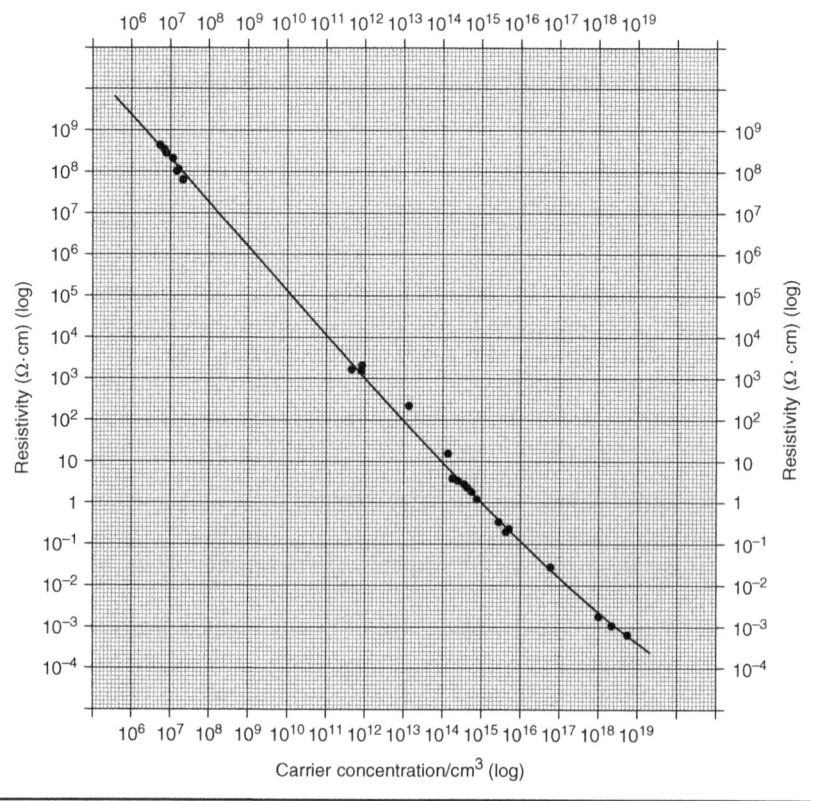

FIGURE 9.17 Gallium arsenide resistivity versus carrier concentration N-type room temperature.

centimeters. Not only was signal lost by absorption when the germanium was heated, but also the system lost focus because of the large index $\Delta N/\Delta T$ for germanium heated above 25°C of $450 \times 10^{-6}/°C$. The use of GaAs in place of Ge in the tank system would improve performance when used in the desert because of the higher use temperature and a lower $\Delta N/\Delta T$ above 25°C of $207 \times 10^{-6}/°C$. Both of the $\Delta N/\Delta T$ values were measured by the author while still at TI.

AMI decided to direct its efforts toward developing a method to produce a plate of GaAs large enough to serve as an avionic window. It was obvious that the best chance for success for AMI was to use its horizontal Bridgman unit. The diameter of the quartz chamber that housed a much larger plate mold would have to be increased. Figure 9.18 shows a photograph of the quartz chamber with the arsenic on the left and the large plate mold containing the gallium on the right. The diameter of the quartz chamber was increased to accommodate a plate mold 12 in long and 6 in wide. Glen Whaley of AMI had routinely fabricated chambers with 8-in

FIGURE 9.18 Quartz chamber loaded for GaAs plate growth.

diameter daily at that time for 15 years. Even some chambers with 10-in diameter had been produced. Larger chambers, a larger-diameter heater, and greater power would be required for scale-up as the need for larger windows arose. AMI estimated growth of plates 8 in × 16 in were possible with this machine. The left photograph of Fig. 9.19 shows an AMI plate 4 in × 10 in × 0.5 in after removal from the chamber. The right photograph shows Dale Welt holding the plate after it has been ground flat and is ready for polishing.

FIGURE 9.19 Rough as grown GaAs plate (right) after being ground flat for polishing.

AMI accepted an order from McDonald Douglas in Kansas City, on a best-effort basis, to provide four GaAs plates 4 in × 10 in × 0.5 in thick. The BN boat was used for all four plates. The two undoped plates were prepared without any difficulty, ground, and polished, and infrared transmission measured 2.5 to 25 μm at six points down the plates with no appreciable variation point to point. The other two plates were to be doped to a resistivity of 1 to 10 Ω·cm to be tested relative to EMI protection. The doping was accomplished by adding to the melt silicon heavily doped Hall samples that had been evaluated. The measured resistivity of the samples determined the Si atoms per gram in each sample. Thus the amount of Si required for a concentration corresponding to the desired plate resistivity was found using Fig. 9.17. The doped samples were weighed and added to the melt at the start. The concentration corresponding to 1 to 10 Ω·cm was low, 10^{13} to 10^{14} per cubic centimeter. The resistivity of both plates after grinding and polishing was measured using the ASTM conductivity method applied by AMI in evaluating germanium bars for the GSA purchases, during the government stockpiling effort and the evaluation of float zoned silicon crystals. Changes in resistivity along the bars were evident on both plates. Silicon was not evenly distributed. Unfortunately, the plates were polycrystalline with grain boundaries running in the direction of growth. Contacts tended to be nonohmic. The readings were different with current in the direction of growth compared to current across the plates. Values taken for the entire plate were unreliable. Silicon was a bad choice as a dopant because of its distribution coefficient, Se or Te would have been better. A discrepancy was found between the measured resistivity and the absorption in the infrared and RF measurements made by McDonald Douglas. To better understand the discrepancy, two other doped plates were made. Strips 2 in wide were cut from the plates, and indium contacts were hot soldered to all four polished edges using an ultrasonic soldering iron. Conductivity was measured at various currents in both directions with and across the grains until they all agreed. The ir transmission at 10.6 μm was measured in several locations in each plate, and the absorption coefficients were calculated and compared to experimental values for different carrier concentrations found in the literature.[21] Figure 9.17 was used to convert the carrier concentrations to resistivity. The results are shown in Fig. 9.20 in a log-log plot that covers 5 orders of magnitude in absorption and resistivity. In the figure, the crosses are the literature values while the circle points are the AMI experimental values from the plates. Figure 9.20 may be used to measure the dopant uniformity in GaAs plates provided the resistivity is 10 Ω or less so that the 10.6-μm absorption is appreciable.

The EMI measurements made on delivered plates to McDonald Douglas indicated the AMI resistivity values based on conductivity were 4 times too high. The AMI results indicated a resistivity rise front to back of 1 to 2 Ω·cm. The new AMI ir measurement method indicated

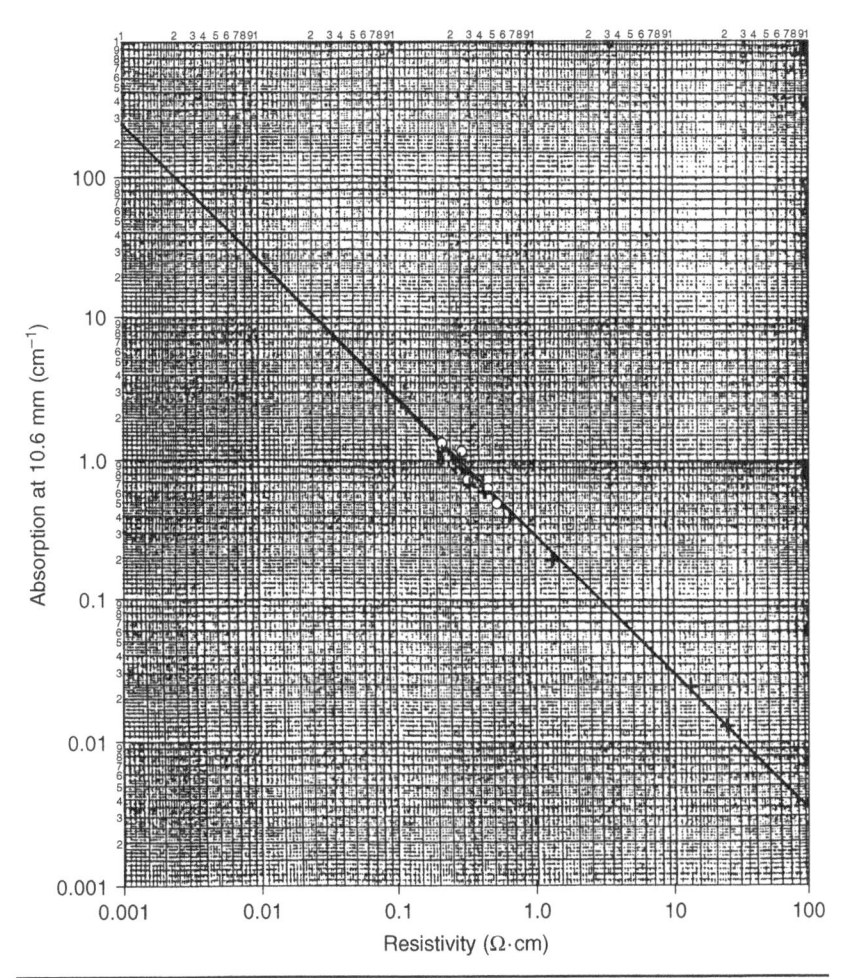

FIGURE 9.20 Free carrier absorption at 10.6 μm in N-type gallium arsenide as a function of resistivity.

a decrease in resistivity front to back of 1 to 0.35 Ω·cm in the 8-in plate length. Such a fall is consistent with an increase in dopant concentration as a melt is slowly frozen down the length of the plate, such as found in the zone refining purification process of high-purity materials.

Classical methods[22,23] may be used to calculate the absorption by infrared, the long wavelengths of microwave and radio frequencies. Such methods have been used before by the author[18] in dealing with the depth of mechanical damage in GaAs. The classical expressions all show the absorption process increasing with the wavelength squared. However, for GaAs the value is reported[24] to be 3 not 2; the effect is much stronger than predicted with classical expressions. When our measured values were reexamined, we found our calculated power value was 2.8, not quite 3.

Using an expression by Moss[25] that allows for changing the wavelength dependency, we were able to estimate values of about 67 percent of observed values in contrast to only 40 percent for classical expressions. The conclusion reached was that a GaAs window doped to 10 Ω·cm would provide sufficient EMI protection with little effect on the 8- to 12-μm transmission.

To demonstrate the merits of forming large plates using the horizontal Bridgman, AMI made one last plate 12 in × 6 in × 1 in, undoped and high-resistivity. The beginning of transmission started at about 0.8 μm and was fully transmitting by 1.5 μm. Transmission at the Yag 1.064 μm was 40 percent uncorrected; with antireflection coating it would be 73 percent. The calculated absorption coefficient was 0.48 cm^{-1}. The plate would be a very good window 1.5 to 14 μm and still useful for the Yag laser. Transmission values, measured at six locations, of 2.5 to 25 μm were identical with no absorption over the range except for a small intrinsic absorption occurring at 13 μm.

AMI decided to attack the other point of criticism directed at GaAs— resistance to rain erosion. Earlier, it was mentioned that silicon had the best physical properties for an infrared window material next to sapphire. One criticism leveled against the material was a weak (0.8 cm^{-1} absorption coefficient) broad intrinsic band occurring at about 9 μm. Silicon containing oxygen as an impurity places an additional absorption band peaked at 9 μm due to the presence of oxygen as an impurity. The SiO absorption may add an additional amount of absorption, reaching a total value of 8 cm^{-1} at the oxygen saturation level of 2 × 10^{18} atoms. However, as pointed out earlier, silicon may be made oxygen-free by vacuum float zoning, and thus only the weak intrinsic band would not be much of an absorption problem for silicon in thin pieces, such as 1 to 2 mm. AMI decided that a GaAs window could be protected from rain erosion damage if it was coated with a silicon layer. Also, common semiconductor techniques could be used to provide conducting grid patterns in the surface useful for EMI protection or even for window deicing. The concept of protecting GaAs windows with a layer of silicon is contained in U.S. Patent 5,194,985 granted to AMI in 1993. Samples were prepared by Bill Herrmann, Jr., of OEC in Dallas. The samples were coated with a 3-mil silicon film. The tests results reported by Des Gibson of Pilkington in the United Kingdom were as follows:

	Damage Threshold (Velocity to Show Damage)
Untreated GaAs	130 m/s (291 mi/h)
Silicon-protected GaAs	210 m/s (470 mi/h) + 162% improvement hardness
Ratio of measured silicon/GaAs	$\frac{950}{575}$ = +167% hardness

The improvement appears to be related to their relative hardness.

John Hall of the U.S. Army Night Vision Laboratory at Ft. Belvoir became a driving force for the use of GaAs in place of germanium in the infrared systems in the Army tanks. He organized a meeting to consider nongermanium optical materials for 8- to 12-μm FLIR production. Of particular importance were the optical requirements for the Horizontal Integration GEN II FLIR program scheduled to supply thousands of M1-Abrams, M8-AGS, and M2 Bradley tanks with second-generation FLIRs. The meeting was held at Ft. Belvoir in May 1994. Attendees included TI (GaAs), Eagle Picher (Ge), Morton International (ZnSe-ZnS), Exotic Materials (Ge), and AMI (GaAs).

Presentations were made describing materials produced and advances made recently. The reader has already learned what AMI presented.[25] The TI presentation by Paul Klocek was very extensive, demonstrating the amount of effort expended and progress made over the last several years. TI enthusiasm for the change to GaAs seemed lacking. The situation in optical Ge and Si production by Dennis Thomas of Eagle Picher was positive. The Morton presentation by Dennis McAllister for ZnSe and ZnS production covered all the widely used CVD materials. Design and fabrication were discussed. Hall talked about new designs using GaAs including some he had already made and tested. His conclusion was that GaAs will prove to be very useful for Army FLIR applications and can dramatically improve the ability of the HTI Gen II FLIR to meet range and bore sight performance in high-temperature environments by reducing the total amount of Ge used. John Hall's words were correct, but resistance at high levels of decision authority led to continued use of Ge optics in all HTI Gen II FLIR systems.[26]

Hall was successful in codeveloping with personnel from Hughes El Segundo an HTI-compatible, GaAs-based imager which is still used by the Night Vision Lab for special testing. Hall eventually left the Night Vision Lab and joined the private sector. He continues to find applications for GaAs optics in the new dual-band "Gen III" sensor systems. Hall has already delivered prototype systems to the Night Vision Lab that have been used for very successful demonstrations and data collection since 2006.[27]

Meanwhile, AMI found great interest on the part of designers and system engineers at Lockheed Martin in Orlando to use GaAs in their upgrade of the TADS PNVS helicopter system. Working with senior optical designer Al Lyon and optics production manager Bob Icovazzi of LMCO, AMI began supplying small amounts of GaAs for prototype designs using the AMI units. Finally, in late 1994, AMI on a best-efforts basis agreed to attempt to grow a single-crystal plate in which the flat faces would have the desired crystalline orientation, <100>, as determined by a seed placed in the melt. The purpose was to produce an outside surface less susceptible to rain impact damage. Obviously, use of the horizontal Bridgman was

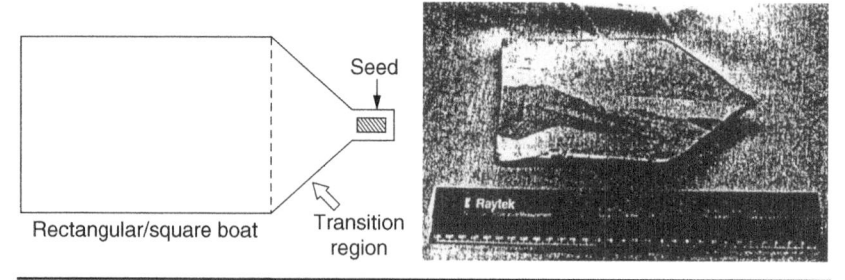

FIGURE 9.21 Boat with seed used to form GaAs face-oriented plate.

the only chance for success. Placement of the seed relative to the melt was critical. The first two attempts failed while the third was partially successful. The left drawing in Fig. 9.21 shows the design of the quartz boat used. The right photograph shows the plate after it was removed, ground, and polished. Note the first half from the seed is almost all single with a twin line extending from the neck, causing the lower half to be poly. A sample cut from the single area evaluated by X-ray failed to confirm the <100> orientation. The conclusion reached was that the horizontal Bridgman in present form would not perform the desired oriented single-crystal growth. Fundamentally, the modification of the equipment and procedures would lead to eventual success.

As the LMCO design advanced, the shape of the required window changed to that of a dome, not flat, and the size increased. The requirement was right at the edge of our capability. Glen Whaley designed and fabricated a quartz chamber we all felt would produce a GaAs blank suitable for fabrication of the dome-shaped window. We recognized that there was risk involved. Because of the importance of this program, we looked for a better solution.

Because of the good relationship Greg Whaley had built up with Bob Ochrym, head of sales at Litton/Airtron, the largest supplier of GaAs in the United States, they agreed to supply AMI with an extremely large crystal. It had twinned during production so the price was modest. The crystal had an 8-in diameter and weighed 25 kg. The arrangement included the head of Airtron, Dave Miller, and head of production, Larry Urick. AMI ground and polished the crystal and evaluated its quality. The crystal then went to Exotic Materials for fabrication and coating.

AMI has continued its relationship with the big producers and still serves as a supplier of GaAs to Lockheed Martin for its Arrowhead helicopter program. All material is 100 percent checked for quality by AMI. Required blanks are polished, ir transmission is measured, and tested by laser transmission for the presence of precipitants.

References

1. Marshal Sparks, "Theoretical Studies of High-Power Infrared Window Materials," Contract DAHC15-73-C-0127, 1973.
2. Stanley K. Dickinson, "Infrared Laser Window Materials Property Data for ZnSe, KCl, NaCl, CaF_2, SrF_2, BaF_2," *Air force Cambridge Res. Lab. Projects* 5620, 3326 (1975).
3. Charles S. Sahagian and Carl A. Pitha, "Compendium on High Power Infrared Laser Window Materials," LQ-10 Program, AFCRL-72-0170, 1972.
4. A. R. Hilton, "Single Crystal Cadmium Telluride High Energy IR Laser Windows," Contract No. F33615-71-C-1762, 1972.
5. W. A. Tiller, "Theoretical Analysis of Requirements for Crystal Growth from Solutions," *J. Crystal Growth* 2, 69 (1968).
6. D. de Nobel, "Phase Equilibria Semiconducting Properties of Cadmium Telluride," *Phillips Res. Rep.* 14, 361–399 (1959).
7. L. R. Shiozawa, J. M. Jost, and D. A. Roberts, "The Application of Vapor Deposition to Semiconductor Materials for Use as High Power Infrared Windows," Contract No. F33615-71-C-1777, Gould Laboratories, 1973.
8. A. Ray Hilton and Charlie Jones, *J. Electrochem. Soc.* 113, 472 (1966).
9. F. A. Harrigan and T. F. Deutsch, Raytheon Contract DAAH 01–70-C-1251, 1970.
10. D. H. Jundt, M. M. Fejer, and R. L. Beyer, "Growth and Optical Properties of Single-Crystal Sapphire Fibers," *SPIE* 1048, 39 (1989).
11. A. Ray Hilton and James McCord, "High Purity Single Crystal Silicon Fibers for Near Infrared Applications," *SPIE* 1048, 32 (1989).
12. Carl Pitha (Ed.), *Conference on High Power Infrared Laser Window Materials*, Air Force Cambridge Laboratory, November 1972.
13. Carl Pitha (Ed.), *Conference on High Power Infrared Laser Window Materials*, Air Force Cambridge Laboratory, November 1973.
14. R. W. Tutison and R. L. Gentilman, "Current and Emerging Materials for LWIR External Windows," *SPIE* 968, 65 (1988).
15. G. R. Cronin and R. W. Haisty, "The Preparation of Semi-Insulating Gallium Arsenide by Chromium Doping," *J. Electrochem. Soc.* 111, 874 (1964).
16. Charlie E. Jones and A. Ray Hilton, "Optical Absorption in Chromium Doped, High Resistivity GaAs in the 0.6 to 1.5 ev Range," *J. Electrochem.Soc.* 118 (1966).
17. Charlie E. Jones and A. Ray Hilton, "The Depth of Mechanical Damage in Gallium Arsenide," *J. Electrochem. Soc.* 112 (1965).
18. Harold Hafner and George Cronin, "Development of Gallium Arsenide Infrared Windows" Contract F33615-74-C-1066-AFML, February 1975.
19. M. J. Brau, L. E. Stone, and W. W. Boucher, "Gallium Arsenide Infrared Windows for High Speed Airborne Applications," *SPIE* 297, 44 (1981).
20. Kelley M. Burke and Russell E. Kremer, "Development in Gallium Arsenide Crystal Growth Technology," *J. Monolithic Technol.*, July 1988.
21. D. Bois, P. Leysal, and C. Schiller, "Free Carrier Absorption at 10.6 µm in GaAs," *J. Electronic Mat.* 5, 275–286 (1976).
22. Margaret Kohin, Steven Wein, Jennifer Traylor, Richard Chase, and Judith Chapman, "Analysis and Design of Transparent Conductive Coatings and Filters," *Opt. Engg.* 32, 911 (1993).
23. B. Jensen, "The Quantum Extension of the Drude-Zener Theory in Polar Semiconductors," *Handbook of Optical Constants of Solids*, Edward Palik (Ed.), Academic Press, New York, 1985, Chapter 9, p. 169.
24. H. Y. Fan, "Effects of Free Carriers on Optical Properties," *Semiconductors and Semimetals, Optical Properties of III-V Components*, Academic Press, New York, 1967, vol. 3, Chapter 9.
25. T. S. Moss, *Optical Properties of Semiconductors*, Academic Press, New York, 1959, p. 30.
26. A. Ray Hilton, Sr., "Doped Gallium Arsenide External Windows," *SPIE* 2286, 91 (1994).
27. J. Hall, J. Vizgaittis, J. Miller, and Dan Berube, "Third Generation FLIR Demonstrator," *SPIE* 6940 OS (2008).

CHAPTER 10
Early Work at Texas Instruments

10.1 First Job

After finishing his Ph.D in 1959, the author was with the drilling and production research division of Humble Oil Company (now Exxon) in Houston, Texas. The problem assignment was to develop a process to turn loose unconsolidated oil sand surrounding the bottom of shallow wells into porous permeable rock using epoxy resins. Loose sand flowing into the well damaged the pumps and was a disposal problem. During the year at Humble Oil, while the author was working with Horace Spain, a suitable process was developed as described in U.S. Patent 3,100,527. The process was used successfully in the field.

10.2 Infrared Applications to Materials

After one year at Humble Oil, a visit by the author to Texas Instruments led to a move to Dallas in June 1960. The Materials Research Laboratory was headed by Tom Burkhalter, a former chemistry professor at Texas A&M and a family friend. A new fuel cell program was just starting, and the job to build a fuel cell operating in the lab by fall was given to the author. The program manager furnished two books, one from the British describing 7 years of research to develop a high-temperature molten salt fuel cell and the other from the Dutch describing 9 years of a similar effort. Both groups had stopped their programs, citing sound technical reasons why the high-temperature molten salt fuel cell was impractical. Nevertheless, the program manager insisted on following their same path. When asked why he thought we would succeed while they failed, he replied that we at TI solve material problems others can't. The answer was, "Baloney."

Through lots of hard work by program personnel, the fall goal was met. The cell used a porous magnesium oxide disk made from hot pressed powders as the membrane. A ceramic process developed by Pete Johnson, a Ph.D. ceramicist, produced the disk. Silver coatings on each side served as the fuel electrode and the oxidizing electrode. The electrolyte filling the pores in the MgO disk was sodium-lithium-carbonate. The cell body was stainless steel. Fuel used was methane or hydrogen with O_2-CO_2 in the other chamber. To prove the cell worked, a child's toy electric airplane was mounted in the lab with the propeller powered by the fuel cell continuously. Methane (CH_4) is the most stable organic molecule from a thermodynamics standpoint. To generate reasonable output while utilizing hydrocarbon fuels, 800°C operation was required. Components would not last over time at such high temperatures. The replacement of the MgO disk to reduce internal resistance and increase output in an improved "screen cell" version was conceived and tested by the author as described in U.S. Patent 3,251,718. It was obvious that without a different fuel cell approach, the program would last only as long as the government funding was available, much like the conditions that exist today.

In 1961 at author request, the first on and first off fuel cell program, was transferred to the Central Analytical Chemistry Facility under Phil Kane. The job opening was for the application of infrared spectroscopy techniques to semiconductor materials. The instrument available for use was the Perkin Elmer 13 U, a research-type instrument at that time. It was modified so that absolute reflection from the surfaces of semiconductor materials could be measured with polarized light or plain as a function of angle of incidence and monochromatic wavelength. The photoconductivity of semiconductor materials could be measured at different wavelengths and correlated with impurity analysis. The instrument also served as the source of monochromatic light for the infrared refractometer attachment used for index measurements as discussed earlier.

The first material evaluated was beta silicon carbide film grown in a process at TI. The question was, Is it a good infrared optical material? The infrared evaluation of SiC films showed the answer was no. Another investigation concerned the effect on the optical and electrical properties of the oxygen impurity in silicon. The magnitude of the absorption coefficient at 9 μm was correlated with the absolute concentration as measured by other means. Another materials research program active at that time under Louis Bailey was finding new thermoelectric materials. One studied very extensively was silicon telluride, Si_2Te_3. The material had a layered mica structure and was transparent in the infrared. It was while studying this material that Werner Beyen of TI learned that the physics branch of the Office of Naval Research in Washington was interested in funding work for the development of new infrared optical materials. Werner Beyen once had worked for the Office of Naval Research. The author was given the assignment by Tom Burkhalter to develop a proposal to respond to this request. In the TI library a copy of the 1959 proceedings of the Russian Ioffe Institute[1] was found that covered papers describing amorphous semiconductors. The institute in Leningrad (St. Petersburg again now) was headed by

Boris Kolomiets. After a reading of their papers, it seemed obvious that a glass could be formed out of Si_2Te_3 if we only added arsenic as a third element. While the author was working with Maurice Brau, the ternary glass forming composition region for Si-As-Te glasses was established.[2] As mentioned earlier in this book, these results were the basis for the proposal and resulted in the many years of support by DARPA through the Office of Naval Research for the chalcogenide infrared glass efforts at TI. The discussion need not be repeated again here.

10.3 Optical Interference and Film Thickness

Many times the success or failure of a technical program depends upon the ability to measure accurately the value of a particular system parameter. The measurement technique may be totally new to the activity of the company or may require development for the particular application. This new capability or ability may become very valuable to the company's technical effort in a short time. Such was the case at TI in the early 1960s as, similar to other semiconductor device producers, they began growing pure silicon layers on highly doped silicon substrates and then grew dielectric films to protect the electronic devices. The two new techniques were the measurement of the thickness of epitaxial layers by the infrared scan of reflected light and the thin dielectric film thickness measurement using ellipsometry. In this chapter, the general principles behind both techniques are discussed along with how they were modified for their applications to two semiconductor problems.

Optical interference occurs when light is reflected from a film-covered surface. As shown in Fig. 10.1a, light intensity I_0 strikes the surface of a film and a fraction of the light is reflected I_1 at the incident angle. The reflected ray I_1 has a phase angle θ_1. The remaining part of I_0 is refracted into the film, partially reflected at the film-surface interface and partially transmitted at the film-air interface as the refracted beam I_2. The phase angle of I_2 (θ_2) is different from that of I_1 (θ_1) because of the phase lag that occurs as the light travels the extra distance through the film and back to the same point as I_1. In Fig. 10.1b, the path difference between light rays I_1 and I_2 is the optical distance $X = 2d \cos \theta_1$, where d is the film thickness and θ_1 is the angle of refraction calculated from Snell's law:

$$n_0 \sin \theta_0 = n_1 \sin \theta_1$$

The refractive indexes n_1 and n_2 are measurements of the velocity of light in medium 1 and medium 2 relative to the velocity of light c in a vacuum:

$$n_1 = \frac{c}{v_1} = \frac{\lambda}{\lambda_1}$$

where v_1 is velocity of light in medium 1, λ_1 is the wavelength in medium 1, and λ is wavelength in air.

Chapter Ten

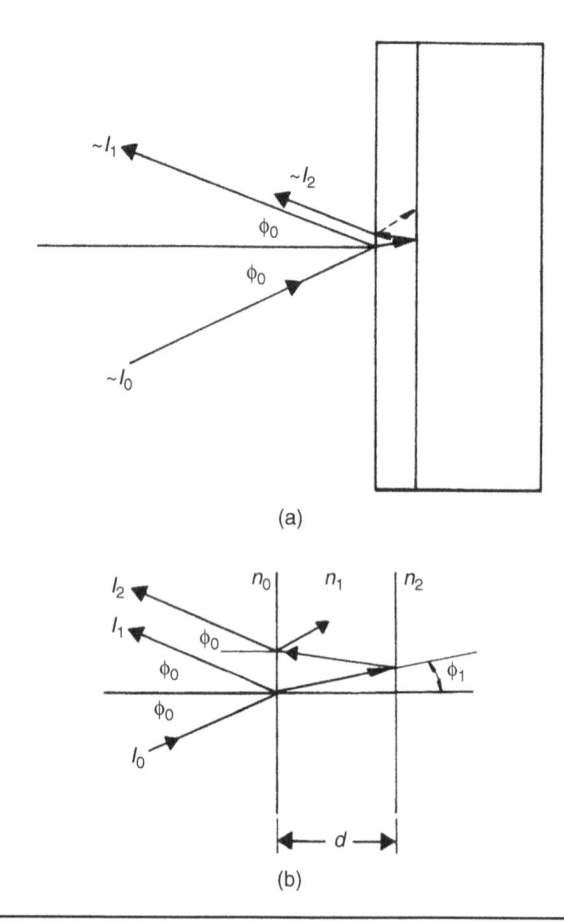

FIGURE 10.1 Optical interference due to reflection of light from a film-covered surface.

The phase lag δ between I_1 and I_2 can be calculated from

$$\delta = \frac{2\pi}{\lambda_1 \times d\cos\theta_1} = \frac{2\pi}{\lambda \times 2n_1 d\cos\theta_1}$$

The two rays I_1 and I_2 combine to produce a resultant light ray of the same frequency but of different intensity and phase angle than either I_1 or I_2. The resultant reflected ray changes in intensity from maximum to minimum values as the value of λ changes. Maximum intensity occurs when $\delta = \pi$, 3π, 5π, or $2\pi(m + \frac{1}{2})$, where m is the order of interference.

10.4 The Infrared Scan Technique for Epitaxial Film Thickness

The technique is applicable because the near pure epitaxial silicon layer is grown on a heavily doped silicon substrate. The refractive indices for layer and substrate may be very close in the short-wavelength

(< 1-μm) region but become quite different in the infrared as the wavelength increases. The same statement is true for other semiconductor processes such as Ge on Ge or GaAs on GaAs. For N-type silicon, the absorption by conduction electrons increases by the square of the wavelength.[3-5] The result is strong reflection at the layer substrate interface without which optical interference would not occur. Reflectivity R at the interface between two dielectric materials with index values of N_2 and N_1 is given by

$$R = \frac{(N_2 - N_1)^2}{(N_2 + N_1)^2}$$

However, for a material with strong absorption the index becomes a complex number $N = N - ik$, where k is the imaginary part; the extinction coefficient calculated from $\alpha = 4\pi k/\lambda$ with α in cm^{-1} units is the linear absorption coefficient calculated from transmission measured at the wavelength λ. In general, working with transparent materials, k is so small it is ignored. But in this case, calculation of R at the silicon-dielectric interface with absorbing silicon becomes more complicated:

$$R_{12} = \frac{(n_2 - n_1)^2 + k_2^2}{(n_2 + n_1)^2 + k_2^2}$$

Figure 10.2 shows measured free carrier absorption coefficient value for N-type silicon as a function of resistivity and wavelength. Notice that the wavelength extends from 1 to 1000 μm (1 mm), and the resistivity range is 0.01 to 19.5 Ω·cm. Values reported in the literature by

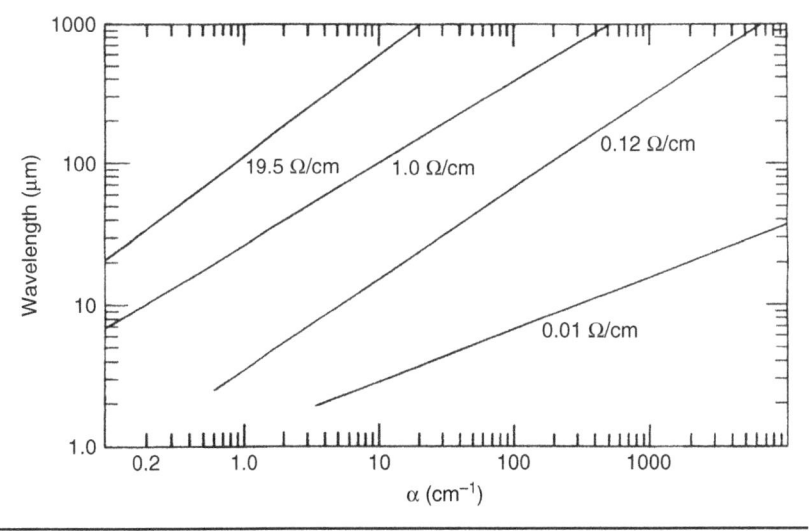

FIGURE 10.2 Absorption coefficient as a function of wavelength for N-type silicon at room temperature.

Spitzer and Fan[5] fall almost exactly on the TI measured values. These measurements were made using the TI Perkin Elmer 301 grating spectrophotometer, standard with a Golay cell detector or using the TI liquid helium cooled, gallium doped germanium bolometer. A study was run by TI and PE to compare the sensitivity of the Golay cell to the TI bolometer[6] that demonstrated the TI bolometer improved S/N by a factor of 3. For every wavelength point, k values may be calculated. In application of the scan technique, change in intensity of the peaks is not the main concern. Only the wavelength location of each peak is important.

Optical interference occurs, and as the infrared spectrophotometer scans from a short wavelength to a longer one, a pattern of maxima and minima occurs with the amplitudes depending in part upon the ratio of the resistivity of layer to substrate. Often the resistivity ratio of layer to substrate was 10 to 20. The scans were generated using the double-beam infrared instruments available at that time. Most were optical null instruments where the reference beam was matched to the intensity of the sample beam by moving a wedge-shaped object into the reference beam until the two beams were equal. The wedge was linked to the recorder to read 1 to 100 percent For this application, a reflectance attachment was mounted in the sample beam. The angle of incidence was fixed at 20° to 30°. An example of such a scan is shown in Fig. 10.3. The wavelength of the scan covers from 12 to over 30 μm. Notice the spacing of patterns is closer together in the shorter wavelengths than in the longer because the order of interference (integer m) is decreasing as λ approaches the same order of magnitude as d. Many of these instruments made by Beckman Instruments were purchased by TI and put in service in the production

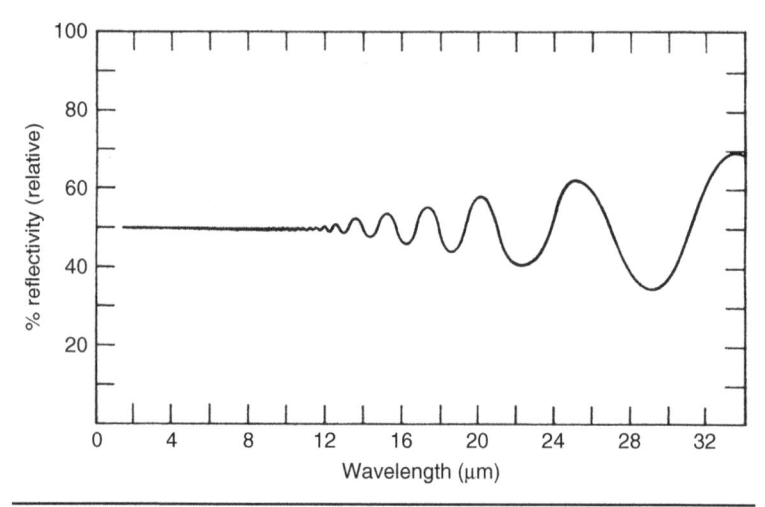

FIGURE 10.3 Typical infrared scan epitaxial layer interference pattern.

area. Personnel were trained in operation and in calculating the results. The simplified interpretation is that the wavelength locations of maxima or minima are separated by 2π degrees. Using the wavelengths for adjacent maxima peaks, the general equation becomes

$$\delta = 2n_1 d \cos \theta_1 = (\lambda_2^{-1} - \lambda_1^{-1})^{-1}$$

Or using cm^{-1} to express wavelengths.

$$2n_1 d \cos \theta_1 = (v_1 - v_2)^{-1}$$

In production areas, charts were made for the operator where all they need to do is read the wavelengths of the peaks. To increase accuracy, they may count a series of peaks and calculate an average wavelength separation.

The method has aspects that have to be taken into account to increase accuracy. One is the steepness of the gradient change in resistivity at the layer-substrate interface. Figure 10.4 illustrates the

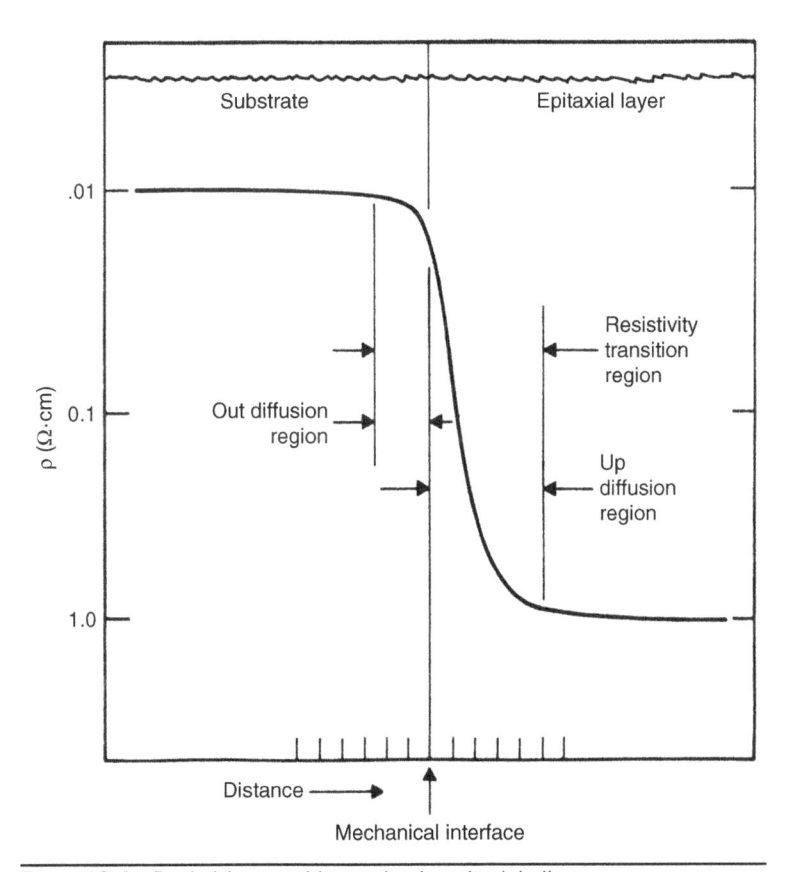

FIGURE 10.4 Resistivity transition region in epitaxial slices.

appearance of the interface profile between film and substrate. As mentioned before, a great deal of technical effort was expended throughout the semiconductor world to improve accuracy and reliability; for example, see Albert and Combs[7] or Schumann.[8] For very thin layer (< 5 µm), the first order of optical interference occurs in the near infrared where the optical difference between layer and substrate is very small. For this reason, the infrared scan technique does not work well for thin epitaxial layers. Layers of 1-µm thickness are almost impossible to measure because their thickness allows less than one order of optical interference where the index of film and substrate is almost the same.

However, the ellipsometer technique has been widely used in the visible region of the spectrum to measure films less than one wavelength in thickness. The need to accurately measure the thickness of very thin epitaxial films looked like a good problem to be solved by the application of ellipsometry in the infrared.

10.5 Elliptical Polarization of Light on Reflection

Light emitted from a hot surface shows no sign of polarization when viewed through a polarizer rotated about its axis. The light is said to be unpolarized. However, when the light is viewed after it has been reflected from a surface at an oblique angle, the intensity is found to vary as the polarizer is rotated. The degree of polarization depends upon the angle of incidence, the optical constants of the surface, and, if present, the thickness of the film on the surface and the optical constants of that film. The optical analysis of the degree of elliptical polarization of reflected light and the use of the measured results to characterize the surface is called ellipsometry. In actual fact, ellipsometry is based on optical interference except in this method angles are measured rather than light intensity. In the opinion of many, including this author, ellipsometry offers the most accurate method for measuring the thickness of transparent films. The method is general in nature, is based on fundamental optical quantities, and is applicable using light wavelengths from the uv to the far infrared. Probably the most important application has been in the semiconductor industry for the control and evaluation of dielectric insulating film deposition processes.

In ellipsometry, it is convenient to consider unpolarized light to be composed of one-half light polarized with its electric vector in the plane of incidence (*P* light) and one-half light polarized with its electric vector normal to the plane of incidence (*N* light). That is, *I unpolarized* = ½I_P and ½I_N.

Figure 10.5 shows the change in reflectivity for *N* light and *P* light as a function of angle of incidence. Notice the reflectivity of *P* light goes to zero at an angle referred to as the polarizing angle $\phi = \tan^{-1} N$. The bottom drawing illustrates the change in phase shift angle for *P* and

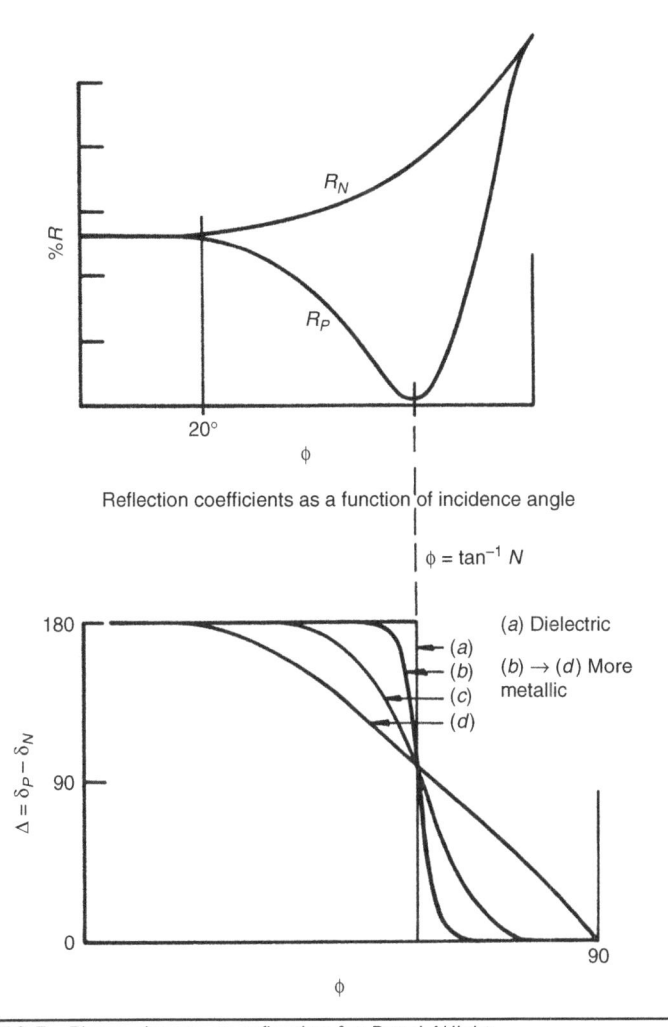

FIGURE 10.5 Phase change on reflection for P and N light.

N light as the angle of incidence changes. The resultant phase shift Δ is shown to be affected as the surface nature changes from dielectric to metallic.

Figure 10.6 illustrates the combined change in the resultant reflected N and P light in intensity and phase angle. Two resultant angles Ψ and Δ are used to describe the elliptical polarization of polarized light. The definitions are as follows:

$$\text{The azimuth angle} \quad \Psi = \tan^{-1} rp/rn$$
$$\text{The phase shift angle} \quad \Delta = \delta_p - \delta_N$$

These two angles are measured by the ellipsometer.

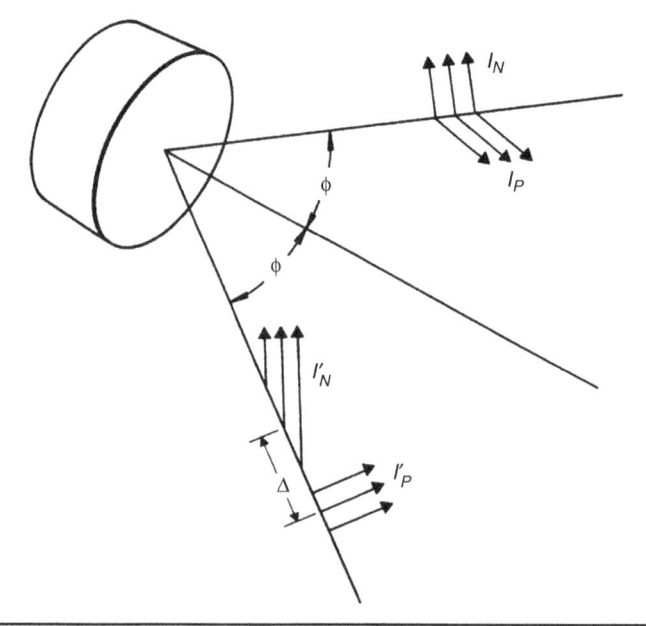

FIGURE 10.6 Intensity and phase shift for P and N light reflection.

The elliptical polarization is easy to visualize for those familiar with Lissajous patterns on an oscilloscope. The patterns are depicted in Fig. 10.7. By analogy, the P light corresponds to the ac voltage applied to the horizontal deflection plates, and the N light to that connected to the vertical plates. The phase angle Δ represents the phase

$\Delta = 0$ $0 < \Delta < \frac{\pi}{2}$ $\Delta = \frac{\pi}{2}$ $\frac{\pi}{2} < \Delta < \pi$ $\Delta = \pi$

$\Delta = \pi$ $\pi < \Delta < \frac{3\pi}{2}$ $\Delta = \frac{3\pi}{2}$ $\frac{3\pi}{2} < \Delta < 2\pi$ $\Delta = 2\pi$

For, $A_P = A_N$
$\Delta = \delta_P - \delta_N$

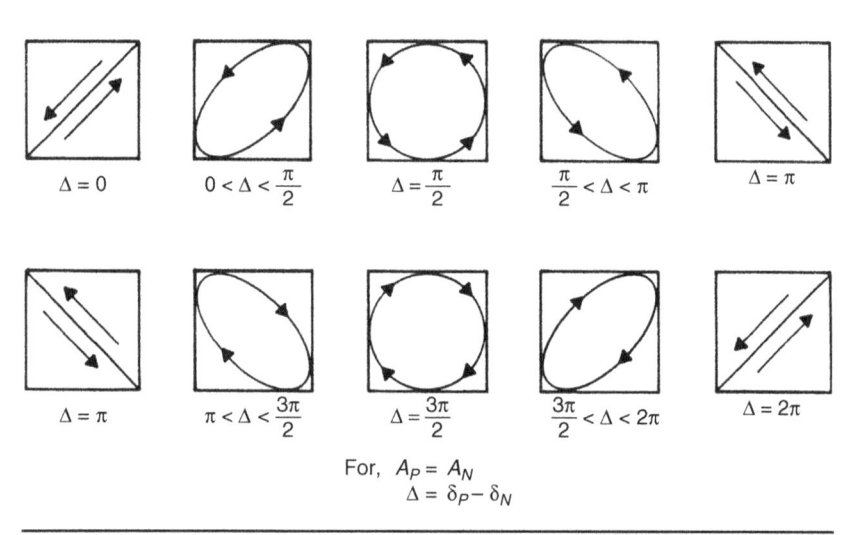

FIGURE 10.7 Oscilloscope Lissajous patterns used to represent elliptical light polarization.

difference between the two applied voltages. The amplitudes A represent light intensity and are shown equal. The patterns shown for different phase values pass through different elliptical forms as well as circular polarization. In reflected light, circular polarization rarely occurs because of the inequality of the reflected intensities. The inequality of the amplitudes may be simulated by making the gain settings of the horizontal and vertical voltage inputs different.

10.6 Measuring the Elliptical Polarization Angles ψ and Δ

Figure 10.8 shows the essential components of a manually operated ellipsometer. A source of parallel monochromatic light passes through the first polarizer P, the quarter wave plate Q, and strikes the surface to be studied at angle of incidence ϕ. The reflected ray leaves the surface at angle ϕ, passes through the analyzer A and is detected. The light energy may be detected by the eye for visible light measurements or a photon sensing device sensitive at the wavelength of measurement. There are several procedures and conventions used in making the ellipsometer measurement. The convention used at TI was to fix the angle of the quarter wave Q to –45 degrees from the angle of incidence. The first polarizer P is then moved relative to the quarter wave plate to put a phase shift in the light equal but opposite to that the light undergoes on reflection. In that way the resultant reflected light becomes plane polarized. This process is called phase compensation. The analyzer polarizer A is then rotated to be perpendicular to the plane of the reflected light resulting in a null or zero

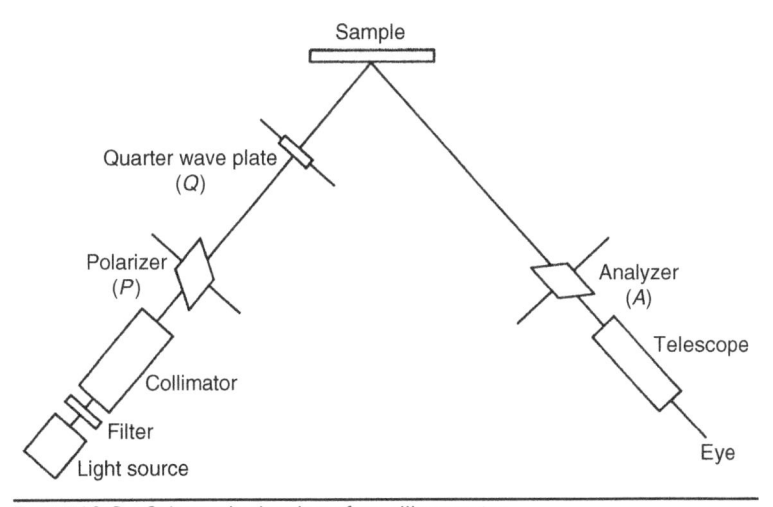

FIGURE 10.8 Schematic drawing of an ellipsometer.

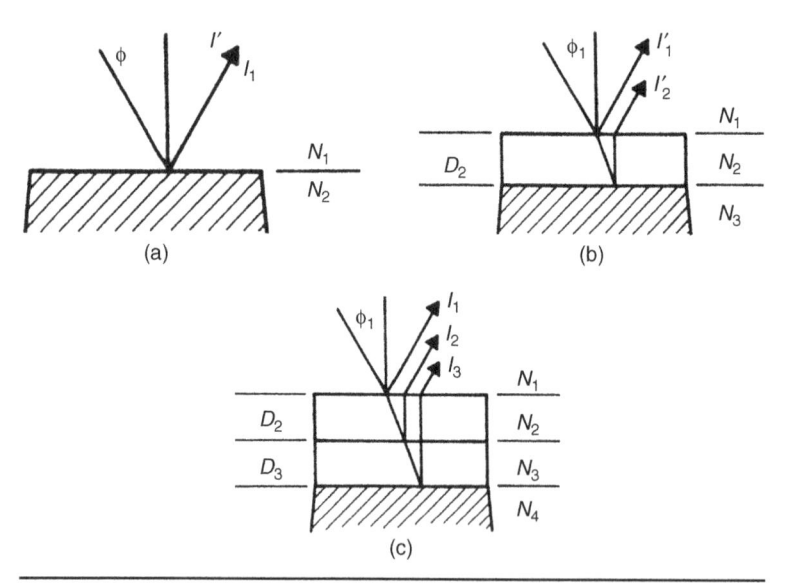

FIGURE 10.9 Intensities and refractive indices required in ellipsometry calculations for (a) single surface, (b) single layer, and (c) double layer.

light intensity. The angles of P and A are read from their very accurate vernier scales. Using this convention, $\psi = A$ and $\Delta = 2P - 90$.

Calculation of curve sets for silicon covered with dielectric films of varying refractive indices was very difficult in the early days of applying ellipsometry because at this stage computers were limited to punch cards or paper tapes and had very small memories needed for calculations. The first deck used at TI to produce silicon curves came from Frank McCrackin of the National Bureau of Standards (NBS). Referring to Fig. 10.9a, the calculations become more difficult because now the values of R_{P12} and R_{N12} must be calculated instead of a single R_{12} as in the scan procedure.

$$R_{P12} = \frac{N_2 \cos\theta_1 - N_1 \cos\theta_2}{N_2 \cos\theta_1 + N_1 \cos\theta_2}$$

$$R_{N12} = \frac{N_2 \cos\theta_2 - N_1 \cos\theta_1}{N_2 \cos\theta_2 + N_1 \cos\theta_1}$$

The calculation gets even more difficult if the film or substrate has a nonzero k or if more than one film is involved. The number of reflection coefficients that must be used in the calculations becomes quite large, as illustrated in Fig. 10.9b which indicates the need to calculate R_{P12}, R_{P23}, R_{N12}, and R_{N23} for D_2. Figure 10.9c indicates the need to calculate R_{P12}, R_{p23}, R_{P34}, R_{N12}, R_{N23}, and R_{N34} for D_2 and D_3. In final form for all three cases, the resultant expression relating all the terms to the measured Ψ and Δ leads to a quadratic equation that may be solved for film thickness. Obviously, computer programs were essential to the full utilization of this technique in the development of semiconductor film deposition processes. R. J. Archer,[9] a pioneer in the application of ellipsometry, used the reverse process to produce this Ψ and Δ set of general curves for nonabsorbing films on silicon shown in Fig. 10.10. The curves are calculated for a 70° angle of incidence using mercury green light (5461 Å). The film index range is very large with the useful range of about 1.2 to 5. A single measured point may be used to determine both the thickness of the film and its index of refraction. Archer was able to correlate index and density of oxide films on silicon made in different ways, with steam, anodizing, or silane decomposition.

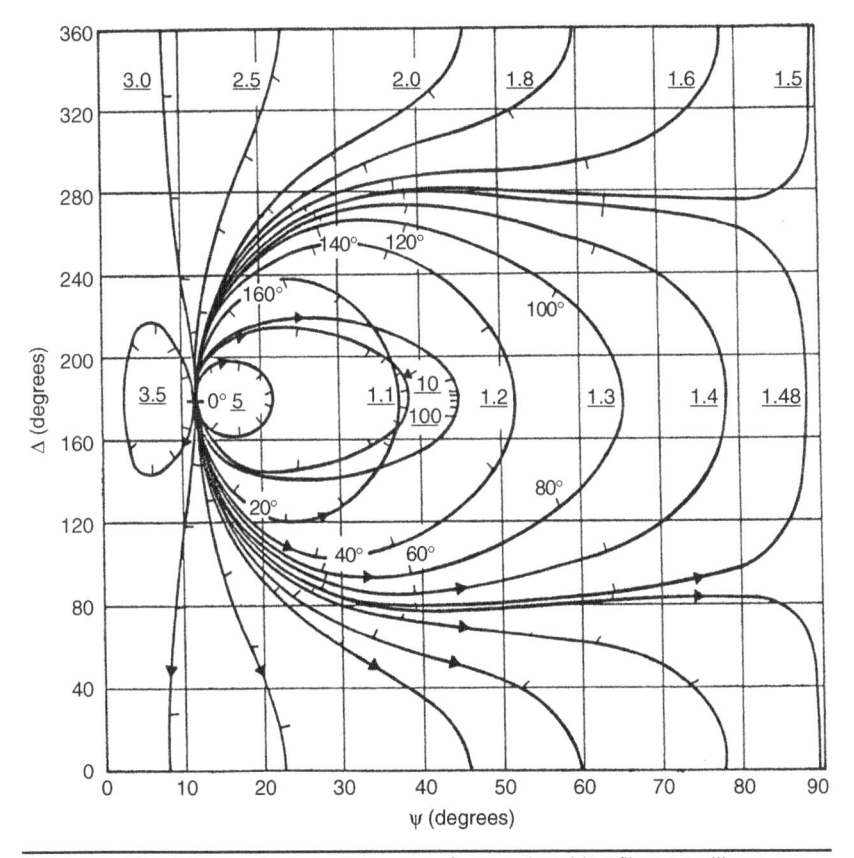

FIGURE 10.10 Ellipsometer ψ and Δ curves for nonabsorbing films on silicon.

258　Chapter Ten

His results are presented in the following:

Method of Film Formation	Index of Refraction	Density (g/cm³)
Oxygen	1.450	2.23
Steam	1.475	2.32
Anodization	1.362	1.80
Silane decomposition	1.430	2.14

A more specific set of curves that would be useful in the TI production area is shown in Fig. 10.11. The index range of the films is 1.4 to 1.6 covering most films in use on silicon devices. Similar curves were calculated for germanium and gallium arsenide substrates.

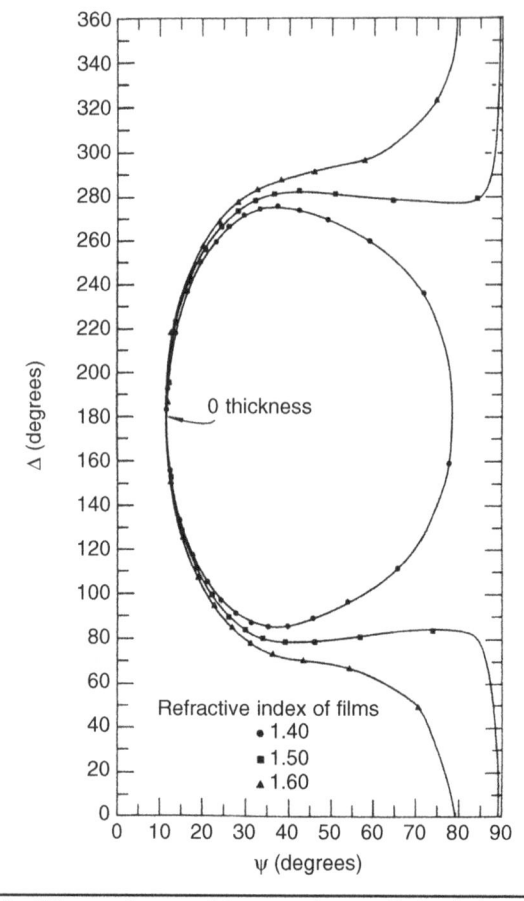

Figure 10.11 Ellipsometer ψ and Δ curves for specific low-index nonabsorbing films on silicon.

10.7 Ellipsometers Used at TI

The author's colleague, Charlie Jones, discovered that there was a research-grade Gaertner ellipsometer in the storage of the basement of the TI CRL building not being used. The instrument was capable of great accuracy and precision. Charlie Jones mastered the instrument and began working closely with the semiconductor process people, aiding them in the development of their insulating film deposition procedures. At the same time, there was an organized task force in CRL helping to develop techniques and instruments to increase quality and yields in the process area. One of the first achievements was to construct color charts for films of different thickness on silicon. The thickness values were measured with the Gaertner ellipsometer. It soon became apparent that daily use of ellipsometry in the process area was needed.

However, the only TI instrument was very expensive and difficult to operate, unsuited for production line use. The author, working with personnel in the CRL model shop, designed and built a much simpler instrument constructed using readily available components. The light source selected was a HeNe laser. The beam was passed through a 30-mil hole in a metal disk to provide a small spot on the sample. Polarizers and quarter wave plates that could be used with the 0.6328-μm laser light were purchased and mounted on the instrument along with suitable vernier scales. A heavy metal base was constructed to support each instrument. The light intensity was measured using a simple CdSe detector with the output measured with a voltmeter and taken as a measure of light intensity. The set up was as before with Q at $-45°$ and angle of incidence $70°$. Several were made for use in the production areas. The cost was small in comparison to that of the Gaertner research instrument. Al Reinberg of the CRL physics laboratory, a member of the CRL task force, trained the technicians in their use and put them in operation in several front-end production units. Posters with Ψ and Δ values were produced for the process areas covering the various insulating dielectric films in use. The application of ellipsometry in the process area was a major step forward in process control.

10.8 Infrared Ellipsometry

The infrared scan technique was unreliable for measuring some epitaxial layers < 5 μm in thickness. Some devices required a layer thickness of 100 μm. In this case, interference patterns occur at longer wavelengths if at all, using instruments available. Also absorption within the thicker layer could sometimes become a problem. If the silicon substrate is high-resistivity, the optical constants do not change until long wavelengths are reached. To demonstrate that far infrared

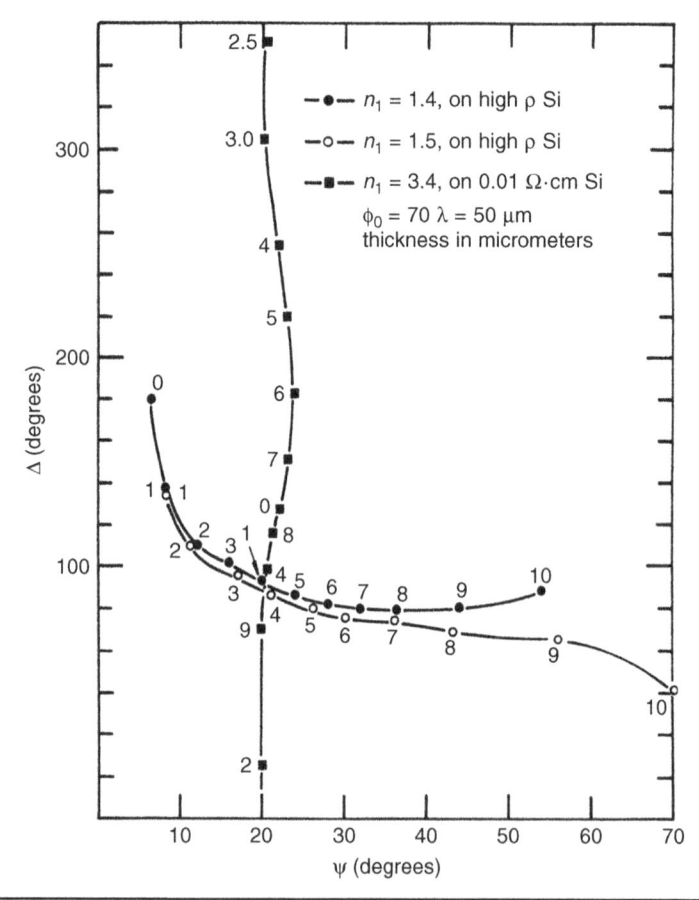

FIGURE 10.12 Ellipsometer curves for films on silicon measured in the far infrared.

ellipsometry may be useful in solving some of these problems, the ψ versus Δ curves shown in Fig. 10.12 were calculated for 1.4 to 1.5 index SiO_2 on high-resistivity Si substrates measured using 50-μm light at a 70° incidence angle. Also included are curves for very thin Si layers on 0.01 Ω·cm silicon substrates. As pointed out earlier, only one wavelength is involved and only angles are measured. The far infrared application should be more accurate and more precise and work well for very thin film layers.

The next step is to build an infrared ellipsometer operating in the far infrared. A far infrared light source, a quarter wave plate transparent in the far infrared, infrared polarizers, and a suitable infrared detector were required. No far infrared monochromatic source was readily available, one had to be made. A unique reflection system was devised that takes advantage of the strong reflection bands (Restrahlen) of ionic crystal to isolate the energy emitted from a globar into a very narrow band of light peaked at around 55 μm. Materials

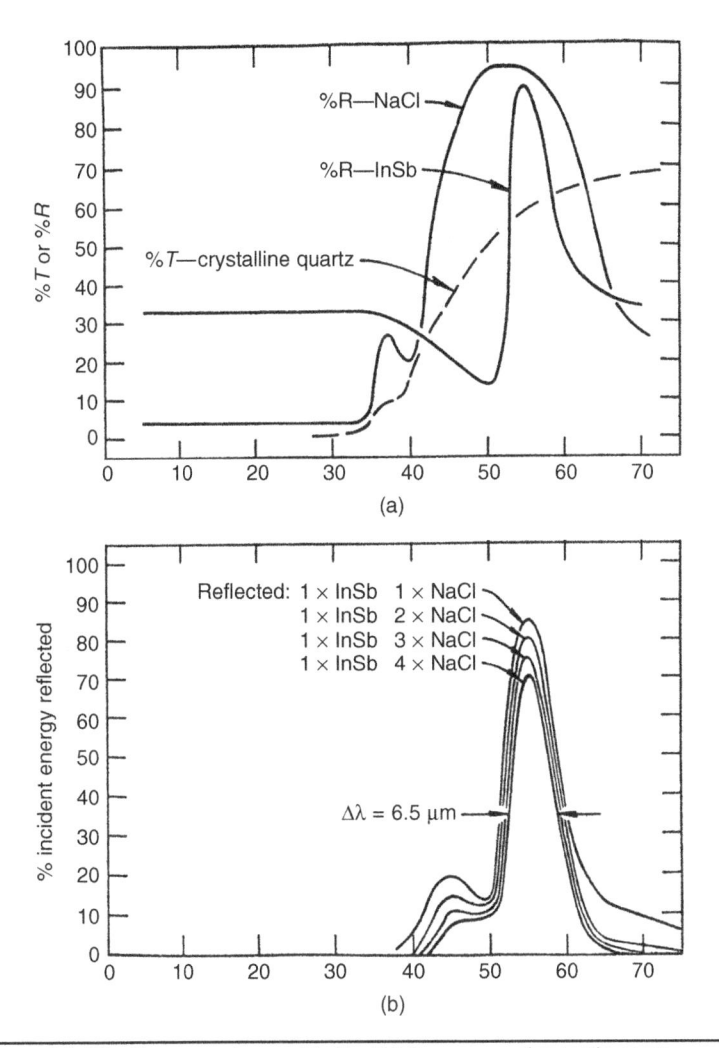

FIGURE 10.13 Components used in the 54.6-μm monochromatic source.

selected for use are shown in Fig. 10.13a. The absolute reflectivities of the two materials, NaCl and InSb, are shown as well as the transmission of crystalline quartz. Crystalline quartz was the material chosen from which a far infrared quarter wave plate could be fabricated. Crystalline quartz transmits at 55 μm and it is birefractive. Quartz stops transmitting in the NIR due to water absorption and its Restrahlen band peaks around 10 μm. Transmission for quartz begins to recover at about 45 μm. Its two refractive indices were known. The selection of NaCl was made because of its high reflection peaked at 55 μm and low reflection on both sides of the peak. But the band was too broad. That is why InSb was chosen, to narrow the resultant peak. Figure 10.13b illustrates how the resultant narrow band is produced by multiple reflections centered at 54.6 μm. The peak reflectivity of

262 Chapter Ten

NaCl is 0.95. Four reflections become a peak of 0.81. The peak for InSb is 0.9, resulting in a calculated energy peak value of 73 percent of incident light. The change in reflected light as a function of the number of reflections is shown. Light is reflected off InSb only once to peak or narrow the reflected light. Also note in Fig. 10.13a the transmission of crystalline quartz is substantial at this wavelength, as shown. Wire grid polarizers would be ideal to use in this instrument. However, they were not available. A photomask technique was used to produce polarizers with 1000 lines per inch and 5000 lines per inch of aluminum wires on high-purity silicon disks. A spacing of $\lambda/3.5$ is required for 99 percent polarization. For a λ of 54.6 µm, spacing of 1600 lines per inch is indicated. Measured polarization for these two at 54.6 µm was 55 percent for 1000 lines per inch and 95 to 98 percent for the 5000 lines per inch. But transmission was low, only 25 to 27 percent.

A schematic drawing of the TI far infrared refractometer is shown in Fig. 10.14. The instrument was built on a Gaertner L118 base. Light

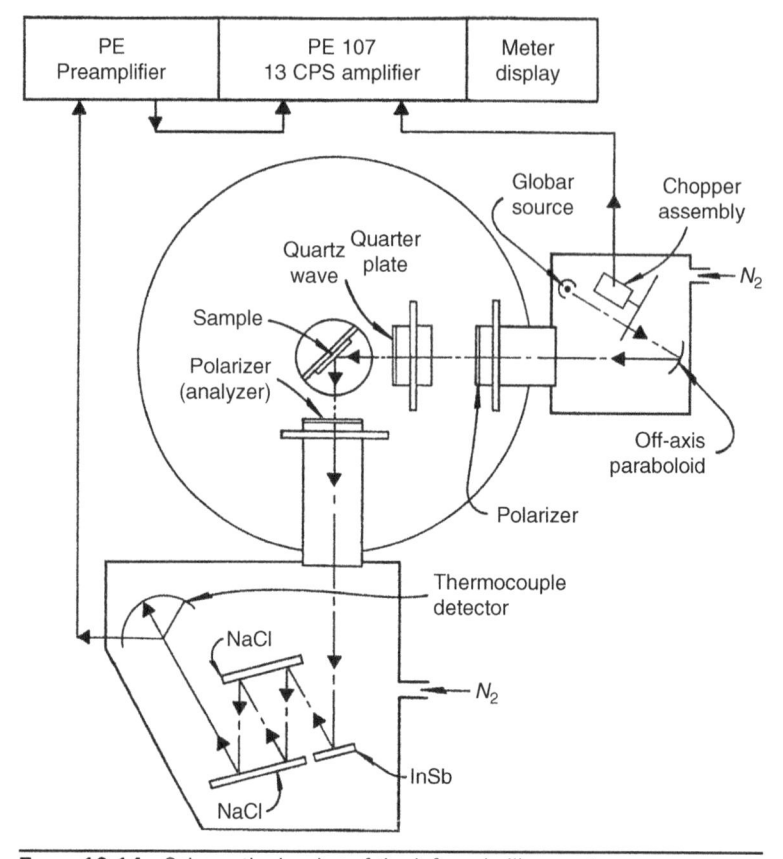

Figure 10.14 Schematic drawing of the infrared ellipsometer.

from the globar source passes first through the 13 cps chopper blades connected to the PE amplifier, strikes an off-axis parabola mirror, and forms a globar image on the sample holder. The parabola mirror is used to collect as much energy as possible from the globar source for the measurements. The beam passes through the P polarizer and the Q plate, reflects off the sample, passes through the A polarizer, and reflects off the InSb and NaCl 4 times which turns it into 54.6-μm energy. Notice the reflections occur in an enclosure that is flushed with dry nitrogen. The purpose was to protect the NaCl plates from moisture. Also airtight chambers were constructed on each arm to minimize infrared absorption by atmospheric moisture. The final energy is collected by an ellipsoid mirror focused on a PE thermocouple detector connected to the PE 13 cps preamplifier/amplifier and displayed on a high-sensitivity voltmeter as a digital signal. The instrument was operated exactly as the manually operated visible light instrument except the eye was not used to detect the null. The system was definitely energy-limited, but readings were surprisingly reproducible. The instrument was built only to demonstrate the applicability of the method to epitaxial film thickness measurements, so tests were required.

The Si substrate and Si layer resistivities were 0.013 and 2 Ω·cm, respectively. The optical constants were calculated from fundamentals for the 54.6-μm wavelength using measured carrier concentrations and carrier mobilities. A set of Ψ and Δ curves was calculated for 1 to 10 μm in thickness with 0.1-μm increments. The angle of incidence was 50°. In the test, 40N on N^+ silicon epitaxial layers were tested. When the measured values were compared to those obtained by the scan technique, agreement was ±6.7 percent. Reproducibility of the ellipsometer readings was ±4.1 percent. The second test involved 40 epitaxial layers of Ge, P on P^+ again 1 to 10 μm in thickness, substrate resistivity of 0.001 Ω·cm and layers of 1 to 2 Ω·cm. Again, the optical constants were calculated from fundamentals and Ψ and Δ curves calculated for 54.6 μm and using a 50° incidence angle. Agreement with the ir scan values was ±9.6 percent with reproducibility of ±2.8 percent. These TI results were reported[10] as support for using far infrared ellipsometry in epitaxial film measurements. Two years later,[11] results of using the TI far infrared ellipsometer to measure the optical constants of 0.016 Ω·cm N-type silicon substrates were reported. The measured n and k values (D) were only in fair agreement to three sets (A, B, C) calculated using three variations of the recognized method of Lyden[12] based on experimentally measured carrier concentration and carrier mobility. Then using the four sets of n and k, four Ψ and Δ curves based on 54.6 μm and 50° angle of incidence were calculated and plotted for thickness up to 10 μm. Each was labeled A, B, C, or D to indicate method of calculation. Then the layer thickness of 16 of the substrates with 1 to 2 Ω·cm Si layers was measured using the TI infrared ellipsometer and their Ψ-Δ values were plotted.

264 Chapter Ten

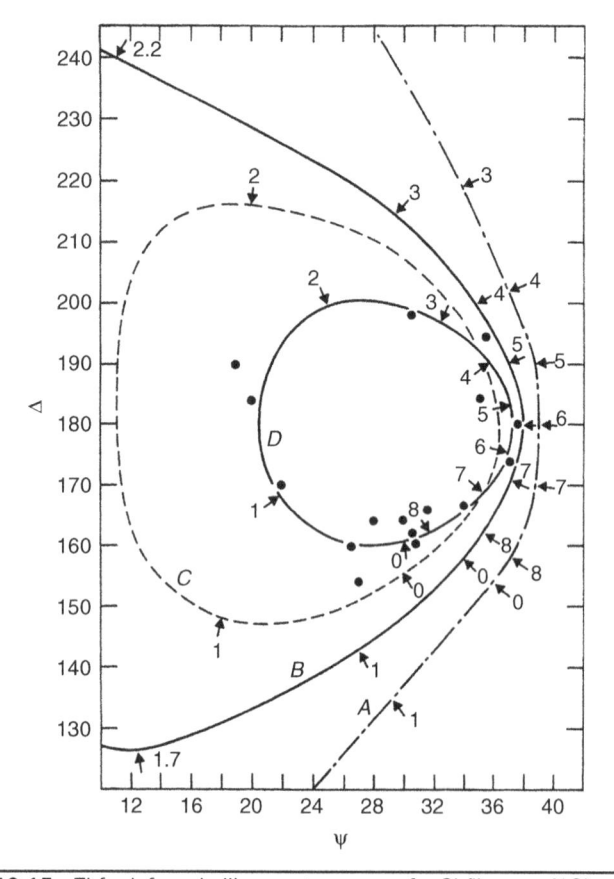

Figure 10.15 TI far infrared ellipsometer curves for Si films on N Si.

The curves and plot points are shown in Fig.10.15. It is obvious that the measured thickness points fit best on curve *D*, the one made using the n and k measured values using the ellipsometer. The conclusion reached from these results was that with a much improved instrument, the method would become more accurate and more direct than the ir scan technique. The instrument and its use to measure film thickness are covered in U.S. Patent 3,425,201: A. R. Hilton and C. E. Jones "Method and Apparatus for Measuring the Thickness of Films by Means of Elliptical Polarization of Reflected Infrared Radiation" (1964).

10.9 The TI Automatic Ellipsometer System

Spurred on by the success of the far infrared ellipsometry effort, a proposal was prepared and presented to the TI "IDEA" Program headed by Fred Bucy, senior vice president for the TI semiconductor

Early Work at Texas Instruments

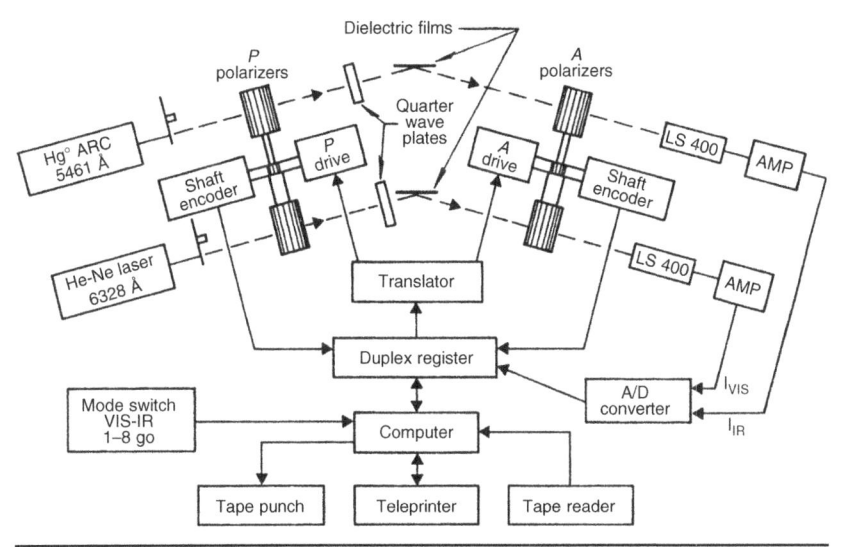

FIGURE 10.16 Schematic diagram of the TI automatic ellipsometer system.

operations. The proposal was to build an instrument with its own stand-alone computer used for control of the instrument, all calculations from ellipsometer-derived data, and print out all answers. The system would not require the use of skilled technical operators, but would be simple enough for anyone to use and obtain answers in a few seconds. Both a visible laser light source and a far infrared laser light source would be provided for two separate channels of operation in one instrument, allowing more-complex film thickness problems to be solved. Cam Allen, a very experienced TI engineer, and Dub Westfall, his technician, were recruited to design and build the system shown in Fig. 10.16.

In place of manual movement, the P polarizer and the A analyzer are moved when pulses are supplied to stepping motors. The mechanical arrangement is such that one pulse supplied to the motor produces a rotation of $0.25°$. The stepping motors are coupled to shaft encoders that supply digitized information to the computer through duplex registers such that the absolute angle of P or A at any position is calculated to $\pm 0.01°$. The one set of drive motors and shaft encoders is used to produce null measurements for the two separate ellipsometers. The two separate ellipsometers have their own light sources, polarizers, quarter wave plates, and detectors. They operate at the same angle of incidence. Selection of one or the other requires only a flip of a two-way switch that selects the proper amplifier for the A to D converter combination. The light sources are a HeNe laser which emits at 0.6328 μm. The second is a Hg arc filtered for the 0.5461-μm line. Both are detected using a TI LS 400 photo device. The computer reads the light intensity and coded angle readings directly through the duplex registers. The information is then transferred to the main program, written in Fortran, for solution of the selected ellipsometry problem. One should again be reminded how much computers and software programs have changed since 1969. The stand-alone computer was housed in a cabinet. It was a Hewlett-Packard 2115A, with only 10,000 memory locations. The input was paper tape into a reader. The ellipsometry calculations are very involved. Computations in computers then used Fortran. To have enough space to operate the required control and calculations, Cam Allen had to go into the memory and strip out all the Fortran subroutines not required for the instrument. An idea of the equipment we had to work with in those times is shown in Fig. 10.17, a photograph of the TI automatic system including computer and printer. The sample area is shown with its protection cover in place in the photograph. You can imagine such a unit built today would probably be placed on a desk.

The original intent was to have two instruments, one visible light and one far infrared. The first far infrared laser light source chosen was a water vapor laser. One was ordered, but the company could not deliver and after 6 months of effort the order was canceled. Next, with the help of members of the CRL physics group, Laird Shearer, Francois Padovani, and Doug Sinclair, the author visited with Prof. Javan of MIT to discuss the construction of a pure He laser he had

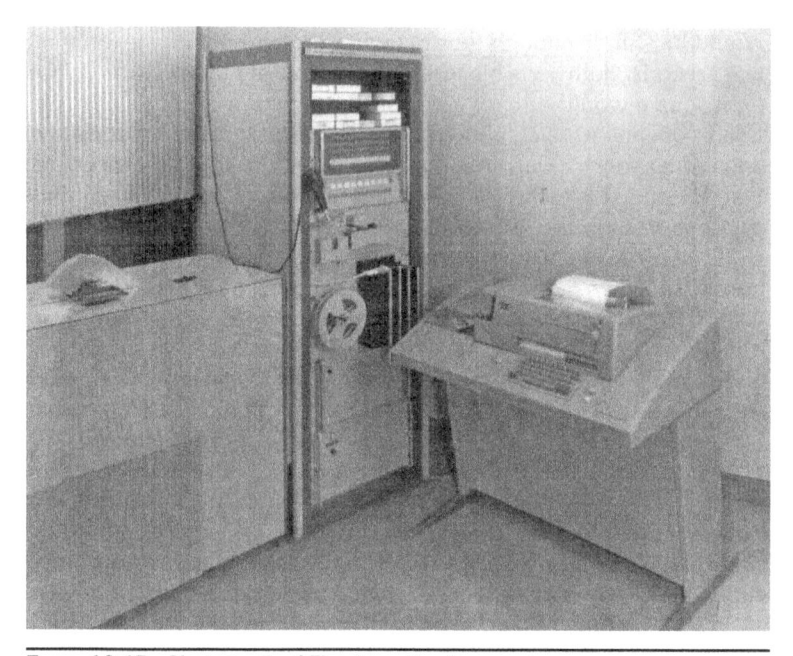

FIGURE 10.17 Photograph of TI automatic ellipsometer.

developed that emitted at 95 µm. Along with help from members of the glass shop, Fred Kennedy, Ron Child, and Glen Whaley, the construction of the laser was undertaken. Some may notice the name Glen Whaley. This was the first time the author worked closely with Glen. It was this experience that led to hiring him as the first employee of AMI some 18 years later. The plasma tube required very skilled fabricators.

The laser was constructed without any problem and the output measured. Making it perform for days and weeks became a serious problem. The He plasma at only 100-µm pressure was very destructive to the plasma tube filaments, sputtering tungsten on the walls covering He and reducing the pressure below operation levels. After months of effort, we could not delay finishing the ellipsometer system any longer. The second light source was chosen to be the filtered mercury arc light. We would abandon the goal for a far infrared laser and put the system in operation as soon as possible. Still, having two light sources at times would be an advantage for some problems. A second set of data could be useful in solving complex problems or measuring optical constants at two different wavelengths. Sometimes the order of optical interference needs to be confirmed. All the software information was contained on a single paper tape. The tape reader and punch were utilized to produce the tape.

The accuracy of the ellipsometer rests on the ability to measure absolute values of Ψ and Δ. All angles are referenced to the plane of incidence of the instrument. The calibration program employs a sequence of events and measurements designed to have the instrument align itself to ensure the angular relationships between angle of incidence, P, A, and Q are correct. Null measurements were obtained in 20 s and printed out in 10 s for a reading in a total of 30 s. A mode switch provided eight ellipsometer program selections: 1 and 2 for SiO_2 on Si, 3 for SiO_2 on Ge, 4 for SiO_2 on GaAs, 5 surface optical constants, 6 and 7 double-layer problems, and 8 an iteration method for N_2 and D_2. The Fortran programs were provided by the CRL computer group of Eric Jones with Charles Ratliff and Gene Daniels.

The operating program called RAD detects the chopped light signal with its reference signal and computes an intensity number. The program Hunt searches for the null as the polarizer is moved in small steps in only one direction, as shown in the upper diagram in Fig. 10.18. After finding a value close to the null, the Scan program takes over and moves away from the null, stops, and while moving in the same previous direction takes nine evenly spaced intensity points, four on each side and one near the peak intensity. The data are moved into place and fit to a quadratic equation. The calculated null is related as an absolute angle value for P or A. Reproducibility of angles P and A values is ±0.01° as shown in Table 10.1 taken on one sample.

Chapter Ten

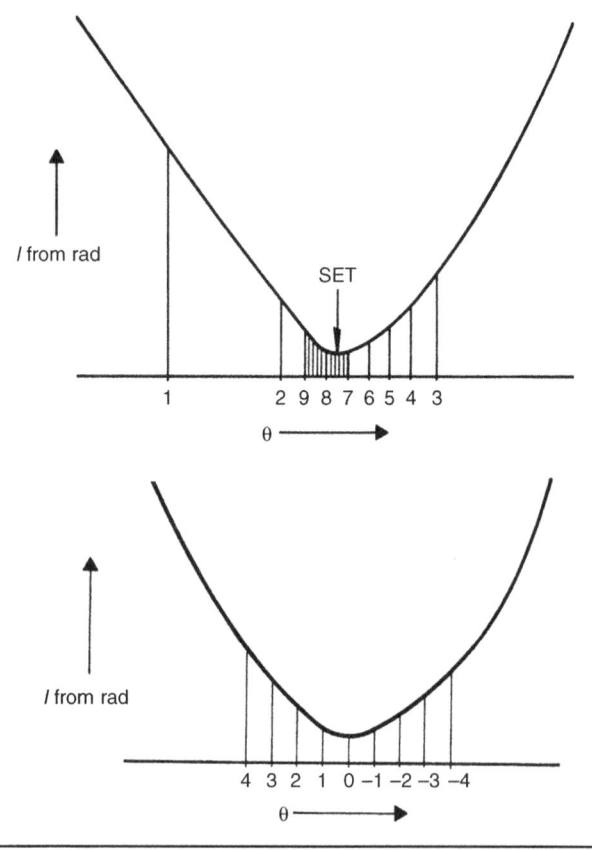

FIGURE 10.18 Schematic drawing of how null angles are measured.

	P		A
P_1	40.33	A_1	31.89
P_2	40.30	A_2	31.91
P_3	40.29	A_3	31.91
P_4	40.30	A_4	31.91
	Average 40.305		Average 31.905
	Deviation ±0.01		Deviation ± 0.01
	$\Delta = 170.61$		$\Psi = 31.91$

TABLE 10.1 SiO_2 on Silicon

The existence of thin oxide layers on active surfaces affects the measured values of surface optical constants obtained using ellipsometry. Also in semiconductor processing "native" oxides are sometimes a problem. The instrument is so sensitive that results vary slightly even with room humidity. The measured growth of an oxide film in room air on a fresh silicon wafer just removed from a reactor with hydrogen in its atmosphere is presented below, illustrating the problem and the sensitivity of the ellipsometer.

\multicolumn{2}{c	}{Relative Humidity 55%}
Time (min) from Reactor	Oxide Thickness (Å)
1	2.23
5	5.42
10	5.54
20	5.86
30	7.28
40	8.46
50	9.72

10.10 Summary

The major portion of this chapter was devoted to discussing optical interference and its use to solve film thickness measurement problems in the semiconductor industry. First, working with instruments available, the infrared scan technique was developed to the point that the thickness of semiconductor materials grown on conducting substrates could be accurately measured in the process areas. As the technology changed and the growth of dielectric insulating layers became part of the process, visible light ellipsometry was put in use for thickness values and process development. Later, it became apparent that very thin epitaxial layers were required for high-frequency devices and the scan technique could not produce accurate values. A somewhat crude infrared ellipsometer was built and used to demonstrate the potential for measuring very thin epitaxial layers using a far infrared (55-µm) light source. The successful demonstration led to a TI-funded program to build a completely automatic computer-controlled ellipsometer system equipped with a visible laser and a far infrared laser light source—an instrument simple to operate and fast to measure with printed answer output. The instrument aligned itself to ensure accuracy. Unfortunately, at that time no source of a far infrared laser with a reasonable lifetime could be found. Nevertheless, the instrument was moved to the production area, equipped with two visible light sources, and used to supply valuable information to process areas for a number of years.

References

1. "The Structure of Glass," Proceedings of the Third All Union Conference on the Glassy State, Leningrad 1959, Consultants Bureau Enterprises Incorporated, vol. 2, New York, 1960.
2. A. Ray Hilton and M. Brau, "New High Temperature Infrared Transmitting Glasses," *Infrared Phys.* 3, 69 (1963).
3. T. S. Moss, *Optical Properties of Semiconductors*, Academic Press, New York, 1959.
4. A. H. Kohn, *Phys. Rev.* 97, 1647 (1955).
5. W. Spitzer and H. Y. Fan, *Phys. Rev.* 105, 268 (1957).
6. Charlie E. Jones, Jr., A. Ray Hilton, J. B. Jamrel, Jr., and C. C. Helms, "The Cooled Germanium Bolometer as a Far Infrared Detector," *Appl. Opt.* 4, 683 (1965).
7. M. P. Albert and J. F. Combs, "Thickness Measurement of Epitaxial Films by the Infrared Interference Method," *J. Electrochem. Soc.* 109, 709 (1962).
8. P. A. Schumann, "The Infrared Interference Method of Measuring Epitaxial Layer Thickness," *J. Electrochem. Soc.* 116, 409 (1966).
9. R. J. Archer, *J. Opt. Soc.* 52 970-77 (1962).
10. A. R. Hilton and C. E. Jones, "Measurement of Epitaxial Film Thickness Using an Infrared Ellipsometer," *J. Electrochem. Soc.* 113, 472 (1966).
11. A. R. Hilton and C. E. Jones, "Optical Properties by Far Infrared Ellipsometry," *J. Electrochem. Soc.* 115, 106 (1968).
12. Henry A. Lyden, *Phys. Rev.* 134, A1106 (1964).

Index

A

Absorption edge wavelength location, 49
Absorption method, 231
Absorption values, 12–13
Agema 210 camera, 137, 188, 190, 208
Agema 210 MWIR images, 190, 208
Air Force Production Development Program, 19
Alkali halides, 14, 24
Allen, Cam, 265, 266
Aluminum oxide, 126
AMI, 20, 89, 122, 225
AMI 10-m drum assembly, 200
AMI Beckman UV 5240 spectrophotometer, 98
AMI C1, 25–27, 150, 163, 166, 168
AMI C2, 164, 166, 168
AMI chalcogenide glass extrusion unit, 177
AMI closed compounding casting glass process, 84–86
AMI computer-controlled infrared refractometer, 105
AMI DARPA-funded large plate process, 215–222
AMI fiber drawing process, 158–168
AMI fiber drawing tower, 159, 160
AMI glass clad fibers, 194
AMI glass thermal expansion apparatus, 90
AMI infrared refractometer, 148
AMI IR fiber, 168–174
AMI lens molding unit, 135
AMI modified Temscal coating chamber, 196
AMI molding process, 146–148
AMI silicon vacuum float zone chambers, 223
AMI striae scope, 115
AMI three-chamber method, 217
AMI vacuum float zone machine, 223, 224
AMI ZYGO interferometer, 124
Amorphous Materials (AMI), 20, 89, 122, 225
Amtir 1, 20–21, 85, 97, 116, 118, 129, 149, 150, 227
Amtir 1V, 107, 149
Amtir 2, 149, 150
Amtir 3, 118, 150
Amtir 4, 98, 99, 150
Amtir 4 absorption edge transmission, 4
Amtir 4 rupture modulus, 111
Amtir 5, 117, 143, 149, 150
Amtir 6, 150
Angle θ, 121
Anneal point, 92
ANTAS, 20
Antimony, 75
Antireflection coatings, 126–129
Applications. *See* Glass processes for other applications
Arc-melter reactor, 61
Arc-source mass spectroscopy, 73
Archer, R. J., 257
Arsenic-germanium-selenium glass, 11
Arsenic selenide glass, 11
Arsenic trisulfide glass, 4, 191, 194–196
Arsenic trisulfide glass compounding process, 195

Index

As, 46
As-Se-Te, 168
As-Se-Te C1 glass fiber, 164
As-Se-Te system, 155
Asarco copper refining electroplating tank house, 72
Asarco selenium, 156, 157
Aspheric surfaces, 132. *See also* Unconventional lens fabrication
As_2S_3, 168
As_2S_3 glass, 104, 118
$As_{40}S_{60}$, 118
Atomic refraction values, 44
Author (Hilton), 18, 19–20, 79, 85, 100, 134, 212, 215, 216, 230, 231, 245, 246, 264, 266
Automatic ellipsometer, 264–269
Avionics window, 235, 236
Avogadro's number, 43
Azimuth angle, 253–259

B

Bailey, Lou, 246
Bandgap, 1
Barrel "dip probe," 173
Beckman UV 5240 spectrophotometer, 98
Beckman Vis-Nir spectrophotometer, 127
Bell, 123
Bend-to-break radius, 163
Beyen, Werner, 246
BN boat, 231, 234, 238
Boldish, Steve, 235
Bolduc, Roy, 172
Boron nitride (BN) boat, 231, 234, 238
Bottom hole caster, 78
Boule, 74
Brau, Maurice, 247
Bridgeport Rotary Table, 99, 103
Bucy, Fred, 264
Bulk absorption coefficient, 51
Bundle 10-M-1, 202–203
Bundle 10-M-2, 203, 206
Bundle 10-M-3, 203, 206
Bundle 10-M-4, 197, 203–204, 206–209
Burke, Kelly, 232
Burkhalter, Tom, 245, 246

C

C1, 25–27, 150, 163, 166, 168
C2, 164, 166, 168
Cadmium telluride, 211–222
Carr, Ed, 126, 196
CBC America IR Zoom, 139, 141
CdTe, 227
Chada, Suneet, 172
Chalcogen-chalcogen bonds, 49
Chalcogenide glass:
  chemical bonding, 48–60
  chemically inert, 114
  density, 41–42
  electrical properties, 47
  electronic conduction, 47, 71
  extrusion, 174–178
  historical overview, 17–21
  IR fibers, 153–158
  mass spectrometric investigation of bonding, 55–57
  molar refraction, 42–44
  physical strength, 47
  softening point, 48
  softening point/hardness, 40–41
  thermal coefficient of expansion, 41
  transition elements, 60–66
  volatility, 134
  x-ray radial distribution analysis, 57–59
Chalcogenide glass extrusion unit, 177
Characterization of glass properties, 89–118
  electrical resistance, 113–114
  final production procedure, 114–118
  hardness, 108–109
  Poisson's ratio, 110
  resistance to chemical attack, 114
  rupture modulus, 110–112
  shear modulus, 110
  thermal conductivity, 112–113
  thermal expansion/glass transition temperature/softening point, 89–94
  transmission/refractive index, 94–107
  Young's modulus, 110
Chemical bonding, 48–60
Chemical reactivity, 114

Index 273

Child, Ron, 267
Closed compounding casting glass process, 84–86
Closed system reactor, 62
Codman Fiber Optics, 154–159
Cohen, M., 226
Common module, 20
Compounding methods, 73–74
Compounding with reactant purification, 74–77
Computer-controlled infrared refractometer, 105
Continuous solution epitaxial growth design, 213
Conventional lens fabrication, 119–129
  antireflection coatings, 126–129
  dramatic changes, 131. *See also* Unconventional lens fabrication
  lens blank preparation, 119–120
  polishing, 122
  spherical surfaces, 121
  testing, 123–126
Conventional solution epitaxial growth process, 212, 213
Copper, 71
Cronin, George, 216, 222, 230, 231
Crossland, Bob, 231
Crosstalk at bundle ends, 210
Crystalline materials, 71. *See also* Infrared crystalline materials
Crystalline quartz, 261
Crystallites, 11
Czochralski method, 222
Czochralski sealed crystal puller LEC method, 232

D

Daniels, Gene, 267
DARPA-funded large plate process, 215–222
DARPA-ONR program, 18
Davidson, Jim, 205
de Nobel, D., 217
Desert, 235
Diamond point turning, 132–133, 136
Dickinson, Stanley, 211
Differential thermal analysis (DTA), 90
Differential thermal analysis (DTA) plot, 26

Diffractive surfaces, 132. *See also* Unconventional lens fabrication
Dilatometric softening point (T_d), 25, 90
Disordered solid, 1
Drozowicz, Z., 226
Druy, Mark, 172
DTA, 90
DTA plot, 26
Dual-band "Gen III" sensor systems, 241

E

Electrical resistance, 113–114
Electronegativity, 5, 23. *See also* Pauling electronegativity
Ellipsometer, 253, 255, 259
Ellipsometer curves, 260, 264
Ellipsometry, 252. *See also* Infrared ellipsometry
Elliptical polarization angles, 255–258
Elliptical polarization of light, 252–255
Epitaxial film thickness, 248–252
Extinction coefficient, 51
Extrinsic loss, 10–12
Extruded C1 glass rods and plates, 177
Extrusion, 174–178

F

Fantozzi, Lou, 124, 131
Far infrared ellipsometry, 259–264
Far infrared reflection spectra, 9
Far infrared Reststrahlen-like reflection bands, 50
Far infrared Reststrahlen-type reflection, 7, 8
Fiber drawing tower, 159, 160
Fiber ribbons, 182, 183
Final production procedure, 114–118
Finished float zoned paveway-grade silicon single crystal, 224
Flat NaCl windows, 99
Fortran, 266, 267
Foster Miller, 172
Fourier transform infrared (FTIR) instrument, 95

Fourier transform infrared (FTIR) spectrophotometer, 170
Fourier transform instrument, 95
Fraden, Jacob, 174, 175, 176
Fraser, W. A., 3, 17
Frerichs, R., 3, 17
Fresnel reflection coefficient, 12
Fresnel reflection loss, 9, 12, 126, 192, 198
Frit, 75, 79
FTIR instrument, 95
FTIR reflection QC scan, 128, 129

G

GaAs, 227
GaAs plates, 233–242
Gaertner ellipsometer, 259
Galileo Electro-Optics, 153, 159, 162
Gallium arsenide, 118, 230–242
Gallium arsenide resistivity, 236
Ge, 46, 227
Ge-Sb-S system, 48
Ge-Sb-Se glass, 57
$Ge_{33}As_{12}Se_{55}$, 118
Germanium, 118
$Ge_2Sb_{12}Se_{60}$, 118
Gibson, Des, 240
Glass clad fibers, 194
Glass extrusion, 174–178
Glass forming composition diagram, 34, 48
Glass forming region, 32
Glass outer shell, 66
Glass processes for other applications, 153–179
  AMI fiber drawing process, 158–168
  chemical applications, 168–174
  extrusion of chalcogenide glass, 174–178
  IR fibers, 153–158, 168–174
Glass production, 71–87
  compounding methods, 73–74
  compounding with reactant purification, 74–77
  open casting methods, 77–84
  purification, compounding, casting (one closed system), 84–86
  reactants, 71–73
  summary (review), 86–87
Glass properties. *See* Characterization of glass properties
Glass softening point, 31, 32, 39, 41, 48
Glass thermal expansion apparatus, 90
Glass transition temperature (T_g), 25, 90
Glassy quartz, 7
Glatkowski, Paul, 172
Glue layers, 126
Golay cell, 250
Gordy's rule, 52, 54
Graham, Amy, 136

H

Hafner, Harold, 19, 92, 231
Haisty, Bob, 230–231
Hall, John, 241
Hanna, Gail, 196
Hardness, 8, 9, 39, 41, 108–109
Harp, Bob, 232
Harrington, James, 161, 163
Hayes, Don, 110
Heavy metal oxides, 28
Hermann, Bill, Jr., 240
Hertzberger equation, 107
High melting oxides, 126
High-purity arsenic trisulfide glass, 194–196
High-purity vacuum float zone technique, 222
High-temperature chalcogenide glass viscometer, 82
High-temperature manometer, 80
Hilton, A. Ray, Jr., 222
Hilton, A. Ray, Sr., 18, 19–20, 79, 85, 100, 134, 212, 215, 216, 230, 231, 245, 246, 264, 266
Historical overview:
  chalcogenide glass, 17–21
  transmission of light, 2–5
Horizontal Bridgman production of GaAs plates, 233–242

I

I/I_0 ratio, 95
Icovazzi, Bob, 241
Image spoiling test, 116

Imaging bundles. *See* IR imaging bundles
Impurities, 72–73, 222
Impurity atoms, 11
Infrared crystalline materials, 211–243
　cadmium telluride, 211–222
　DARPA-funded large plate process, 215–222
　gallium arsenide, 230–242
　horizontal Bridgman production of GaAs plates, 233–242
　silicon, 225–230
　single-crystal silicon fibers, 228–230
　vacuum float zoned silicon detector material, 222–225
Infrared detector arrays, 131
Infrared ellipsometry, 259–264
Infrared Fiber Systems, 159
Infrared refractometer, 105, 148
Infrared scan technique, 248–252
Infrared transmission *(T)*, 31
Injection molding, 134
Intensity-angle data points, 106
Interatomic bond distances, 53
Interferometer, 116
IR fibers, 153–158, 168–174
IR imaging bundles, 181–210
　leachable bundle method, 181
　1-m length bundles, 184–191
　SBIR phase II program. *See* Navy SBIR IR imaging bundle program
　stacked ribbon method, 181–183
IR reflection of glassy quartz, 7
Isoelectronic, 230
IVA-VA-VIA system, 30, 35

J

Javan, Ali, 266
Jerger, Joe, 3, 17, 155
Johnson, Pete, 245
Joint Strike Fighter, 143
Jones, Charlie, 19, 231, 259, 264
Jones, Eric, 267
Jones, Mitchell, 222
JSF lens, 144

K

Kane, Phil, 246
Kennedy, Fred, 267
Kennedy, Howard, 100
Kinoforms, 132. *See also* Unconventional lens fabrication
Klocek, Paul, 241
Knoop hardness, 8, 9, 39, 41, 108–109
Knudsen cell, 55, 56
Kodak, 211, 212
Kokorina, Valentina, 18
Kolomiets, Boris, 17, 247
Kreidl, Norbert, 19
Kremer, Russell, 232
KRS-5, 227
Kyle, William, 172

L

Laakman, Kathy, 154
Large area mosaic CdTe substrates, 215
Laser calorimetry, 77
Lawson, John, 139
Layered glass, 66
Leachable bundle method, 181
LeBlanc, Rich, 191
Leitz Miniload Hardness Tester, 108
Lens drawing, 132
Lens fabrication. *See* Conventional lens fabrication; Unconventional lens fabrication
Lens molding unit, 135
Light phase cancellation, 78
Liquidus curve, 30
Lissajous patterns, 254
Litrow mount deviation angle method, 100–101
LMCO, 133, 142
Lockheed Martin in Orlando (LMCO), 133, 142
Log viscosity, 93, 94
Lombardo, Gino, 154, 158
Long-wavelength cutoff, 5–10
Lorentz-Lorenz equation, 43
Loretz, Tom, 90, 153, 154, 158, 175, 176, 182, 199, 203

M

Magnesium oxide, 126
Mass spectrometric investigation of bonding, 55–57
Mass spectrum, 56, 57
Materials Advisory Panel, 19
McAllister, Dennis, 241
McCord, James, 202, 222
McCrackin, Frank, 256
McDonald Douglas, 238
McKenzie, Douglas, 18
Melling, Peter, 172, 176
Metallic impurities, 73
Methane, 246
Method A, 91
"Method and Apparatus for Measuring the Thickness of Films by Means of Elliptical Polarization of Reflected Infrared Radiation" (Hilton/Jones), 264
Method B, 91
MgO disk, 246
MIDAC Corporation, 173
Miller, Dave, 242
Mn-Ge-Se, 68
Mn-Ge-Te, 67, 68
Modified Temscal coating chamber, 196
Modlin, Paul, 199, 203
Modulation transfer function (MTF), 78–79, 116, 124
Molar refraction, 42–44
Molded 5.38-in-diameter Amtir 5 JSF lens, 144
Molded 8-mm-diameter lens, 144
Molded 17-mm-diameter lens, 145
Molded vs. DPT optical elements, 136
Molding process, 146–148
Molding units, 146
Monochromator, 104
MTF, 78–79, 116, 124
Multiblocked Amtir 1 planoconvex 1-in lens, 122
Multifiber probe, 172
Multimolding, 147
MWIR Agema 210 camera, 137, 188, 190, 208

N

N light, 252
Navy P3 system, 19
Navy SBIR IR imaging bundle program, 191–210
  AMI glass clad fibers, 194
  50 percent transmission goal, 196–199
  high-purity arsenic trisulfide glass, 194–196
  Navy bundle 10-M-1, 202–203
  Navy bundle 10-M-2, 203, 206
  Navy bundle 10-M-3, 203, 206
  Navy bundle 10-M-4, 197, 203–204, 206–209
  objectives, 192
  1-m C2 imaging bundles, 192–194
  overview of process, 209–210
  10-m imaging bundles, 199–209
Negative impurities, 73
Neslab temperature bath, 102
Ni-Ge-S, 68
Ni-Ge-Se, 68
Ni-Ge-Se composition, 65
Ni-Ge-Te, 68
Ni-Zn-Se, 68
Ni-Zn-Te, 67, 68
Nicolet AVATAR 320, 95
NIR camera, 204, 206, 207
NIR Electrophysics 7290 tube camera, 188
Null angle, 267, 268

O

Ochrym, Bob, 242
1-in-diameter planoconvex sensor lenses, 122
1-m C2 imaging bundles, 192–194
1-m length bundles, 184–191
100-mm FL lens 1, 138
OPD, 116
Open casting methods, 77–84
Open down hole glass casting unit, 83, 84
Optical design, 131–132
Optical homogeneity, 78
Optical interference, 247–248
Optical null instruments, 95

# Index   277

Optical wavefront distortion (OPD), 116
Oscilloscope Lissajous patterns, 254
Ovshinsky, Stanley, 18
Owen, Alan, 18
Oxide materials, 6
Oxygen, 23

P

P, 46
P light, 252
Padovani, Francois, 266
Palm IR camera, 138–143, 188
Panametric 25-hp ultrasonic thickness gauge, 109
Pappis, Jim, 19
Parabola mirror, 263
Patterson, Robert, 18, 91
Pauling, Linus, 5
Pauling electronegativity, 5, 7, 8, 21, 28
Pauling electronegativity elemental values, 23
Paveway grade silicon single crystal, 224
Paveway silicon business, 225
Pearson, A. David, 18
Periodic table, 21–24
Perkin Elmer 13 U spectrophotometer, 36, 99, 102, 246
Perkin Elmer FTIR, 127
Perkin Elmer FTIR transmission scan, 96
Perkin Elmer Paragon 1000, 95
Phase compensation, 255
Phase lag (δ), 248
Phase shift angle, 253–259
Pitha, Carl, 212
Poisson's ratio, 110
Polarizing angle, 252
Polishing, 122
Pollicove, Harvey, 133–134
Polycrystalline, 1
Pour caster, 77
Precision molding, 133–146
Purity, 72, 73

Q

Quartz, 227, 261
Quartz chamber, 236, 237

R

R, 31, 51
RAD, 267
Radiance IT camera, 205, 209
Radius of curvature, 123
Raman effect, 53
Raman spectra, 54, 55
Ranat, R. M., 116
Raoult's law, 55
Ratliff, Charles, 267
Raytek, 168, 170
Raytheon Palm IR, 138–143, 188
Raytheon Palm IR BST uncooled bolometer 8- to 12-µm camera, 137
Raytheon Radiance IT camera, 205, 209
Raytheon zinc selenide, 103
Reactants, 71–73
Reflectivity *(R)*, 31, 51
Refractive index, 12, 13, 42, 51, 94–107, 115, 132
Refractive index change, 148–150
Refractive index measurement procedure, 101
Refractometer, 102
Reinberg, Al, 259
Relative scattering power, 58
Remote spectroscopy, 171
Remspec multifiber FTIR probe, 172, 173
Resistance to chemical attack, 114
Resistance to rain erosion, 240
Reststrahlen band, 5
Reststrahlen-like reflection bands, 9
Reynolds, Dick, 212, 215
Ribbon stacking method, 181–183
Rocking furnace method, 73–74
Rupture modulus, 110–112
Russian Ioffee Institute, 246

S

S, 46
Sahagian, Charles, 212
Sancliff, Inc., 159
Sancliff split coating die, 161
Sapphire, 227
SBIR phase II program. *See* Navy SBIR IR imaging bundle program
Se, 46

Index

Sealed chamber extrusion-pull method, 168–173
Selenide glass, 29
Selenium, 23, 73
Selenium compositions, 68
Selenium glass, 4, 72
Sellmeier equation, 107
Semiconductor-grade quartz, 73
Sensiv, 173
Servo Corporation, 3, 4, 17, 44, 163
Shear modulus, 110
Shearer, Laird, 266
Si, 46, 227
Si-Ge-As-Te glass, 36, 38
Si-P-Te system, 32–35
Silicate glass, 29
Silicon, 225–230, 240
Silicon telluride, 246
Sinclair, Doug, 266
Single-crystal, 1
Single-crystal silicon fibers, 228–230
SiO_2 glass, 28
$SiTe_4$ glass, 59
Slump molding, 133
Small-diameter single-crystal silicon rods, 229
Smith, John, 158
Snell's law, 155, 247
Softening point, 31, 32, 39, 41, 48
Solid, 1
Spain, Horace, 245
Sparks, Marshal, 211
Spectroscopic selection rules for active vibrations, 7
Spherical surfaces, 121. *See also* Conventional lens fabrication
Spherometer, 123
Spindles, 122
Spurlock, Bill, 116
Stacked ribbon method, 181–183
Stevenson, Chuck, 172
Stoichiometry, 48
Strain point, 92
Striae, 12
Striae comparison, 78
Striae scope, 115, 116
Striae scope photograph, 86, 117
Strong absorption, 12–14
Sulfide glass, 29
Sulfur, 23, 73
Sulfur-based glass, 36, 73

Sulfur compositions, 68
Sulfur glass, 11, 72
Surface hardness, 8. *See also* Knoop hardness
Swink, Larry, 79

T

T, 31
T_d, 25, 90
T_g, 25, 90
TADS PNVS, 20
Te, 46
TEA curve, 25
Teal, Gordon, 2
Telluride glass, 29
Tellurium, 23, 51, 73
Tellurium compositions, 67, 68
Tellurium glass, 4
Temscal coating chamber, 196
10-m drum assembly, 200
10-m imaging bundles, 199–209
Tensile strength, 185
Terlow, Jan, 139
Texas Instruments (TI), 245–270
  Air Force Production Development Program, 19
  automatic ellipsometer, 264–269
  crystalline materials, 212–214
  DARPA-ONR program, 18
  ellipsometer, 259
  elliptical polarization angles, 255–258
  elliptical polarization of light, 252–255
  first exploratory program, 29, 59–60
  GaAs, 241
  Galileo Electro-Optics, 153
  infrared applications to materials, 245–247
  infrared ellipsometry, 259–264
  infrared scan technique, 248–252
  large windows for high-energy CO_2 lasers, 74
  Navy P3 system, 19
  optical interference, 247–248
  Perkin Elmer 13 spectrophotometer, 36, 38
  silicon, 225
  slump molding, 133
  vacuum float zone technique, 222
Thermal coefficient of expansion, 41

Index **279**

Thermal comparator, 109, 112, 113
Thermal conductivity, 112–113
Thermal expansion analyzer (TEA) curve, 25
Thermal expansion coefficient, 41
Thermoscan, 174, 176
Thermoscan tympanic membrane thermometer, 175
Thomas, Dennis, 241
Thompson, Bill, 107
Thomson, Mary, 172
Three-chamber method, 217
TI. *See* Texas Instruments (TI)
TI 20 glass, 19, 74–78
TI 1173 glass, 20, 74–78, 83, 153
TI automatic ellipsometer system, 264–269
TI far infrared ellipsometer curves, 264
Ti-Ge-Se, 68
Ti-Ni-Te, 68
TI pour glass casting unit, 77
Ti-Si-Te, 68
Ti-V-Se, 68
Ti-V-Te, 68
Titanium, 60
$Ti_{15}V_{15}Te_{70}$, 63, 64
Two-element 218-mm FL lens, 140
Tyler, Tommy, 175

U

Unconventional lens fabrication, 131–151
  diamond turning, 132–133
  optical design, 131–132
  precision molding, 133–146
  refractive index change, 148–150
  slump molding, 133
  volume production, 146–148
Unpolarized light, 252
Urethane paint, 174
Urick, Larry, 242

V

Vacuum float zoned silicon detector material, 222–225
Vanadium, 60
Vapor growth approach, 217
Vapor species, 55
Vertical Bridgman crystal grower, 232, 233
Viscosity, 80–83, 90, 92–94
Volume expansion, 14
Volume production, 146–148

W

Weirauch, Don, 19
Welt, Dale, 233, 237
Westfall, Dub, 265
Whaley, Glen, 84–85, 156, 236, 242, 267
Wiese, Gary, 188, 192, 193
Woody, Dan, 142
Wyko interferometer, 117

X

X-ray radial distribution analysis, 57–59

Y

Young's modulus, 110

Z

ZEMAX, 132
Zinc selenide, 103, 118
Zirconium, 60
Zn-Ge-Se, 68
Zn-Ge-Te, 67, 68
ZnS, 227
ZnSe, 227
Zoom lens 1, 140
ZYGO evaluation sheet, 124–126
ZYGO interferometer, 124

CPSIA information can be obtained at www.ICGtesting.com
Printed in the USA
LVOW04*2138021014

407068LV00002B/10/P